UNITEXT for Physics

UNITEXT for Physics series publishes textbooks in physics and astronomy, characterized by a didactic style and comprehensiveness. The books are addressed to upper-undergraduate and graduate students, but also to scientists and researchers as important resources for their education, knowledge, and teaching.

Umberto Lombardo · Gianluca Giuliani · Yifei Niu

Quantum Mechanics

From Atoms to Nuclei

Science Press
Beijing

Springer

Umberto Lombardo
University of Catania
Catania, Italy

Gianluca Giuliani
Nuclear Science and Technology
Lanzhou University
Lanzhou, China

Yifei Niu
Nuclear Science and Technology
Lanzhou University
Lanzhou, China

ISSN 2198-7882 ISSN 2198-7890 (electronic)
UNITEXT for Physics
ISBN 978-981-97-1672-2 ISBN 978-981-97-1670-8 (eBook)
https://doi.org/10.1007/978-981-97-1670-8

Jointly published with Science Press
The print edition is not for sale in China (Mainland). Customers from China (Mainland) please order the print book from: Science Press.

This Springer imprint is published by the registered company Springer Nature Singapore Pte Ltd.
The registered company address is: 152 Beach Road, #21-01/04 Gateway East, Singapore 189721, Singapore

To Graziella and Nicolò

Preface

This textbook is intended to provide an introduction to non-relativistic Quantum Mechanics, focussed on a comparative analysis of atomic and nuclear phenomena. The aim is to facilitate students' understanding of how Quantum Mechanics approaches two fields of physics, characterized by significantly different space–time scales.

To begin with, the probabilistic character of Quantum Mechanics is illustrated throughout the breakthrough experiments that led to the principle of quantum superposition states. Afterwards, a self-contained presentation of the quantum mechanical formalism is developed through the analysis of the measurement processes in an ideal experimental setup, in the footsteps of the Julian Schwinger formulation. The quantum states as well as physical observables are then converted into abstract entities, that is, vectors and operators of Hilbert's space. Analogously to Classical Mechanics, the basic observables are identified as generating functions of the space–time symmetry transformations, relying on the relationships between symmetries of quantum systems and conservation laws of the dynamical variables. Likewise, the new physical variables, spin and isospin, are also introduced as generators of the SU(2) symmetry transformations in their respective two-dimensional spaces. In this context, the Hamiltonian is identified as the generator of the time translations and, as a consequence, the Schrödinger equation straightforwardly comes out of describing the time evolution of quantum systems.

The second part of the book is devoted to the presentation of quantum mechanical applications to atomic and nuclear systems, running on two parallel tracks. Firstly, the solutions of the Schrödinger equation are discussed for the simplest bound two-body systems, the hydrogen atom, where proton and electron interact via the long-range Coulomb force, and the deuteron nucleus, where proton and neutron interact via the short-range nuclear force. The investigation of the two-body problem is completed with the solution of the Schrödinger equation for the continuum state, and with the low-energy atomic and nuclear elastic scattering.

Going forward, Quantum Mechanics is applied to complex many-body systems, i.e. many-electron atoms and many-nucleon atomic nuclei. The presentation emphasizes the common property that brings them together in the diversity of their interactions, i.e. both are Fermi systems, subjected to the Pauli principle. This principle, due to the non-distinguishability of identical particles, in fact, is shown to play a major role in the interpretation of the shell structure of atomic electrons as well as nuclear protons and neutrons. However, the many-body effects on the interaction between two particles, due to the surrounding particles, lead to significantly different effective interactions. The Hartree–Fock mean field, which works quite well with atomic systems, proves insufficient when applied to the nuclear systems. As a consequence, the semi-empirical liquid drop model of the nucleus is introduced for finite nuclear systems and extended to the infinite nucleon system, known as nuclear matter. This ideal system, introduced by Hans Bethe in the fifties of the last century, proved to be extremely useful in the physics of heavy-ion collisions and compact stars.

This latter issue has motivated the inclusion in the book contents of homogeneous atomic and nuclear Fermi fluids in the framework of the quantum statistical theory. In particular, the properties of degenerate Fermi fluids, such as high-density electron and nucleon systems, are applied to explain the stability of white dwarfs and neutron stars.

The last part is devoted to the presentation of the collective excitations in atomic gas and nuclei in the framework of the semiclassical treatment of the many-body systems, based on the Wigner transform. The sound propagation in ^{3}He is addressed to the discussion of two extreme mechanisms: the propagation of the first sound as an adiabatic process and the propagation of the zero sound as a collisionless one. The observations show that we are faced with a unique process, evolving from the first sound to the zero sound when reducing temperature to extremely low values. Finally, the collective modes in nuclei, known as giant resonances, are discussed as in-phase oscillations of nucleons or out-of-phase oscillations of neutrons against protons. They are distinguished from each other by the multi-polarity degree (monopole, dipole, quadrupole,...). There is experimental evidence that, moving from lower to higher multi-polarities, the adiabaticity degree of the medium excitations rapidly decreases.

This book was inspired by the lectures on Quantum Mechanics, held by one of the authors (U.L.) at the Department of Physics and Astrophysics of the University of Catania in the period 2000–2015.

Catania, Italy — Umberto Lombardo
Lanzhou, China — Gianluca Giuliani
Lanzhou, China — Yifei Niu
July 2021

Preface to the Second Edition

This book is a re-edition of the book 'Quantum Mechanics: from Atoms to Nuclei, printed by Science Press (Beijing 2022). Apart from the stylistic improvement of the text and the correction of a number of oversights and printing errors, this edition presents the following new subjects (1) a brief introduction to the entangled states; the extension of the Hartree-Fock approximation to the dynamics of many body systems; (2) a review of the presentation of the nuclear giant resonances. Umberto Lombardo (Catania, August 2024).

Acknowledgements

The authors are indebted to several colleagues and friends for valuable comments and suggestions, in particular, M. Baldo and H.-J. Schulze from INFN-Catania, P. Schuck from IPN-Orsay, V. Greco from the University of Catania, I. Bombaci from the University of Pisa, A. Bonaccorso from INFN-Pisa, M. Ruggieri and H. F. Zhang from the Lanzhou University. A thank you is also due to Susan Whitehouse for her careful language revision.

This work was supported by the School of Nuclear Science and Technology of Lanzhou University (China).

Contents

Symbols

a	Scalar
$\overrightarrow{a}$	Vector
$\overrightarrow{a} \cdot \overrightarrow{b}$	Scalar product
$\overrightarrow{a} \times \overrightarrow{b}$	Vector product
$\lvert\psi\rangle$	State vector
$\langle\psi\rvert$	Dual-state vector
O	Operator
$\overrightarrow{O}$	Vector operator
$\lvert\psi\rangle\langle\psi\rvert$	Projection operator
$\{\mathrm{f,g}\} = \frac{\partial f}{\partial p}\frac{\partial g}{\partial q} - \frac{\partial g}{\partial p}\frac{\partial f}{\partial q}$	Poisson bracket
$[\mathbf{A}, \mathbf{B}] \equiv \mathbf{AB} - \mathbf{BA}$	Commutation bracket
$\{\mathbf{A}, \mathbf{B}\}_+ \equiv \mathbf{AB} + \mathbf{BA}$	Anticommutation bracket
$\partial/\partial t$	Partial derivative
d/dt	Total derivative
$\overrightarrow{\nabla} \equiv (\partial/\partial x, \partial/\partial y, \partial/\partial z)$	Nabla (also ∇)
$\nabla^2 \equiv \Delta \equiv (\partial^2/\partial x^2 + \partial^2/\partial y^2 + \partial^2/\partial z^2)$	3D-nabla squared
$\Box \equiv (\partial^2/\partial x^2 + \partial^2/\partial y^2 + \partial^2/\partial z^2 - \partial^2/\partial (ct)^2)$	4D-nabla squared

Chapter 1
Space–Time Symmetries and Classical Observables

Quantum Mechanics, introduced and developed at the beginning of the last century, represents a revolution of the ideas and laws of Classical Mechanics. It prevailed under the push of more and more discoveries on the atomic and subatomic scale. At the same time it has proven to include Classical Mechanics as a limiting case when moving from microscopic to macroscopic scale. This is an aspect of continuity from Classical to Quantum Mechanics, that will be underlined in the following chapters of this textbook. Indeed, some achievements of Classical Mechanics still survive in Quantum Mechanics, even if the context is completely new. Extremely important examples of continuity are the space and time symmetries, in that they are closely related to the conservation of dynamical variables. This relationship, established in the framework of Classical Mechanics, must be extended to Quantum Mechanics, provided that the generators of the symmetry transformations are the same in the two theories. Since these generators are the dynamical variables themselves, we have thus a solid key to identify the quantum dynamical variables.

In this chapter the laws of classical dynamics and their form-invariance under space and time symmetry transformations will be reviewed in the Hamiltonian formalism (see e.g. [LAL]). Afterwards, the conservation theorems of momentum, angular momentum, and energy will be derived from the space and time symmetry in the framework of canonical transformations.

1.1 Hamilton's Equations

In Hamilton's formalism of classical dynamics, the state of a physical system with f degrees of freedom at time t is defined by f generalized coordinates $q_i(t)$ and f generalized momenta $p_i(t)$ $(i = 1, \ldots, f)$. What we mean by state, in the most

---

Supplementary Information The online version contains supplementary material available at https://doi.org/10.1007/978-981-97-1670-8_1.

U. Lombardo et al., *Quantum Mechanics*, UNITEXT for Physics,
https://doi.org/10.1007/978-981-97-1670-8_1

general sense, is the most complete information on the system at time t. Given a function containing the maximum physically significant information on the system, named Hamiltonian, the laws of motion are

$$\dot{q}_i = \frac{\partial H}{\partial p_i}, \qquad \dot{p}_i = -\frac{\partial H}{\partial q_i}. \tag{1.1}$$

This is a set of $2f$ coupled first-order differential equations. The state of the system at any time t is uniquely determined by integrating these equations in the interval $t - t_0$, as far as generalized coordinates and momenta are known at the time t_0. This correspondence between the initial and final state is the essence of *classical determinism*, namely the capability of the classical laws of motion to predict with absolute certainty the state of the system at any time, once it is known at the time t_0. In quantum mechanics there is still determinism, i.e. the existence of laws of motion, but the correspondence between the initial and final state has a probabilistic character, the quantum states are significantly different from classical ones.

Defining the Poisson brackets between two observables $A(q, p)$ and $B(q, p)$ (q and p are the sets of all coordinates and momenta, respectively) as

$$\{A, B\} \equiv \sum_{i=1}^{f} \left(\frac{\partial A}{\partial p_i} \frac{\partial B}{\partial q_i} - \frac{\partial A}{\partial q_i} \frac{\partial B}{\partial p_i} \right),$$

Hamilton's equations, Eq. (1.1), can be written

$$\dot{q}_i = \{H, q_i\}, \qquad \dot{p}_i = \{H, p_i\}. \tag{1.2}$$

We refer to Eq. (1.2) as the canonical equations of motion. Coordinate and momentum are called canonically conjugate variables.

1.2 Space–Time Symmetries and Conservation of Dynamical Variables

During the motion of a mechanical system, the q_i's and the p_i's, generalized coordinates and momenta, vary with time. There exist, however, functions of such variables which do not vary as long as the Hamiltonian satisfies some symmetries. We will refer to these functions as integrals (or constants) of motion, and in particular we will focus on the integrals of motion which are related to the space–time symmetries, namely time is uniform, space is homogeneous and isotropic.

1. Uniformity of Time

We assume that the Hamiltonian does not explicitly depend on time, that is $H(q(t), p(t), t) \equiv H(q(t), p(t))$. By calculating the total time derivative of H, from the equations of motion, Eq. (1.1), one easily gets

$$\frac{\mathrm{d}H}{\mathrm{d}t} = \sum_i \left(\frac{\partial H}{\partial q_i} \dot{q}_i + \frac{\partial H}{\partial p_i} \dot{p}_i \right) = 0.$$

It follows that $H(q(t), p(t)) = E$ is constant at any time t. This constant represents the energy of the system.

2. Homogeneity of Space

To simplify matters, we consider a system of two interacting particles whose Hamiltonian is invariant under space translations and rotations, that is

$$H = \frac{p_1^2}{2m_1} + \frac{p_2^2}{2m_2} + V(|\vec{r}_1 - \vec{r}_2|).$$

Let us consider the time derivative of the total momentum of the two particles $\vec{P} = \vec{p}_1 + \vec{p}_2$ and, as previously done, make use of Hamilton's equations, Eq. (1.1), we find:

$$\frac{\mathrm{d}\vec{P}}{\mathrm{d}t} = \frac{\mathrm{d}}{\mathrm{d}t}(\vec{p}_1 + \vec{p}_2) = -\frac{\partial H}{\partial \vec{r}_1} - \frac{\partial H}{\partial \vec{r}_2} = -\frac{\partial V}{\partial \vec{r}_1} - \frac{\partial V}{\partial \vec{r}_2} = 0,$$

the two space derivatives of the potential having the same magnitude but opposite sign. We conclude that the invariance for translations of the interaction leads to the conservation of total momentum.

3. Isotropy of Space

The Hamiltonian of the two interacting particles is normally invariant not only under space translations but also under rotations, like the one just considered. Let us examine now the total time derivative of the total orbital angular momentum:

$$\begin{aligned}
\frac{\mathrm{d}\vec{L}}{\mathrm{d}t} &= \frac{\mathrm{d}}{\mathrm{d}t}(\vec{r}_1 \times \vec{p}_1 + \vec{r}_2 \times \vec{p}_2) \\
&= \vec{r}_1 \times \frac{\mathrm{d}\vec{p}_1}{\mathrm{d}t} + \frac{\mathrm{d}\vec{r}_1}{\mathrm{d}t} \times \vec{p}_1 + \vec{r}_2 \times \frac{\mathrm{d}\vec{p}_2}{\mathrm{d}t} + \frac{\mathrm{d}\vec{r}_2}{\mathrm{d}t} \times \vec{p}_2 \\
&= \vec{r}_1 \times \left(-\frac{\partial H}{\partial \vec{r}_1} \right) + \frac{\partial H}{\partial \vec{p}_1} \times \vec{p}_1 + \vec{r}_2 \times \left(-\frac{\partial H}{\partial \vec{r}_2} \right) + \frac{\partial H}{\partial \vec{p}_2} \times \vec{p}_2.
\end{aligned} \tag{1.3}$$

By plugging the derivatives of the Hamiltonian

$$-\frac{\partial H}{\partial \vec{r}_i} = -\frac{\partial V}{\partial \vec{r}_i}, \qquad \frac{\partial H}{\partial \vec{p}_i} = \frac{\vec{p}_i}{m_i}, \qquad i = 1, 2$$

into Eq. (1.3), we obtain

$$\frac{d\vec{L}}{dt} = \vec{r}_1 \times \left(-\frac{\partial V}{\partial \vec{r}_1}\right) + \vec{r}_2 \times \left(-\frac{\partial V}{\partial \vec{r}_2}\right).$$

The potential only depends on the magnitude of the distance between the two particles, hence

$$\frac{d\vec{L}}{dt} = -\frac{dV}{dr}\hat{r} \times (\vec{r}_1 - \vec{r}_2) = 0, \tag{1.4}$$

where $\hat{r} = (\vec{r}_1 - \vec{r}_2)/r$, r being the distance between the two particles. Equation (1.4) tells us that the invariance of the interaction under rotations leads to the conservation of the total orbital angular momentum.

The relation between space–time symmetries and conservation of dynamical variables will be addressed within a more general framework in the next sections.

1.3 Canonical Transformations and Space–Time Symmetries

The set of generalized coordinates and momenta defining the states of a physical system is not unique. Different sets can describe the same motion of the system, as long as they fulfill equations with the same form-invariance. The most general transformations between two sets of generalized coordinates and momenta are

$$Q_i = Q_i(q_1, \ldots, q_f, p_1, \ldots, p_f, t),$$
$$P_i = P_i(q_1, \ldots, q_f, p_1, \ldots, p_f, t).$$

If the equations of motion are form-invariant under these transformations, the latter are called canonical transformations [GOL]. Without losing generality, the condition for a transformation to be canonical can be derived by restricting to the class of infinitesimal transformations. In a compact form it can be written as

$$Q = q + \delta q, \qquad P = p + \delta p. \tag{1.5}$$

Let us choose an arbitrary function $G(q, p)$ and consider the trans-formation

$$\delta q = \varepsilon \frac{\partial G}{\partial p} = -\varepsilon\{q, G\}, \qquad \delta p = -\varepsilon \frac{\partial G}{\partial q} = -\varepsilon\{p, G\}, \tag{1.6}$$

where ε is an arbitrary smallness parameter. We now prove that this is an infinitesimal canonical transformation. The conditions for the form-invariance of the equations of motion, Eq. (1.2) are

$$\dot{q} - \{H(q, p), q\} = \dot{Q} - \{H(Q, P), Q\} = 0,$$
$$\dot{p} - \{H(q, p), p\} = \dot{P} - \{H(Q, P), P\} = 0,$$

which, under an infinitesimal transformation, are written as

$$\delta\dot{q} - \{\delta H, q\} - \{H, \delta q\} = 0, \tag{1.7}$$
$$\delta\dot{p} - \{\delta H, p\} - \{H, \delta p\} = 0, \tag{1.8}$$

where δq and δp are given by Eq. (1.6). In turn, we have

$$\delta H = \varepsilon\frac{\delta H}{\delta q}\delta q + \varepsilon\frac{\delta H}{\delta p}\delta p = \varepsilon\{G, H\} \tag{1.9}$$

and hence the form-invariance conditions can be written

$$\{\{q, H\}, G\} + \{\{H, G\}, q\} + \{\{G, q\}, H\} = 0,$$
$$\{\{p, H\}, G\} + \{\{H, G\}, p\} + \{\{G, p\}, H\} = 0, \tag{1.10}$$

which are identically fulfilled, being the well-known Jacobi identity applied to the three variables q, H, G and p, H, G, respectively. Therefore, we may conclude that the infinitesimal transformation, generated by the function $G(q, p)$, is a canonical transformation. The function is indeed called *generator of a canonical transformation*. Now we shall prove that the symmetry transformations, such as space translations, time translations, and space rotations, belong to the class of canonical transformations. This is not a surprise because for those systems with Hamiltonian satisfying the space–time symmetries, the equations of motion have to be form-invariant under those symmetry transformations.

(1) Time Translations: due to the uniformity of time, the Hamiltonian H of an isolated system is a conserved quantity, being explicitly independent of time. Let us assume the canonical transformation with $G = H$ and $\varepsilon = \mathrm{d}t$, $\mathrm{d}t$ being an infinitesimal time interval. Applying Eq. (1.6), we get

$$P_i = p_i - \frac{\partial H}{\partial q_i}\mathrm{d}t = p_i + \dot{p}_i\mathrm{d}t = p_i(t + \mathrm{d}t),$$
$$Q_i = q_i + \frac{\partial H}{\partial p_i}\mathrm{d}t = q_i + \dot{q}_i\mathrm{d}t = q_i(t + \mathrm{d}t),$$

showing that the canonical transformation generated by the Hamiltonian is nothing else than a time translation. We conclude that the Hamiltonian is the generator of an infinitesimal time transformation displacing coordinates and momenta from time t to time $t + \mathrm{d}t$. We will see that the same holds true in quantum mechanics.

(2) Space Translations: due to the homogeneity of space the total momentum of a system $\vec{P} = \sum_i \vec{p}_i$ is a conserved quantity. We consider the transformation with

total momentum as the generator of an infinitesimal translation. We fix the x axis in the direction of the translation and $\varepsilon = \delta x$. Again from Eq. (1.6), we get

$$X_i = x_i + \frac{\partial p_i^x}{\partial p_i^x}\delta x = x_i + \delta x,$$
$$P_i^x = p_i^x - \frac{\partial p_i^x}{\partial x_i}\delta x = p_i^x,$$

showing that the canonical transformation generated by the total momentum is an infinitesimal space translation. As a result, the generator of an infinitesimal translation in a fixed direction is the projection of the total momentum in that direction.

(3) Space Rotations: due to the isotropy of space the total angular momentum is a conserved quantity. We consider an infinitesimal rotation around the z-axis with $\varepsilon G = \delta\phi\hat{z}\cdot\vec{L}$, where $\vec{L} = \sum_i(\vec{r}_i \times \vec{p}_i)$. We have the following equations:

$$\begin{aligned} X_i &= x_i + \delta\phi\frac{\partial}{\partial p_i^x}\Big[\sum_j(\vec{r}_j \times \vec{p}_j)_z\Big] = x_i + \delta\phi\frac{\partial}{\partial p_i^x}(-y_i p_i^x) \\ &= x_i - \delta\phi y_i, \\ Y_i &= y_i + \delta\phi\frac{\partial}{\partial p_i^y}\Big[\sum_j(\vec{r}_j \times \vec{p}_j)_z\Big] = y_i + \delta\phi\frac{\partial}{\partial p_i^y}(x_i p_i^y) \\ &= y_i + \delta\phi x_i, \\ Z_i &= z_i. \end{aligned}$$

The infinitesimal rotation around a generic axis, say $\hat{u}$, is described by a canonical transformation that is generated by the projection of the total angular momentum of the system onto the $\hat{u}$ axis.

Suppose we have a function of coordinates, momenta and time $f(q, p, t)$. The total time derivative is

$$\frac{\mathrm{d}f}{\mathrm{d}t} = \frac{\partial f}{\partial t} + \sum_i\left(\frac{\partial f}{\partial q_i}\dot{q}_i + \frac{\partial f}{\partial p_i}\dot{p}_i\right) \tag{1.11}$$

and, using Hamilton's equations, Eqs. (1.1), it becomes

$$\begin{aligned} \frac{\mathrm{d}f}{\mathrm{d}t} &= \frac{\partial f}{\partial t} + \sum_i\left(\frac{\partial H}{\partial p_i}\frac{\partial f}{\partial q_i} - \frac{\partial H}{\partial q_i}\frac{\partial f}{\partial p_i}\right) \\ &= \frac{\partial f}{\partial t} + \{H, f\}. \end{aligned} \tag{1.12}$$

Two remarkable properties stem from this equation:

(1) if f is an integral of motion, i.e. $\frac{\mathrm{d}f}{\mathrm{d}t} = 0$, and it is not explicitly dependent on time, then

$$\{H, f\} = 0; \tag{1.13}$$

(2) if two quantities f and g are both integrals of motion, their Poisson bracket is an integral of motion too, and hence

$$\{H, \{f, g\}\} = 0, \tag{1.14}$$

(see Problem 6). This result is known as the Poisson theorem.

We consider an infinitesimal canonical transformation generated by the function G. Let the variation of the Hamiltonian under such a transformation be

$$\delta H = \sum_i \left(\frac{\partial H}{\partial p_i} \delta p_i + \frac{\partial H}{\partial q_i} \delta q_i \right).$$

From the canonical transformation, expressed by Eq. (1.5) and Eq. (1.6), we have

$$\delta q_i = Q_i - q_i = \varepsilon \frac{\partial G}{\partial p_i},$$

$$\delta p_i = P_i - p_i = -\varepsilon \frac{\partial G}{\partial q_i}.$$

As a consequence, δH becomes

$$\delta H = \varepsilon \sum_i \left(\frac{\partial H}{\partial q_i} \frac{\partial G}{\partial p_i} - \frac{\partial H}{\partial p_i} \frac{\partial G}{\partial q_i} \right),$$

and using the Poisson bracket between G and H, it is written as

$$\delta H \equiv \epsilon \{G, H\}. \tag{1.15}$$

This result tells us that the Hamiltonian is an invariant under the infinitesimal canonical transformation generated by G if

$$\{G, H\} = 0.$$

On the other hand, we have already seen that G is an integral of the motion if

$$\frac{\mathrm{d}G}{\mathrm{d}t} = \{H, G\} = 0.$$

As for the generators of the space symmetry transformations, we can express the conservation equations by means of the Poisson brackets

$$\{H, \vec{p}\} = 0,$$
$$\{H, \vec{L}\} = 0.$$

The conservation of the dynamic variables is a consequence of the space–time symmetries. We will see that such symmetries are also true in quantum mechanics; therefore, we expect the same conservation relations are also true.

1.4 Notes and References

LAL Landau, L.D., Lifshitz, E.M.: Mechanics. Pergamon Press (1982).
GOL Goldstein, H.: Classical Mechanics. Addison-Wesley, Reading, USA (1965), Chap. 8.

1.5 Problems

- Problem 1. Consider a particle with Hamiltonian $H_0 = \dfrac{p^2}{2m} + \dfrac{kq^2}{2}$ (harmonic oscillator). (a) Derive and solve the equations of motion for fixed energy E. (b) Suppose a perturbation $V(q) = k\lambda q$ is added to H_0. Without solving the new equations of motion, calculate the new angular frequency $\left(\omega = \sqrt{\dfrac{k}{m}}\right)$ and energy.
- Problem 2. Derive and solve the equations of motion of a rigid body rotating around the z-axis with Hamiltonian $H = l_z^2/2\Im$, where l_z is the z-projection of the angular momentum and $\Im$ is the moment of inertia (rigid rotor).
- Problem 3. Calculate and solve the equations of motion of an electron submitted to a static and uniform electromagnetic field (e.m.), described by the four-potential $(\vec{A}, \Phi)$, with magnetic field $\vec{B} = \vec{\nabla} \times \vec{A}$ directed along the z-axis and electric field $\vec{E} = -\nabla\Phi$ orthogonal to the x-axis.
- Problem 4. Using the infinitesimal canonical transformations, prove the Jacobi identity.
- Problem 5. Determine the constants of motion of two particles with mass m interacting through a potential dependent on the distance $r_{12} = |\vec{r}_1 - \vec{r}_2|$.
- Problem 6. Prove that the Poisson bracket between two constants of motion is also a constant of motion.
- Problem 7. Determine the canonical transformation generated by $G(q, p) = qp$.

Chapter 2
Superposition Principle

Classical Mechanics endorses the sharp separation between particle and wave phenomena in the macroscopic world, the former being governed by Newton's laws and the latter by Huygens–Fresnel's laws. With the opening of the microscopic world to experiment, this separation disappeared. On the one hand, the photo-electric and Compton effects are examples of the particle-like behaviour of radiation; on the other hand, interference and diffraction are examples of the wave-like behaviour of particles. In this chapter, we show how quantum mechanics accomplishes the unification of waves and particles through an extension of the superposition principle. A probabilistic definition of state comes out, meaning that a physical system, submitted to measurement, can be detected in various configurations, each with its own probability.

2.1 An Historic Experiment

Consider the Young experiment of light interference from a double-slit. Figure 2.1 (left) depicts the experimental set, consisting of a source S of a beam of light, a screen with two slits, F_1 and F_2, symmetrically placed with respect to the axis SO, and a detector R.

The light waves with a wavelength λ, emitted by the source S, travel towards the screen. The beam, crossing the screen, is split by the two slits into two beams. According to the Huygens–Fresnel principle, the slits behave like two elementary sources of in-phase spherical waves. By placing the detector R at a considerable distance from the slits, we can assume the two waves to reach the detector as plane waves. In the generic point $P(x)$ of the detector, the two waves are no longer in-phase,

Supplementary Information The online version contains supplementary material available at https://doi.org/10.1007/978-981-97-1670-8_2.

U. Lombardo et al., *Quantum Mechanics*, UNITEXT for Physics,
https://doi.org/10.1007/978-981-97-1670-8_2

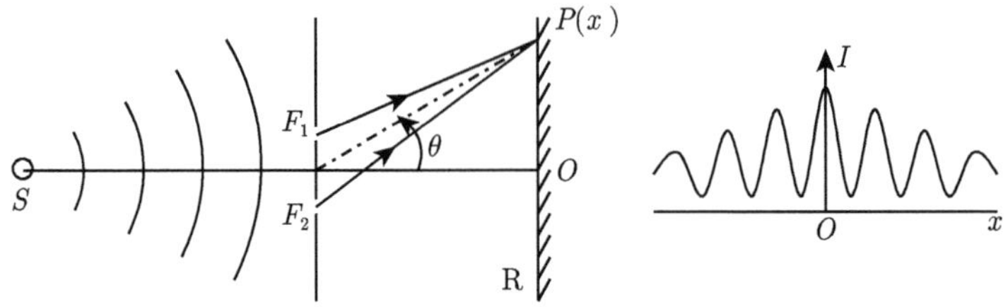

Fig. 2.1 Sketch of the setup of the Young double-slit experiment (left) and distribution of light intensity in the detector (right)

but they keep a phase relation, which depends on the difference between their optical paths, $d_1 = P(x) - F_1$ and $d_2 = P(x) - F_2$. In $P(x)$ at the time t, their respective electric fields (and magnetic fields) take the form of plane waves

$$\vec{E}_1 = \vec{E}_0 e^{i(kd_1 - \omega t)},$$
$$\vec{E}_2 = \vec{E}_0 e^{i(kd_2 - \omega t)},$$

where $k = 2\pi/\lambda$ is the wave vector and ω is the angular frequency. According to the Huygens–Fresnel theory, the resulting electric field in $P(x)$ is the vector sum of the electric fields of the two impinging fields (*superposition principle*), and therefore the light intensity is given by the square modulus of the resulting field

$$I(x) = |\vec{E}_1 + \vec{E}_2|^2 = 2I_0 \left[1 + \cos\left(\frac{2\pi}{\lambda} D \sin\theta\right)\right], \tag{2.1}$$

where $I_0 = |\vec{E}_1|^2 = |\vec{E}_2|^2$ is the intensity of each component and θ is the angle between the $P(x)$ direction and the horizontal line SO, and D is the distance between the two slits, as depicted in Fig. 2.1. For $\theta = 0$, we observe a central peak $I = 4I_0$ at the point O of the detector. The next two peaks are symmetrically located at the angle given by

$$\sin\theta = \frac{\lambda}{D}.$$

Other maxima ($I = 4I_0$) and minima ($I = 0$) follow one another on the left and right sides, as shown in the interference pattern of Fig. 2.1 (right). By decreasing λ, the peaks move closer and closer, and for $D \gg \lambda$, they cannot be resolved any longer. In conclusion, this experiment shows that, for two coherent beams, the total intensity on the detector is not the sum of their respective intensities, because one has to first sum up the two wave amplitudes and then compute the square modulus, as dictated by the superposition principle. This procedure gives rise to the interference term of Eq. (2.1).

2.2 Wave-like Behaviour of Particles

Let us repeat the experiment with the double-slit setup, replacing the light source with a neutron source (such an experiment was performed over two decades ago [ZEI]). To observe the expected interference phenomenon, the momentum of the neutron beam must fulfill a condition, equivalent to the light wavelength, of being the order of the slit size. This comes from the de Broglie hypothesis that associates a matter wave to a particle of wavelength $\lambda = h/p$.[1] Accordingly, the momentum value must fall into the range of low-energy neutrons, i.e. with energy less than one eV large. If the experiment is made with a weak source, i.e. emitting one particle at a time, we avoid any possible interactions between the emitted particles so that the experiment with N particles is equivalent to N independent experiments with a single-particle. As a consequence, we can define the probability $\mathcal{P}(x)$ of detecting a particle at the point $P(x)$ of the detector

$$\mathcal{P}(x) = \lim_{N\to\infty} \frac{N(x)}{N}, \tag{2.2}$$

where $N(x)$ is the number of particles detected in $P(x)$ and N the total number of particles in the experiment. The experiment can be performed with three slit configurations:

- If F_1 is open and F_2 is closed, the neutrons, which cross F_1, give rise to the distribution $\mathcal{P}_1(x)$ centred around $P(x_1)$, image point of F_1;
- If F_1 is closed and F_2 is open, the neutrons, which cross F_2, give rise to the distribution $\mathcal{P}_2(x)$ centred around $P(x_2)$, image point of F_2;
- If F_1 is open and F_2 is open, we observe an interference pattern similar to what happens with light. The experiment performed with neutrons yields the result shown in Fig. 2.2.

From the classical point of view, the neutrons should cross the screen either through slit F_1 or through slit F_2, depending on which slit is in the neutron trajectory, before impinging on the screen, and the probability at $P(x)$ should be the sum $\mathcal{P}_1 + \mathcal{P}_2$. There is no room for interference! In probabilistic language, the two events, i.e. the neutron crossing through F_1 and the neutron crossing through F_2, should be independent of each other. In order to understand the interference, we must assume that the neutron (remember only one neutron at a time!) was not in a single trajectory before impinging the screen but in a superposition of the two trajectories associated with the two slits. As a consequence, in the probabilistic language, the two events are not independent of each other.

To explain Fig. 2.2, we would like to apply the superposition principle with an analogue strategy to the light interference case. In analogy with the wave amplitude,

[1] In the early stage of Quantum Mechanics, L. de Broglie (1924) proposed the existence of matter waves associated to particles to justify electron diffraction [DAV]. Later, this idea was abandoned, but the relation between momentum and wavelength was proved to be correct.

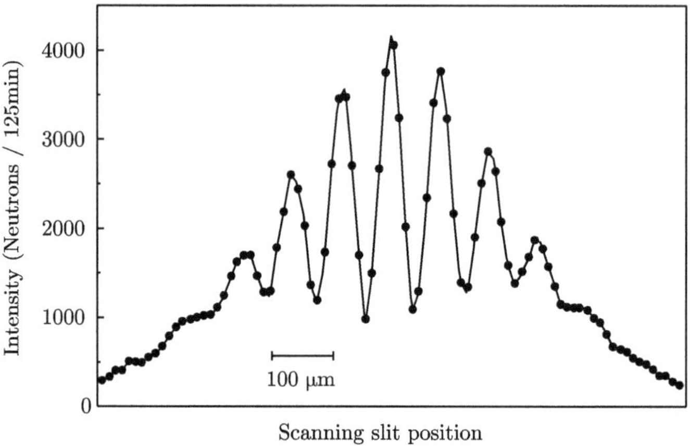

Fig. 2.2 Experimental neutron interference pattern [ZEI]. Courtesy of American Physical Soc

quantum mechanics introduces a theoretical quantity $A(x)$, known as *probability amplitude*, not directly observable, but associated to the observable probability $\mathcal{P}(x)$ of any event, Eq. (2.2), by the relation $\mathcal{P}(x) = |\ \mathrm{A(x)}|^2$. Thus, we introduce the amplitude $A_1(x)$ with probability $|A_1(x)|^2$ that the particle reaches the point $P(x)$ when only F_1 is open, and the amplitude $A_2(x)$ with probability $|A_2(x)|^2$ that the particle reaches the point $P(x)$ when only F_2 is open. Therefore, by applying the superposition principle, the total probability at the point $P(x)$ on the detector, will be

$$\begin{aligned}\mathcal{P}(x) &= |A_1(x) + A_2(x)|^2 \\ &= \mathcal{P}_1(x) + \mathcal{P}_2(x) + \mathcal{P}_{\text{inter}}(x),\end{aligned}$$

when both slits are open. The term $\mathcal{P}_{\text{inter}}(x)$ is responsible for the interference pattern. The probability amplitudes are complex numbers. Their phase difference determines the value of $\mathcal{P}(x)$, maximum or minimum intensity, at $P(x)$ (see Problem 1). This is in keeping with the phase difference of the two electric fields in the light experiment. So far, we have not yet explained how to calculate the two amplitudes $\mathcal{P}_i(x)$. The probability amplitudes of the physical processes on a microscopic scale are the purpose of quantum mechanics.

By taking a closer look, we consider the same experiment with a beam of electrons and we place a strong light source beyond the slits [FEY]. This source emits a Compton light flash when crossed by one electron so that it is possible to determine the number of electrons coming from F_1 and F_2 by counting the number of flashes near each slit. This is possible only if the flash wavelength is significantly smaller than the distance between the two slits. Otherwise, there is no longer any interference and the distribution of the detected electrons is classical: $\mathcal{P} = \mathcal{P}_1 + \mathcal{P}_2$. This is a consequence of the uncertainty principle stated by Heisenberg [HEI], imposing a limit on knowledge of nature.

2.3 Particle-like Behaviour of Waves

We review the first experimental effects showing that, under certain conditions, light exhibits a particle-like behaviour. The first is the photoelectric effect [MIL]: when light falls on the surface of a metallic conductor electrons are emitted. The electron flow does not depend on light intensity, but rather on its frequency, as proven by the existence of a frequency threshold. Einstein interpreted this effect by postulating that light is made of photons with $E = h\nu$, where E and ν are the photon energy and frequency, respectively, and h is the Planck constant [EIN]. According to this hypothesis, photons of the energy $E = h\nu$ are absorbed by electrons, which are emitted only if $h\nu$ exceeds the ionization threshold. The second is the Compton effect [COM]: a monochromatic wave with a wavelength in the X-ray band ($\lambda \sim 1$ Å) hits a graphite target, as shown in Fig. 2.3. After the interaction with an electron of the target, the outgoing X-ray beam splits into two components, one of which is in the forward direction and the other is at an angle θ. The wavelength shift of this latter is equal to

$$\lambda' - \lambda = \frac{h}{mc}(1 - \cos\theta),$$

where m is the electron mass and c the speed of light. Classically, only the forward component is expected to be observed. The secondary beam is interpreted as the output of a collision between the electron and photon according to the Einstein hypothesis. The wavelength shift is easily derived from two-body kinematics (see Problem 1).

The third experiment is the black-body radiation, which furnished the first evidence for the energy quantization. In the modern view, it is interpreted as a typical example of a boson gas. The black-body radiation is a standing electromagnetic (e.m.) field confined in a cavity in equilibrium with conductive walls at temperature T. The electromagnetic field $\vec{A}(\vec{r}, t)$ (it can be either the electric or magnetic field) can be expressed in terms of a superposition of monochromatic plane waves

$$\vec{A}(\vec{r}, t) = \sum_k \vec{A}_k^0 e^{i(\vec{k}\cdot\vec{r} - \omega t)}, \tag{2.3}$$

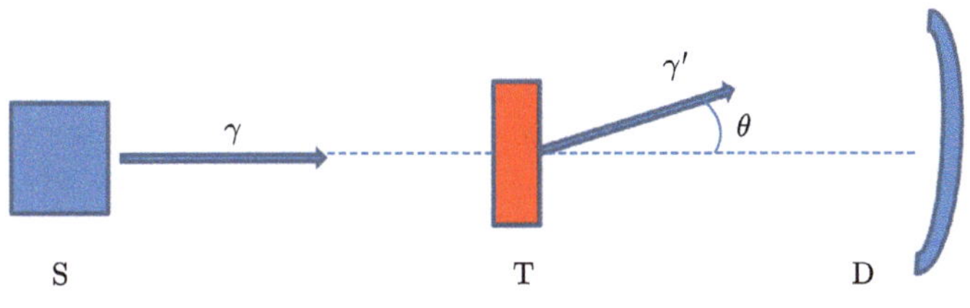

Fig. 2.3 Sketch of the compton scattering

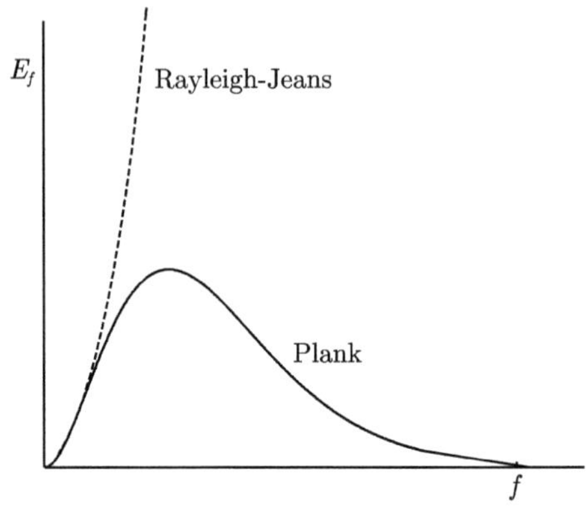

Fig. 2.4 Black-body radiation: spectral energy density versus frequency

where $\vec{A}_k^0$ is the amplitude, $k = 2\pi/\lambda$ is the wave number, and ω is the angular frequency. It can be proven (see Sect. 16.8) that the e.m. field in the cavity is equivalent to a system of decoupled harmonic oscillators and the number of oscillators per unitary volume and frequency is $N(\nu) = 8\pi\nu^2/c^3$. From statistical mechanics it turns out that the average energy density of the electromagnetic field in the cavity is

$$E_\nu = N(\nu) \frac{\displaystyle\int \mathrm{d}\vec{r}\mathrm{d}\vec{p} E(\vec{r}, \vec{p}) \mathrm{e}^{-E(\vec{r},\vec{p})/k_\mathrm{B}T}}{\displaystyle\int \mathrm{d}\vec{r}\mathrm{d}\vec{p}\ \mathrm{e}^{-E(\vec{r},\vec{p})/k_\mathrm{B}T}}, \tag{2.4}$$

where $E(\vec{r}, \vec{p})$ is the energy of the oscillator of frequency ν and the factor $\mathrm{e}^{-E(\vec{r},\vec{p})/k_\mathrm{B}T}$ is the statistical weight. The factor k_B is the Boltzmann constant. Following the postulate of photon gas, we assume that the energy E of the oscillator of frequency ν is quantized with $n = 1, 2, 3, \ldots$ photons associated with it. So, the energy E in Eq. (2.4) must be replaced by $nh\nu$ and the integral by the sum over the number of photons.[2] Then the new average energy, known as the Planck formula, is

$$E_\nu = N(\nu) \frac{\sum_n n \cdot h\nu\, \mathrm{e}^{-n \cdot h\nu/kT}}{\sum_n \mathrm{e}^{-n \cdot h\nu/kT}} = N(\nu) \frac{h\nu}{\mathrm{e}^{\frac{h\nu}{k_\mathrm{B}T}} - 1}. \tag{2.5}$$

The two predictions are plotted in Fig. 2.4. The classical energy density, known as Rayleigh–Jeans law, has a parabolic dependence on frequency, whereas the quantum energy density has the same shape at low frequency, but it goes over into exponential decay at high frequency.

[2] A rigorous derivation is carried out in Sect. 16.8.

2.4 The Stern–Gerlach Experiment

The Stern–Gerlach experiment is of great interest because it is the first experimental proof of electron spin and also because it represents a unique example of how the superposition principle directly leads to the quantum mechanical state of a physical system. A beam of silver atoms emitted by the source S crosses a magnetic field orthogonal to its direction (z-axis) and reaches a detector R, as shown in Fig. 2.5. If the magnetic field $\vec{B}$ is constant, the atoms reach the central point O of the detector without deflecting. If, instead, the field is inhomogeneous, its non-vanishing gradient couples to the magnetic moment of the silver atoms generating the force

$$F_z = -\frac{\partial}{\partial z}(-\vec{m} \cdot \vec{B}) = m\frac{\partial B_z}{\partial z}\cos\theta = m_z\frac{\partial B_z}{\partial z}, \tag{2.6}$$

which splits and deflects the beam into two components. Surprisingly, the atoms concentrate only at the points A and B of the detector (see Fig. 2.6) (right), instead of distributing over the whole area between A and B as expected from the dependence of the magnetic force on the deflection angle θ. The quantum interpretation is that the non-polarized atoms, entering the interaction area, are forced to polarize either parallel or anti-parallel to the magnetic field direction. In other words, the interaction with the magnetic field induces a transition of the beam from a non-polarized state to two and only two polarization states. By replicating the experiment with a source of partially polarized atoms, the final state does not change, i.e. either polarized parallel or anti-parallel to the magnetic field, but the number of atoms detected in A and B is different. In conclusion, the polarization state of the silver atoms is described by quantum mechanics as a superposition of two detection states. We shall turn back to the quantum states in the next chapter.

Before closing this section, we want to explain why the Stern–Gerlach experiment is evidence of electron spin. The silver atom, which is a paramagnetic atom (i.e. it has a permanent magnetic moment), has 47 electrons, 46 of which form a spherically symmetric cloud with no net angular momentum. The 47th unpaired electron, known as valence electron, is responsible for the total angular momentum of the whole atom.

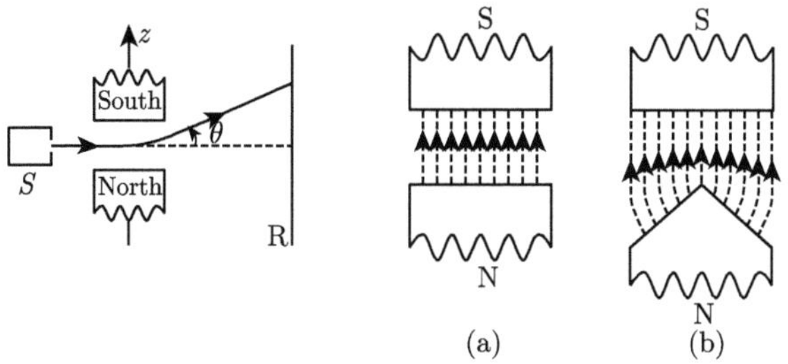

Fig. 2.5 Sketch of the setup for the Stern–Gerlach experiment (left), homogeneous magnetic field (left (**a**)) and inhomogeneous magnetic field (right (**b**))

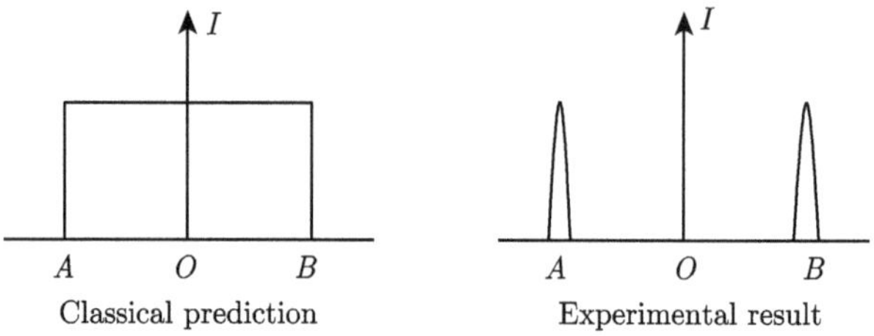

Fig. 2.6 Intensity distributions of the atoms on the detector. Left: classical outcome. Right: experimental result

Since it occupies an s-orbital ($L = 0$), the total orbital angular momentum is zero. But, in the Stern and Gerlach experiment the silver atoms interact with a magnetic field, so they must possess another type of angular momentum other than the orbital one. This is the spin already predicted by the Dirac relativistic electron theory [DIR].

2.5 Notes and References

DAV The Davisson and Germer's experiment (Nature **119** (1927)) of the electron diffraction from a lattice was one of the first tests of the wave-like behaviour of particles.

EIN Einstein, A.: Annalen der Physik, **17**, 132 (1905).

MIL The Einstein interpretation of the photoelectric effect with the hypothesis of photon (1905) was confirmed experimentally. R.A. Millikan (Phys. Rev. **4**, 73(1914)).

COM Compton, A.: Phys. Rev. **21**, 483 (1923).

ZEI Zeilinger A., et al.: Rev. Mod. Phys. **60**, 1067 (1998) .

FEY Feynman, R., Leighton, R.B., Sands, M.: Lectures on Physics. Volume III: Quantum Mechanics, The New Millennium edn. Addison-Wesley (1965).

HEI Heisenberg, W.: Zeitschrift fur Physik **43**, 172 (1927).

DIR Dirac, P.A.M.: Proc. Royal Soc. (London) A **117**, 610 (1928).

2.6 Problems

- Problem 1. Compton scattering: A photon with frequency ν and energy $E_\nu = h\nu$ collides with an electron at rest. The photon transfers a fraction of its energy to the electron and then scatters out with energy $E_{\nu'} = h\nu'$ ($\nu' < \nu$) and frequency ν' at the angle θ_ν with respect to the initial direction, and the electron scatters with energy E_e at the angle θ_e (see figure). Determine the wavelength shift as a function of the two angles (Hint: Apply the energy and momentum conservation from the initial to the final scattering states).

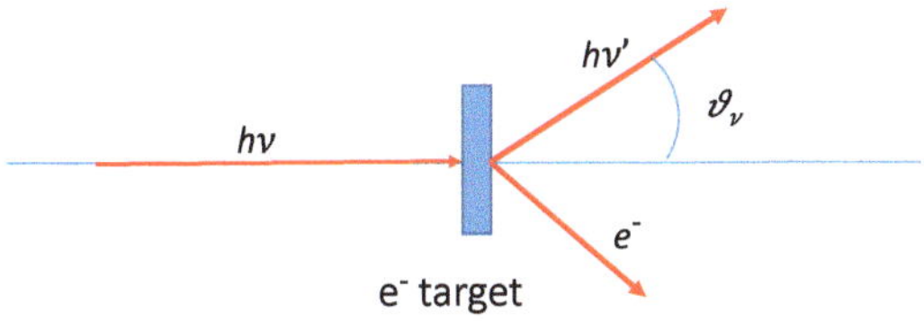

- Problem 2. Discuss the two extreme limits $h\nu \ll k_B T$ and $h\nu \gg k_B T$ of the Planck formula.
- Problem 3. In a Stern–Gerlach (SG) experiment, a non-polarized silver beam with intensity I_0 splits into two beams: one corresponding to parallel (up) spin to $z-$axis, one corresponding to anti-parallel (down) spin to the $z-$axis. Their intensities are $I(up) = I(down) = \frac{I_0}{2}$. Suppose one is suppressed, whereas the other one crosses a second SG whose magnetic field B has the same magnitude of the previous one and an orientation along the $z'-$axis, which is rotated of an angle θ with respect to the $z-$axis. How to describe the quantum state of the second beam and which are the outcomes from the second detector? [Hint: apply the superposition principle].
- Problem 4. Let us consider the double-slit experiment with photons. Suppose that, after crossing the double-slit device, one of the two photon beams gets the phase shift δ with respect to the other. Calculate the new interference pattern in the detector and discuss the physical interpretation of the interference for the two values $\delta = \pi$ and $\delta = \frac{\pi}{2}$.
- Problem 5. The interference from double-slit is observed with light wavelength $\lambda = 4000$ Angstrom. In the case of electrons calculate the energy needed to observe analogous interference. What is the temperature of the source to supply the electrons with that energy? (Hint: Use the Maxwell–Boltzmann energy distribution).
- Problem 6. A beam of spin 1/2 particles, polarized in the state $s_z = +1/2$, crosses a Stern–Gerlach apparatus with magnetic field oriented in the plane x-z. Determine the possible field directions, knowing that the beam splits into two components with intensity I and $3I$, respectively.

Chapter 3
States and Dynamical Variables

In the previous chapter, we resorted to the Huygens–Fresnel superposition principle to explain the discoveries of microscopic world and ended up with the quantum state as a superposition of detection states, such as the two polarization states in the Stern–Gerlach experiment. In this chapter, we shall show how the quantum–mechanical formalism of states and dynamical variables naturally arises from the analysis of measurements of observables made with experimental setups [SCW]. An experimental set-up, either double slit or Stern–Gerlach and many others, can be idealized as that shown in Fig. 3.1. This is made up of a source S, which prepares a beam of particles, a suitable target T, with which the particles interact, and an array of detectors R, which intercept the particles, coming out of the target, in all possible final states. The target induces a transition of the system from the initial to the final state. This transition is described by operators of the Hilbert space, which represent the dynamical variables measured in the experiment, e.g., the spin in Stern–Gerlach experiment. The states and observables are described by vectors and operators, respectively, defined in a multidimensional linear space, known as Hilbert space.[1] The operator character of the physical variables in quantum mechanics reflects the sharp limits imposed by nature on their simultaneous measurement. Indeed, dynamical variables expressed by non-commuting operators, cannot be measured simultaneously, at variance with the classical dynamical variables, where they are expressed by c-numbers.

[1] We are including the finite-dimensional spaces in the class of Hilbert spaces which, in fact, are the infinitely-dimensional spaces of square-summable functions.

Supplementary Information The online version contains supplementary material available at https://doi.org/10.1007/978-981-97-1670-8_3.

U. Lombardo et al., *Quantum Mechanics*, UNITEXT for Physics,
https://doi.org/10.1007/978-981-97-1670-8_3

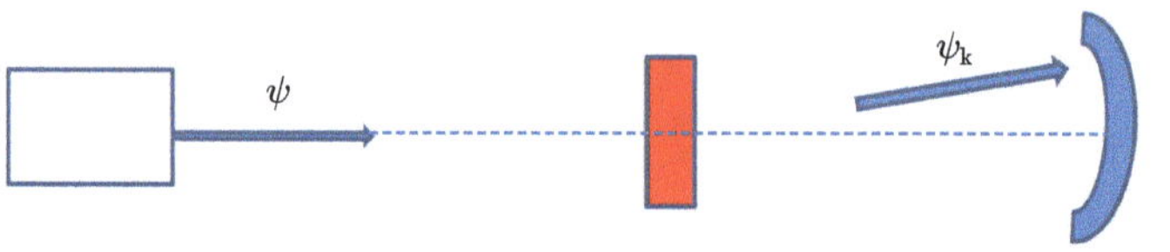

Fig. 3.1 A schematic experimental setup

3.1 States of a Quantum System as Vectors of Hilbert Space

Consider an experiment, e.g. that of Stern–Gerlach. The source of a given kind of particle is supposed to emit one particle at a time (weak source) so that the experiment made with N particles is equivalent to N experiments replicated with only one particle at a time under the same experimental conditions. The detectors' array is made of D detectors, to each of which corresponds a value O_k $(k = 1, D)$ of the observable O under consideration, e.g., the deflection angle in the Stern–Gerlach experiment. Each particle coming from the source, after interacting with the target, reaches one of the detectors. At the end of the experiment $\mathcal{N}_k$ of the N particles are counted in the kth detector. If N is quite large, as is normally the case, the frequency $\mathcal{N}_k/\mathcal{N}$ of the particles detected by each detector can be extrapolated with the probability

$$\mathcal{P}_k = \lim_{\mathcal{N}\to\infty} \frac{\mathcal{N}_k}{\mathcal{N}}, \tag{3.1}$$

namely the probability that a particle, prepared by the source in a certain state, is detected by the kth detector. There is not a one-to-one correspondence between initial and final states of the particle, as is the case in classical mechanics. Suppose we describe the detection states by vectors of a D-dimensional linear vector space $\mathcal{D}$, denoted with the Dirac ket $|\psi_k\rangle$ $(k = 1, \ldots, D)$.[2] Two requirements are dictated by the experiment:

- the two events, one particle in detector k and the other one in detector $k'(\neq k)$, are mutually exclusive, hence the two corresponding states are represented by orthogonal vectors, i.e., the scalar product $\langle\psi_{k'}|\psi_k\rangle = 0$;
- the number of detectors are assumed to exhaust all possible values of the observable O, hence the states $|\psi_k\rangle$, $(k = 1, D)$ form a complete set of vectors, which means that any state of $\mathcal{D}$ space can be expressed as a linear combination of this set. For convenience the norm of each state can be assumed to be equal to 1, i.e., $\langle\psi_k|\psi_k\rangle = 1$.

Turning back to our ideal experiment, the question is how to describe the state in which the source prepares the particles. We expect it to be related to the set

[2] Hereafter the Dirac notation [DRC] of vectors with ket and bra will be adopted (see Appendix A).

of probabilities $\mathcal{P}_k$, $(k = 1, D)$ which is the experimental output, Eq. (3.1). As we learned from the interference experiment of particles in Chap. 2, we have to assume that such $\mathcal{P}_k$ is the square-modulus of probability amplitude that a particle, prepared by the source in a given state $|\psi\rangle$, is detected in the state $|\psi_k\rangle$. Geometrically, this amplitude is the projection of the state vector $|\psi\rangle$ onto the state vector $|\psi_k\rangle$. Since the $|\psi_k\rangle$ vectors form a complete set of states in $\mathcal{D}$ space, the state of the source can be written as a linear combination of the set of detected states

$$|\psi\rangle = \sum_{k=1}^{D} |\psi_k\rangle A_k,$$

'in accordance with the superposition principle. The superposition coefficients $A_k = \langle\psi_k|\psi\rangle$ are the probability amplitudes of detecting the system in the $|\psi_k\rangle$ state and

$$\mathcal{P}_k = |A_k|^2 = |\langle\psi_k|\psi\rangle|^2, \tag{3.2}$$

are the corresponding probabilities. Since the sum of all probabilities must be equal to 1, the norm of the state of the source must be $\langle\psi|\psi\rangle = 1$. This requirement imposes a severe restriction on the space where the state vectors are defined, i.e., it must only represent vectors with finite norm, as previously mentioned. Moreover, the norm determines the state up to a phase factor $e^{i\phi}$, where ϕ is a real arbitrary number. In other words, two states which differ for a phase factor are physically identical. Formally we can write

$$\sum_k \mathcal{P}_k = \sum_k \langle\psi|\psi_k\rangle\langle\psi_k|\psi\rangle = 1, \tag{3.3}$$

$$\sum_k |\psi_k\rangle\langle\psi_k| = 1, \tag{3.4}$$

where $\mathbf{P}_k = |\psi_k\rangle\langle\psi_k|$ is an example of operator in the Dirac notation. In general, the operator $\mathbf{O}$ is defined as a correspondence between vectors of $\mathcal{D}$ space, meaning that it transforms one state into another, symbolically $\mathbf{O}|\psi\rangle = |\phi\rangle$. In the particular case of $\mathcal{P}_k$, the correspondence law between $|\psi\rangle$ and $|\psi_k\rangle$ is

$$\mathbf{P}_k|\psi\rangle = (|\psi_k\rangle\langle\psi_k|)|\psi\rangle = |\psi_k\rangle(\langle\psi_k|\psi\rangle).$$

In our introduction to the formalism of quantum mechanics, the projection operators have been assigned a special task, i.e. that of generating the operators representing physical observables, as is illustrated in the next section.

Equation (3.4), expressing the completeness of the set of detected states, is said to be a resolution of the identity operator **1**. This equation reflects the property of the D state vectors $|\psi_k\rangle$ to be the only states in which a particle can be detected. Equation (3.3) entails the condition $\langle\psi|\psi\rangle = 1$. It follows that the quantum states are represented by vectors of $\mathcal{D}$ space, but only the direction of these vectors has physical significance.[3] In addition, the correspondence between states and probabilities is up to a phase factor $e^{i\phi}$.

3.2 Observables as Operators in Hilbert Space

In our ideal experimental setup the particle, emitted by the source, interacts with a device, e.g. the magnetic field in the Stern–Gerlach experiment, sensitive to the observable O to be measured, and then it is detected in a state with a definite value O_n of the observable. After iterating N times the experiment, the value O_n is measured with probability $\mathcal{P}_n$ in each of the D detectors. Due to its probabilistic nature, the result of our measurement is analyzed in terms of the statistical tools, in particular the average value (or expectation value) and the root-mean-square

$$\langle O\rangle = \sum_n O_n \mathcal{P}_n, \tag{3.5}$$

$$\langle \Delta O^2\rangle = \sum_n (O_n - \langle O\rangle)^2 \mathcal{P}_n, \tag{3.6}$$

respectively. Using the new formalism, we can recast the two statistical quantities as follows

$$\langle O\rangle = \sum_n O_n \mathcal{P}_n = \sum_n O_n |A_n|^2 = \langle\psi|\sum_n O_n \mathbf{P}_n|\psi\rangle,$$

$$\begin{aligned}\langle(\Delta O)^2\rangle &= \sum_n (O_n - \langle O\rangle)^2 \mathcal{P}_n = \langle\psi|\left(\sum_n O_n \mathbf{P}_n - \langle O\rangle\right)^2|\psi\rangle \\ &= \langle O^2\rangle - \langle O\rangle^2,\end{aligned}$$

respectively. As a result, the operator describing the observable O is

$$\mathbf{O} = \sum_n O_n \mathbf{P}_n, \tag{3.7}$$

This expression, known as *spectral representation*, leads the definition of the observable back to a measurement operation of the set of its values. We conclude that quantum mechanics describes the physical observables O with operators defined in $\mathcal{D}$ space.

---

[3] The implications of this general condition are discussed in Ref. [GOT].

3.3 General Properties of Quantum Observables

Once an operator $\mathbf{O}$ has been associated with a physical variable O, the measurement results O_n and the corresponding vectors $|n\rangle$ acquire the meaning of eigenvalues and eigenvectors, respectively, of the operator $\mathbf{O}$. Indeed, applying $\mathbf{O}$, written in terms of its spectral representation, Eq. (3.7), to one of the states $|n\rangle$ $(n = 1, D)$ one obtains

$$\mathbf{O}|m\rangle = \sum_n O_n|n\rangle\langle n|m\rangle = \sum_n O_n\delta_{nm}|n\rangle = |m\rangle O_m,$$

and hence

$$\mathbf{O}|O_m\rangle = |O_m\rangle O_m,$$

which we recognize as the eigenvalue equation of the operator $\mathbf{O}$ (see Appendix A.4). Since the measurement results of the observable O are real numbers, the corresponding operator $\mathbf{O}$ must be Hermitian (or self-adjoint), $\mathbf{O} = \mathbf{O}^\dagger$. An additional requirement for the operator to represent a physical variable is that its eigenstates form a complete set of states in $\mathcal{D}$ space. Indeed this requirement ensures that the physical variable can be measured in any state of $\mathcal{D}$. In other words, if this were not the case, there would be states inaccessible to measures of $\mathbf{O}$. The operator character of the physical variables in quantum mechanics has physical implications which are in strong contrast with their classical counterparts. Let us indeed consider two operators $\mathbf{A}$ and $\mathbf{B}$, describing two observables A and B, respectively, and assume that they have a complete set of simultaneous eigenstates, which means that they are simultaneously measurable, then the two eigenvalue equations

$$\begin{aligned}\mathbf{A}|A_nB_m\rangle &= |A_nB_m\rangle A_n,\\ \mathbf{B}|A_nB_m\rangle &= |A_nB_m\rangle B_m\end{aligned}$$

imply that the two operators commute. Indeed, after multiplying the first equation times $\mathbf{B}$ and the second one times $\mathbf{A}$ and then, subtracting one from the other, we obtain

$$[\mathbf{B}, \mathbf{A}]|A_nB_m\rangle = |A_nB_m\rangle(B_mA_n - A_nB_m) = 0$$

i.e., the commutator of the two operators $[\mathbf{B}, \mathbf{A}] = \mathbf{BA} - \mathbf{AB}$ is equal to zero. Generally, the opposite is true, i.e., if two operators do not commute the corresponding variables cannot be simultaneously measured. We shall see further that basic variables such as position and momentum as well as the components of angular momentum do not commute.

So far, we have considered only states and operators defined in linear multidimensional spaces. But some observables have a set of infinite, enumerable or not, eigenstates, that requires the introduction of infinitely-dimensional spaces. These spaces are named *Hilbert spaces*. We shall see that there are no particular conceptual difficulties in extending the considerations made so far to Hilbert spaces.

3.4 Unitary Transformations

From Chap. 1 we learned that Hamilton's equations are form-invariant under the canonical transformations. A class of these transformations are the so-called symmetry transformations, i.e., for the transformations corresponding to symmetries of the system. The interest for the symmetry transformations is that their generating functions are dynamical variables. Such a scenario of classical mechanics can be transferred, *mutatis mutandis*, into quantum mechanics, as we will see in the following chapters.

In quantum mechanics the role played by the canonical transformations is replaced by the unitary operators. Indeed, suppose a physical symmetry, e.g. space–time symmetry, is not violated by the observable **O** submitted to measurement. In this case, its probability amplitude must be form-invariant under the given symmetry transformation. We define the probability amplitude of measuring the eigenvalue O_k $(k = 1, D)$ as $A_k = \langle \psi_k \mid \psi \rangle$, and then the probability amplitude after the transformation will be

$$A'_k = \langle \psi'_k \mid \psi' \rangle = \langle \psi_k | \mathbf{U}^\dagger \mathbf{U} | \psi \rangle, \tag{3.8}$$

where **U** is the operator carrying out the transformation. The form-invariance $A_k = A'_k$ requires

$$\begin{aligned} \langle \psi_k \mid \psi \rangle &= \langle \psi_k \, | \mathbf{U}^\dagger \mathbf{U} | \, \psi \rangle, \\ \langle \psi'_k \mid \psi' \rangle &= \langle \psi'_k \, | (\mathbf{U}^\dagger)^{-1} (\mathbf{U})^{-1} | \psi' \rangle. \end{aligned} \tag{3.9}$$

It follows that the operator **U** is unitary

$$\mathbf{U}^\dagger \mathbf{U} = \mathbf{U}\mathbf{U}^\dagger = \mathbf{1}. \tag{3.10}$$

The inverse operator $\mathbf{U}^{-1}$ stems from the existence of the inverse transformation. The various symmetry transformations will be discussed in the following chapters, starting from their infinitesimal expression. In general, the latter can be written as

$$\mathbf{U}(\delta\alpha) = \mathbf{1} + \frac{\mathrm{i}}{\hbar} \delta\alpha \, \mathbf{g}. \tag{3.11}$$

The imaginary unit i is introduced to make **g** a hermitian operator in view of its interpretation as a physical observable. The meaning of the smallness parameter α as well as the operator **g** changes according to the type of transformation under consideration.

3.5 Complement to Sect. 3.1: Entangled States

In Sect. 3.1, we have described the measurement process in Quantum Mechanics. Before the observable **O** of a quantum system is measured, its state is in a superposition of states. After measurement the system is detected in an eigenstate of **O**. We consider now a system, made of two interacting particles, e.g. in a spin state $S = 0$[4], that is

$$|\psi\rangle = \sqrt{\frac{1}{2}}(|\uparrow\rangle_1 |\downarrow\rangle_2 - |\uparrow\rangle_2 |\downarrow\rangle_1) \cdot \tag{3.12}$$

In any measurement, if we find one particle in the state with spin *up*, then the other one will be found in the state with spin *down* and vice versa. Such a state is called *entangled state*. This is a tricky phenomenon since it happens instantaneously regardless of how far apart the two particles are. Therefore, it apparently violates the local causality hypothesis [EPR], despite being an experimental outcome. The above mentioned case is not the kind of correlation we now are interested in, because this has a classical analogue. Instead, we are interested in pure quantum correlations, a special case of which is that of two non-interacting identical particles. Due to the fact that the two identical particles are not distinguishable, the configuration with a particle in the u state and another particle in the v state (often these are position states) is then

$$|\psi\rangle = \sqrt{\frac{1}{2}}(|u\rangle_1 |v\rangle_2 \pm |u\rangle_2 |v\rangle_1), \tag{3.13}$$

where sign $+$ is for bosons and $-$ for fermions (see Chap. 5.4). Here the correlation between the two particles is not due to the conservation of a global variable (e.g. $S = 0$) but due to the indistinguishable two particles, which is a purely quantum feature. In the language of quantum informatics, the two superposition states are named *qbits*. Common examples of qubits are particle states of spin up and spin down, two-level excitation states of atomic electrons, vertical and horizontal polarization states of photons. In all cases the two events, A and B, represent exclusive quantum states, that is <A|B> = 0.

The entanglement has proved to be not only a subject of investigation on the Quantum Mechanics foundations, but also a subject of research and development on information, computing and other applications. Below we present an application of entanglement, called quantum *teleportation*, which is a technique for transferring quantum information between two observers as far apart from each other as you like. This transfer, although instantaneous, does not violate relativity because it is not a transfer of matter or energy, but only quantum states, as explained below. Let us consider the experiment depicted in Fig. 3.2. The source S emits two correlated photons from the decay of a single photon in the superposition of two orthogonal spin polarization states: vertical polarization V and horizontal polarization (H).

[4] The spin observable will be introduced in Sect. 6.6

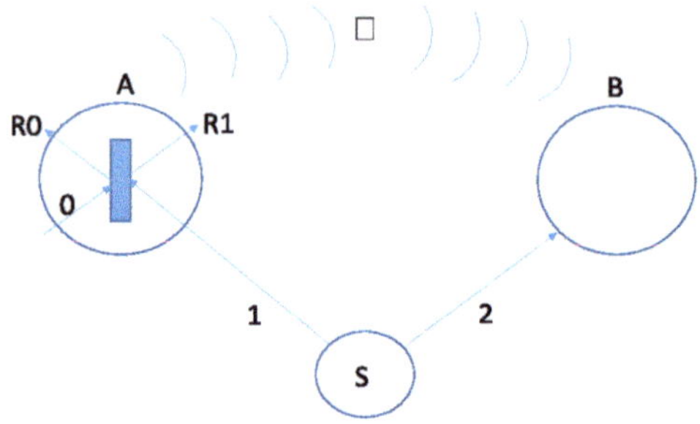

Fig. 3.2 The schematic experimental setup of teleportation (see [TELE])

$$|\psi\rangle_{12} = \sqrt{\frac{1}{2}}(|V\rangle_1|H\rangle_2 - |H\rangle_2|V\rangle_1)\cdot \tag{3.14}$$

The two polarization states have the same probability amplitude. The photon 1 reaches the observer A and the photon 2 reaches the observer B, the latter being located far away from A. A new photon 0 reaches the observer A in a state, which is the superposition of the two orthogonal polarization states

$$|\psi\rangle_0 = \alpha|V\rangle + \beta|H\rangle, \tag{3.15}$$

where $|\alpha|^2 + |\beta|^2 = 1$. This state is to be teleported to the observer B. To such a purpose the photon 0 and the photon 1 are sent together to a beam splitter shown in Fig. 3.2, which entangles the two particles in the state

$$|\psi\rangle_{01} = \sqrt{\frac{1}{2}}(|H\rangle_0|V\rangle_1 - |H\rangle_1|O\rangle_0) \tag{3.16}$$

The teleportation mechanism proceeds as follows: for any photon detected in the state $|H\rangle$ by the detector R0 a photon is detected in coincidence in the state $|V\rangle$ by the detector R1, due to the entanglement between 0 and 1 and simultaneously and instantaneously the photon 2 reaches the observer B in the state $|H\rangle$, being entangled to the photon 2 with orthogonal polarization. After a large number of events, a number of photons proportional to α reaches B in the state$|V\rangle$ and number proportional to β reaches B in the state $|H\rangle$. Since other entangled states could be detected, the observer A must inform B via a classical communication device (e.g. telephone) the state set by the beam splitter (see Ref. [TELE]). In conclusion, the initial state $|\psi\rangle_0$ has been teleported in B, no matter how B is far away from$A$.

3.6 Notes and References

DRC Dirac, P.A.M.: The Principles of Quantum Mechanics. Clarendon Press, Chap. 2.

SCW This kind of introduction to the quantum–mechanical formalism by reference to the Stern–Gerlach experiment was first presented by J. Schwinger in *Quantum Mechanics: Symbolism of Atomic Measurements*. Springer (2011).

GOT Gottfried, K., Yan, T.-M.: Quantum Mechanics: Fundamentals. Springer (2003). Chap. 7.

EPR Einstein, A., Podolskwy, B., Rosen, N.: Phys. Rev. **47**, 777 (1935).

TELE Bouwmeester, D., Pan, J.-W., Mattle, K., Weinfurter, H., Zeilinger, A.: Nature **390**, 575 (1997).

3.7 Problems

- Problem 1. Prove the following identity $[\mathbf{A}, [\mathbf{B}, \mathbf{C}]] + [\mathbf{C}, [\mathbf{A}, \mathbf{B}]] + [\mathbf{B}, [\mathbf{C}, \mathbf{A}]] = \mathbf{0}$
- Problem 2. Prove that the commutator between two observables is form-invariant under unitary transformations.
- Problem 3. Let $f(\mathbf{A}) = \mathbf{0}$ be an equation, whose left-hand side is a polynomial f function of the operator $\mathbf{A}$, i.e. $f(\mathbf{A}) = \sum_{n=1}^{N} c_n \mathbf{A}^n$. Show that the eigenvalues A_α of $\mathbf{A}$ are the roots of the equation $f(A) = 0$.
- Problem 4. The operator $\mathbf{P}$ is such that $\mathbf{P}|\psi\rangle = -|\psi\rangle$. Show that its eigenvalues are ± 1.
- Problem 5. Let $\mathbf{A}$ be a dynamical variable and $\bar{\mathbf{A}} = U^\dagger \mathbf{A} U$ the dynamical variable obtained by a unitary transformation $\mathbf{U}$. Determine the relation between eigenvalues and eigenvectors of the two dynamical variables.
- Problem 6. Let $[\phi_k, k = 1, D]$ be a complete set of orthonormal states of a D-dimensional space and $\mathbf{O}$ a dynamical variable in that space. Show that the eigenvalue problem for $\mathbf{O}$ can be recast in the form of diagonalization of the matrix with elements $O_{ij} = \langle \phi_i | \mathbf{O} | \phi_j \rangle$.

Chapter 4
Space Translations and Momentum

This is the first of three chapters devoted to introducing the basic dynamical variables, i.e. momentum, energy and angular momentum as generators of space–time symmetry transformations. Firstly, the position variable will be postulated as physical observable, as empirically proven, e.g., in the two-slit experiment. Then, the momentum will be introduced as a generator of the space translations and its properties discussed, including the eigenvalue equations. Finally, the Heisenberg uncertainty principle will be deduced, as a consequence of the non-commutativity between the position and momentum operators.

4.1 Wave Function and Position Operator

Let us suppose we measure the position of a quantum system, prepared in the state $|\psi\rangle$, repeating the measurement a considerable number of times. Let us call $\mathcal{P}_{\vec{r}}$ the probability of finding the system in the position state $|\vec{r}\,\rangle$. From the experimental viewpoint, it is legitimate to assume that the system can be localized in space and any position of space $\vec{r}$ is a possible measurement result (Localization Postulate). In other words, the position is a dynamic variable with a continuous spectrum of real values, likewise the corresponding classical variable. Hence, the superposition principle dictates that the state $|\psi\rangle$ is to be written as

$$|\psi\rangle = \int \mathrm{d}^3\vec{r}\,|\vec{r}\,\rangle\langle\vec{r}\,|\psi\rangle = \int \mathrm{d}^3\vec{r}\,\psi(\vec{r})\,|\vec{r}\,\rangle, \tag{4.1}$$

Supplementary Information The online version contains supplementary material available at https://doi.org/10.1007/978-981-97-1670-8_4.

U. Lombardo et al., *Quantum Mechanics*, UNITEXT for Physics,
https://doi.org/10.1007/978-981-97-1670-8_4

where $\psi(\vec{r}) = \langle \vec{r}\,|\psi\rangle$ is the amplitude associated with the probability $\mathcal{P}(\vec{r}) = |\langle \vec{r}\,|\psi\rangle|^2$. The calculation of the function $\psi(\vec{r})$, named wave function, is one of the main tasks of quantum mechanics.

As we learned in Chap. 2, the measurement results of the particle locating enable us to build up a Hermitian operator describing the variable position through its spectral representation and its eigenvalue equation

$$\vec{\mathbf{r}} = \int \mathrm{d}^3\vec{r}\,|\vec{r}\rangle\langle\vec{r}|\vec{r}, \tag{4.2}$$

$$\vec{\mathbf{r}}\,|\vec{r}\,'\rangle = |\vec{r}\,'\rangle\,\vec{r}\,', \tag{4.3}$$

respectively. The eigenvalues form a *continuous spectrum*, meaning that it extends on the continuum of real numbers. The position operator is a vector operator, whose components $\mathbf{x}, \mathbf{y}, \mathbf{z}$, being simultaneously measurable, commute, i.e.

$$[\mathbf{x}, \mathbf{y}] = [\mathbf{x}, \mathbf{z}] = [\mathbf{y}, \mathbf{z}] = 0. \tag{4.4}$$

As a consequence, the kets $|\vec{r}\,'\rangle = |x', y', z'\rangle$ are simultaneous eigenstates of $\mathbf{x}, \mathbf{y}, \mathbf{z}$, respectively, and fulfil the eigenvalue equations:

$$\begin{aligned}
\mathbf{x}|x', y', z'\rangle &= |x', y', z'\rangle x', \\
\mathbf{y}|x', y', z'\rangle &= |x', y', z'\rangle y', \\
\mathbf{z}|x', y', z'\rangle &= |x', y', z'\rangle z'.
\end{aligned}$$

The eigenstates of the position operator form a complete set and the eigenstates belonging to different eigenvalues are orthonormal, i.e. $\langle \vec{r}\,'|\vec{r}\rangle = 0$ if $\vec{r}\,' \neq \vec{r}$. We now examine the case $\vec{r}\,' = \vec{r}$. According to Eq. (4.1), the wave function $\psi(\vec{r})$ can be written as

$$\psi(\vec{r}) = \int \mathrm{d}^3\vec{r}\,'\,\langle\vec{r}\,|\vec{r}\,'\rangle\,\psi(\vec{r}\,') = \psi(\vec{r})\int \mathrm{d}^3\vec{r}\,'\,\langle\vec{r}\,|\vec{r}\,'\rangle.$$

The wave function $\psi(\vec{r})$ has been displaced outside the integral, because only its value at $\vec{r} = \vec{r}\,'$ contributes, owing to the orthogonality condition, so that the remaining integral must be equal to one. For this integral to be finite the integrand must be ∞ only at that point, where it is not vanishing. These properties define the Dirac δ-function

$$\delta(\vec{r} - \vec{r}\,') = \begin{cases} 0 & \text{if } \vec{r} \neq \vec{r}\,' \\ +\infty & \text{if } \vec{r} = \vec{r}\,'. \end{cases}$$

In fact, it is not a real function, but a functional[1]. With the introduction of the Dirac δ-function, we end up with

$$\int d^3\vec{r}\,\langle\vec{r}\,|\vec{r}\,'\rangle = \int_{-\infty}^{+\infty} d^3\vec{r}\,\delta(\vec{r}-\vec{r}\,') = 1. \tag{4.5}$$

This equation represents the extension of the orthonormality property to eigenstates of dynamical variables with continuous spectrum.

4.2 Space Translations

Let us imagine doing a rigid translation, which displaces the position of a point-like system (it could also be the center of mass of an extended system) from $\vec{r}$ to $\vec{r}+\vec{a}$, the displacement $\vec{a}$ being arbitrary. Thus, if the initial wave function of the system is $\psi(\vec{r})$, then the wave function of the translated system will be $\psi_{\vec{a}}(\vec{r}) = \psi(\vec{r}-\vec{a})$. The translations establish a one-to-one correspondence between any two states in space. In turn, this correspondence defines the translation operator $\mathbf{U}_{\vec{a}}$

$$\mathbf{U}_{\vec{a}}|\psi\rangle = |\psi_{\vec{a}}\rangle, \tag{4.6}$$

$$\mathbf{U}_{\vec{a}}|\vec{r}\,\rangle = |\vec{r}+\vec{a}\,\rangle. \tag{4.7}$$

Physically, the translations form a class of *equivalent observers*, meaning that, by virtue of the homogeneity of space, all observers belonging to this class report the same results from the same experiment. For quantum systems these results are given by the transition probabilities, so that

$$|\langle\psi_{\vec{a}}|\phi_{\vec{a}}\rangle|^2 = |\langle\psi|\phi\rangle|^2. \tag{4.8}$$

This condition is fulfilled if $\mathbf{U}_{\vec{a}}^{\dagger}\mathbf{U}_{\vec{a}} = \mathbf{1}$.[2] Moreover, we can also define the inverse operator $\mathbf{U}_{\vec{a}}^{-1} = \mathbf{U}_{-\vec{a}}$, corresponding to the inverse translation. As a consequence, $\mathbf{U}_{\vec{a}}$ is a unitary operator

$$\mathbf{U}_{\vec{a}}^{\dagger}\,\mathbf{U}_{\vec{a}} = \mathbf{U}_{\vec{a}}\,\mathbf{U}_{\vec{a}}^{\dagger}. \tag{4.9}$$

From the unitarity of translation operator we can deduce the transformation law of the position operator. Indeed, an infinitesimal translation shifts the position eigen-

[1] The properties of Dirac δ-function and its possible realizations in terms of real functions are discussed in Appendix A.

[2] In fact, the equality of the amplitudes is requested [WIG].

value from $\vec{r}$ to $\vec{r}+\delta\vec{r}$, which is also an eigenvalue position, belonging to the eigenvector $\mathbf{U}_{\delta r}|\vec{r}\,\rangle$, hence

$$\vec{\mathbf{r}}\,\mathbf{U}_{\delta r}|\vec{r}\rangle = (\vec{r}+\delta\vec{r})\mathbf{U}_{\delta r}|\vec{r}\rangle.$$

By multiplying on the left side by $\mathbf{U}^{\dagger}$, we get

$$\mathbf{U}^{\dagger}{}_{\delta\vec{r}}\,\vec{\mathbf{r}}\,\mathbf{U}_{\delta\vec{r}} = \vec{\mathbf{r}} + \delta\vec{r}. \tag{4.10}$$

From this equation we learn how the position operator transforms under space translations.

An important property of the translation operators stems from the fact that the result of two successive translations is independent of the order in which they are carried out. The consequence for the corresponding operators is that they commute

$$\mathbf{U}_{\vec{a}}\mathbf{U}_{\vec{a}'} \;=\; \mathbf{U}_{\vec{a}'}\mathbf{U}_{\vec{a}}. \tag{4.11}$$

Thus, the class of translation operators form an Abelian group. As we will see in the next section, this property leads to the commutation rules of momentum, which, in turn, lead to the simultaneous measurability of all three components. Conversely, we anticipate that other transformations, e.g. the rotations, do not form an Abelian group, that entails a limitation to measurability of angular momentum.

4.3 Momentum as a Generator of Infinitesimal Translations

In analogy with what we did in Chap. 1, we can limit ourselves to infinitesimal translations with step $\delta\vec{a}_j = \delta\vec{r}$ $(\delta r^2 \ll \delta r)$. Hence the translation unitary operator can be written as

$$\mathbf{U}_{\delta\vec{r}} = \mathbf{1} - \mathrm{i}\vec{\mathbf{g}}\cdot\delta\vec{r}, \tag{4.12}$$

which has to be considered as the first-order expansion in $\delta\vec{r}$. The imaginary unit i has been introduced to make $\vec{\mathbf{g}}$ a Hermitian operator, as it can be easily deduced from the unitarity condition.

According to Eq. (4.7), we write

$$\mathbf{U}_{\delta\vec{r}}\,|\vec{r}\,\rangle = |\vec{r}+\delta\vec{r}\,\rangle, \tag{4.13}$$

$$\mathbf{U}_{\delta\vec{r}} = \mathbf{1} - \mathrm{i}(\mathbf{g}_x\delta x + \mathbf{g}_y\delta y + \mathbf{g}_z\delta z). \tag{4.14}$$

The second equation stems from the fact that the translation $\delta\vec{r}$ is the combination of three independent translations, e.g. δx, δy, δz. By exploiting the correspondence between the infinitesimal translation unitary operators in quantum mechanics and the infinitesimal canonical translations in classical mechanics, we identify the generator $\vec{\mathbf{g}}$ of the former as the quantum dynamical variable corresponding to the classical momentum. We observe that $\mathbf{U}_{\delta\vec{r}}$ is dimensionless, while the scalar product $\vec{\mathbf{g}} \cdot \delta\vec{r}$ has the dimension of an action. It follows that $\vec{\mathbf{g}}$ is to have the dimension of a momentum divided by a universal constant with the dimension of an action. The de Broglie relation $\frac{2\pi}{\lambda} = \frac{p}{\hbar}$, where p is the momentum of a particle and λ the associated wavelength, suggests that this constant must be identified with the Planck constant h, or $\hbar = h/2\pi$ merely for convenience. Thus, Eq. (4.7) assumes the ultimate form

$$\mathbf{U}_{\delta\vec{r}} = \mathbf{1} - \frac{\mathrm{i}}{\hbar}\vec{\mathbf{p}} \cdot \mathbf{d}\vec{\mathbf{r}}. \tag{4.15}$$

Momentum, as an operator, establishes a correspondence between space state vectors. This correspondence can be determined by comparing the translated wave function

$$\psi(\vec{r} - \delta\vec{r}) = \psi(\vec{r}) - \frac{\mathrm{i}}{\hbar}\delta\vec{r} \cdot \langle\vec{r}\,|\vec{\mathbf{p}}\,|\psi\rangle, \tag{4.16}$$

with that calculated from Taylor series expansion

$$\psi(\vec{r} - \delta\vec{r}) = \psi(\vec{r}) - \delta\vec{r} \cdot \nabla\psi. \tag{4.17}$$

From the comparison that correspondence turns out to be

$$\langle\vec{r}|\vec{\mathbf{p}}|\psi\rangle = \frac{\hbar}{\mathrm{i}}\nabla\psi(\vec{r}) \qquad \left(\text{formally } \;\vec{\mathbf{p}} \to \frac{\hbar}{\mathrm{i}}\nabla\right). \tag{4.18}$$

As previously mentioned, the space translations form an Abelian group, which implies that any two infinitesimal translation operators commute in the same way as the corresponding generators. Hence, the three components of the generator $\vec{\mathbf{p}}$ of the infinitesimal translation $\delta\vec{r}$ commute

$$[\mathbf{p}_x, \mathbf{p}_y] = [\mathbf{p}_y, \mathbf{p}_z] = [\mathbf{p}_z, \mathbf{p}_x] = 0, \tag{4.19}$$

and so they are simultaneously measurable. Another consequence of the Abelian character of the translations is that the operator of a finite translation $\vec{a}$ of a system can be easily obtained. The translation of $\vec{a}$ can be thought of as a succession of infinitesimal translations $\delta\vec{a}_j = \frac{\vec{a}}{N}$ with $j = 1, \ldots, N$, thus

$$\mathbf{U}_{\delta\vec{a}} = \mathbf{1} - \frac{\mathrm{i}}{\hbar}\frac{\vec{\mathbf{p}} \cdot \vec{a}}{N}.$$

In the limit $N \to \infty$, we obtain

$$\mathbf{U}_{\vec{a}} = \lim_{N\to\infty}\left(\mathbf{1} - \frac{\mathrm{i}}{\hbar}\frac{\vec{\mathbf{p}}\cdot\vec{a}}{N}\right)^N = \mathrm{e}^{-\frac{\mathrm{i}}{\hbar}\vec{a}\cdot\vec{\mathbf{p}}}. \tag{4.20}$$

4.4 Free Particle in a Box

The spectrum of possible values of momentum is obtained by solving the eigenvalue equation

$$\vec{\mathbf{p}}|\psi_{\vec{p}'}\rangle = |\psi_{\vec{p}'}\rangle\vec{p}\,'. \tag{4.21}$$

Actually, this is a compact way of writing the equations for the three components of the vector operator $\vec{\mathbf{p}} \equiv (p_x, p_y, p_z)$. As expected from the Hermitian character of this operator, its eigenvalues are real numbers and the eigenvectors belonging to different eigenvalues are orthogonal. Projecting into the state vectors $|\vec{r}\rangle$[3] and using Eq. (4.18) the eigenvalue equation transforms into a first-order homogeneous differential equation for the function $\psi_{\vec{p}'}(\vec{r}) = \langle\vec{r}|\psi_{\vec{p}'}\rangle$

$$(-\mathrm{i}\hbar)\vec{\nabla}\psi_{\vec{p}'}(\vec{r}\,) = \vec{p}\,'\psi_{\vec{p}'}(\vec{r}). \tag{4.22}$$

For any real value of momentum p' the solution is a plane wave

$$\psi_{\vec{p}'}(\vec{r}) = \frac{1}{(2\pi\hbar)^{3/2}}\mathrm{e}^{\frac{\mathrm{i}}{\hbar}\vec{p}\,'\cdot\vec{r}}, \tag{4.23}$$

where the constant normalizes the wave function to the Dirac δ-function that can be written in the form (see Appendix A.6)

$$\langle\vec{p}\,'|\vec{p}\,''\rangle = \left(\frac{1}{2\pi\hbar}\right)^3\int \mathrm{d}^3\vec{r}\,\mathrm{e}^{\frac{\mathrm{i}}{\hbar}(\vec{p}\,'-\vec{p}\,'')\cdot\vec{r}}. \tag{4.24}$$

Summarizing, the eigenfunctions of momentum are plane waves and the eigenvalue spectrum is a continuous spectrum. As plane waves, they describe free particles with fixed values of momentum in much the same way as monochromatic electromagnetic waves with fixed wavelength λ. It is worthwhile pointing out that the plane wave with fixed momentum $\vec{p}\,'$ describes a fully delocalized particle, as shown by the density probability $|\psi_{\vec{p}'}(\vec{r})|^2$ = const in any position, namely it does not contain any information on the position of the particle. We shall come back to this point in

[3] The wave function $\psi_{\vec{p}}(\vec{r})$ is a momentum state analyzed on the basis of position states, named *representation*, like a vector in real space which is analyzed by means of its coordinates in a reference frame. But other representations are possible provided that their states form a complete set of states.

the last section. However, the plane wave describes an ideal state. As a matter of fact, a real particle is always in a superposition of momentum eigenstates, known as wave packet

$$\psi(\vec{r}\,) = \int \mathrm{d}^3\vec{p}\, c_{\vec{p}}\, \mathrm{e}^{\frac{\mathrm{i}}{\hbar}\vec{r}\cdot\vec{p}}, \tag{4.25}$$

which, in fact, localizes the particle in a region of space, whose size depends on the weight function $c_{\vec{p}}$. In the extreme situation that the weight-function is independent of momentum, namely $c_{\vec{p}} = c_0$, the wave function becomes

$$\psi(\vec{r}) = c_0 \int \mathrm{d}^3\vec{p}\, \mathrm{e}^{\frac{\mathrm{i}}{\hbar}\vec{r}\cdot\vec{p}} = c_0\,\delta(\vec{r}), \tag{4.26}$$

meaning that the particle is located at $\vec{r} = 0$, but its momentum is completely indeterminate.

An important observable, function of momentum, is the kinetic energy $\boldsymbol{K} = \dfrac{\boldsymbol{p}^2}{2m}$, which is the energy of a free particle with mass m. As $\boldsymbol{K}$ is a function of momentum $\boldsymbol{p}$, its eigenstates are also eigenstates of momentum with eigenvalues equal to $\epsilon_p = p^2/2m$. In the coordinates, its eigenvalue equation is

$$-\frac{\hbar^2}{2m}\nabla^2\psi_{\vec{p}\,'}(\vec{r}) = \frac{p'^{\,2}}{2m}\psi_{\vec{p}\,'}(\vec{r}). \tag{4.27}$$

Instead of studying this problem for a particle in free space, it is more interesting by far to study the case in which it is confined in a finite region of space. This situation, indeed, frequently occurs as, e.g., in the case of diluted quantum fluids in a finite volume. In such cases Eq. (4.27) has to be solved with boundary conditions. Let us start with one-dimensional case of a particle confined on the segment $|q| < L$,

$$-\,\hbar^2\nabla^2\psi(q) = p^2\psi(q), \tag{4.28}$$

This is a homogeneous second-order differential equation with general solution

$$\psi(q) = u\,\cos(kq) + v\,\sin(kq), \tag{4.29}$$

where $k = \dfrac{p}{\hbar}$ and u and v are integration constants. From the boundary conditions, i.e. $\psi(\pm L) = 0$, two equations are obtained.

$$\begin{aligned} u\,\cos(kL) + v\,\sin(kL) &= 0, \\ u\,\cos(kL) - v\,\sin(kL) &= 0, \end{aligned}$$

which fix the integration constants. The full solution turns out to be

$$\psi_n(q) = \text{const} \times \cos(k_n q), \qquad k_n = \frac{(2n+1)\pi}{2L}, \qquad n = 1, 3, \ldots \text{(odd)} \tag{4.30}$$

$$\psi_n(q) = \text{const} \times \sin(k_n q), \qquad k_n = \frac{n\pi}{L}, \qquad n = 2, 4, \ldots \text{(even)} \tag{4.31}$$

The associated eigenvalues are

$$E_n = \frac{\hbar^2 k_n^2}{2m}. \tag{4.32}$$

The remarkable result is that, for the first time, we meet the quantization of momentum and energy. The extension from one-dimensional to three-dimensional motion in the volume $V = L^3$ is trivially the combination of three independent motions, hence energy eigenfunctions and eigenvalues are written

$$\psi_n(\vec{r}) = \psi_{n_x}(x)\,\psi_{n_y}(y)\psi_{n_z}(z), \tag{4.33}$$

$$E_n = \frac{\hbar^2}{2m}(k_{n_x}^2 + k_{n_y}^2 + k_{n_z}^2), \tag{4.34}$$

respectively, where $\vec{n} \equiv (n_x, n_y, n_z)$ components fulfill the quantization conditions in Eqs. (4.31) and (4.32) and $n^2 = n_x^2 + n_y^2 + n_z^2$. Any particle wave function $\psi(\vec{r})$, inside the volume V, can be expressed as superposition of the eigenvectors $|\vec{p}_n\rangle \equiv |p_{n_x}, p_{n_y}, p_{n_z}\rangle$

$$\psi(\vec{r}) = \sum_n \langle \vec{r}|\vec{p}_n\rangle \psi(\vec{p}_n). \tag{4.35}$$

In several physical systems V is very large so that for any continuous function $f(\vec{p})$, one can approximate

$$\sum_n f(\vec{p}_n) = \frac{V}{(2\pi\hbar)^3} \sum_n \Delta p_x \Delta p_y \Delta p_z f(\vec{p}) \approx \frac{V}{(2\pi\hbar)^3} \int \mathrm{d}^3\vec{p}\, f(\vec{p}). \tag{4.36}$$

This formula is exact for $V \to \infty$, in which case the wave function $\psi(\vec{r})$ of Eq. (4.35) becomes the Fourier transform of $\psi(\vec{p})$.

The quantization is a consequence of the particle confinement, typical of a bound state. This is one of the main features of quantum mechanics. The behaviour of particles reminds us of classical oscillations in elastic media (e.g. string, gas, lattice) with wavelength $\lambda_n = 2\pi/k_n$. Extrapolating to infinite volume $L \to \infty$ the discrete spectrum goes over into continuum, either for particles or waves, so that $p_n = \hbar k_n \to p$, $\lambda_n \to \lambda$, and we recover the de Broglie relation

$$p = \frac{h}{\lambda}, \tag{4.37}$$

already introduced in Chap. 1 discussing on the wave–particle dualism.

4.5 Heisenberg Uncertainty Relations

In classical mechanics position and momentum define the state of a particle. In quantum mechanics a particle with definite momentum is fully delocalized in space, as it has been shown in Sect. 4.3. Conversely, a particle localized in space has a fully indeterminate momentum. In intermediate situations, it happens that a partial delocalization of the particle is accompanied by a partial indetermination of its momentum. The relation between the two uncertainties is the well-known Heisenberg uncertainty principle. In this section we shall show that this principle stems from the non-commutativity between the position and momentum operators, and that a lower limit to simultaneous measurements of the two observables stems from their commutation rules. In accordance with Eq. (4.10) the position transforms under translations as

$$\mathbf{U}^{\dagger}{}_{\delta\vec{r}}\,\vec{\mathbf{r}}\,\mathbf{U}_{\delta\vec{r}} = \vec{\mathbf{r}} + \mathbf{1}\delta\vec{r}. \tag{4.38}$$

From the explicit expression of the infinitesimal translation operator we also have

$$\mathbf{U}^{\dagger}{}_{\delta\vec{r}}\,\vec{\mathbf{r}}\,\mathbf{U}_{\delta\vec{r}} = \vec{\mathbf{r}} - \frac{\mathrm{i}}{\hbar}[\vec{\mathbf{r}}, \delta\vec{r}\cdot\vec{\mathbf{p}}\,].$$

Comparing the right hand side of the two equations, Eq. (4.38) and Eq. (4.39), we get the commutation rules

$$[\vec{\mathbf{r}}, \vec{\mathbf{p}}\cdot\delta\vec{r}] = \mathrm{i}\hbar\mathbf{1}\delta\vec{r}. \tag{4.39}$$

After getting rid of the infinitesimal displacement $\delta\vec{r}$, we end up with the following commutation rules:

$$[\mathbf{x}, \mathbf{p}_x] = [\mathbf{y}, \mathbf{p}_y] = [\mathbf{z}, \mathbf{p}_z] = \mathrm{i}\hbar, \tag{4.40}$$

and

$$[\mathbf{x}, \mathbf{p}_y] = [\mathbf{x}, \mathbf{p}_z] = [\mathbf{y}, \mathbf{p}_x] = \cdots = 0. \tag{4.41}$$

As a result, the components of position and momentum in the same direction are not compatible observables. But, it does not mean that the components of position and momentum of a particle can not be measured simultaneously. The Heisenberg principle tells us that there is a degree of uncertainty in such measurements. Going further, we prove that Heisenberg principle is a consequence of the commutation rules of position and momentum.

Let us consider the one-dimensional case of a particle occupying the state $|\psi\rangle$ normalized to one. We denote with q_0 and p_0 its average values of position and momentum, respectively, and define the following auxiliary position and momentum states:

$$|\psi_q\rangle = (\mathbf{q} - q_0)|\psi\rangle,$$
$$|\psi_p\rangle = (\mathbf{p} - p_0)|\psi\rangle.$$

From the commutation rules between position and momentum, we find

$$2\mathrm{i}\mathrm{Im}(\langle\psi_q|\psi_p\rangle) = \langle\psi_q|\psi_p\rangle - \langle\psi_p|\psi_q\rangle = \langle\psi|[\mathbf{q} - q_0, \mathbf{p} - p_0]|\psi\rangle = \mathrm{i}\hbar,$$

and then

$$\mathrm{Im}(\langle\psi_q|\psi_p\rangle) = \left(\frac{\hbar}{2}\right)^2.$$

Applying the Schwartz inequality, we get

$$\mathrm{Im}(\langle\psi_q|\psi_p\rangle) = \left(\frac{\hbar}{2}\right)^2 \leq |\langle\psi_q|\psi_p\rangle|,$$

and then

$$|\psi_q| \cdot |\psi_q| \;\geq\; \langle\psi_q|\psi_p\rangle \;\geq\; \mathrm{Im}(\langle\psi_q|\psi_p\rangle) = \left(\frac{\hbar}{2}\right)^2. \tag{4.42}$$

It is easy to show that the modulus $|\psi_q|$ and the modulus $|\psi_p|$ are the standard deviations of position and momentum

$$\langle\psi_q|\psi_q\rangle = \int \mathrm{d}q'(q' - q_0)^2|\langle q'|\psi\rangle|^2 \equiv (\Delta q)^2,$$
$$\langle\psi_p|\psi_p\rangle = \int \mathrm{d}p'(p' - p_0)^2|\langle p'|\psi\rangle|^2 \equiv (\Delta p)^2,$$

respectively. By plugging these definitions into Eq. (4.42), we recover the Heisenberg principle

$$\Delta p \Delta q \geq \frac{\hbar}{2}, \tag{4.43}$$

which establishes the limit imposed by nature on the simultaneous measurement of position and momentum. There are two extreme cases already mentioned before: the exact determination of the position of the system, $\Delta q = 0$, leads to an absolute indetermination of momentum, i.e. $\Delta p = \infty$; vice versa, the exact determination of its momentum, $\Delta p = 0$, leads to an absolute indetermination of position, i.e. $\Delta q = \infty$. The latter case is met with a plane wave, describing the idealized situation of a system absolutely delocalized in space, its momentum being completely determined. Finally, we observe that any uncertainty disappears at the classical limit $\hbar \to 0$.

An interesting application of the Heisenberg principle concerns the stability of atoms. The Rutherford model depicts the hydrogen atom as a negative charge, i.e. the electron, interacting by mutual Coulomb attraction, with the nucleus, made of a single proton with positive charge. According to the classical electromagnetism, the electron, while orbiting around the proton, gradually slows down emitting electromagnetic waves. As a result, orbiting closer and closer, it eventually falls down on the proton with zero energy. The Heisenberg principle prevents such an occurrence. Indeed, the electron inside the nucleus would have a well-determined position as well as momentum, the same as the proton, i.e. at rest in the center of the nucleus, that means $\Delta q = \Delta p = 0$. What really happens is that the electron occupies an orbit outside the nucleus in which its total energy, kinetic plus potential, has the minimum non-vanishing value compatible with the uncertainty relation. Thus, assuming the electron to be in a state corresponding to the minimum indetermination $\Delta q \Delta p = \frac{\hbar}{2}$, one may estimate the radius of the hydrogen atom. The electron moves with a kinetic energy $\frac{\hat{p}^2}{2m}$ in the Coulomb field of the nucleus. Since the Coulomb potential has a spherical symmetry, we have to look for the minimum energy state with spherical symmetry. Moreover, in order to prevent the electron from leaving the atom, the average value of its momentum has to be zero. Under these conditions, we calculate the average value of the total energy of the electron

$$\langle H \rangle = \langle \psi | \frac{\boldsymbol{p}^2}{2m} | \psi \rangle - \langle \psi | \frac{e^2}{\boldsymbol{r}} | \psi \rangle, \tag{4.44}$$

where $-\frac{e^2}{\boldsymbol{r}}$ is the electron–proton Coulomb potential energy. Making use of the approximation $\langle 1/\boldsymbol{r} \rangle \approx \frac{1}{r_0}$, where r_0 is the radius of the electron orbit, and $\langle r^2 \rangle \approx {r_0}^2$ from Eq. (4.43) we get $\langle p^2 \rangle \approx \frac{\hbar^2}{r_0^2}$[4]

$$\langle H \rangle = \frac{\hbar^2}{2mr_0^2} - \frac{e^2}{r_0}. \tag{4.45}$$

The minimization of the average energy $\langle H \rangle$ yields

$$r_0 = \frac{\hbar^2}{me^2}. \tag{4.46}$$

Regardless of the approximations made, it is, however, surprising to have found that r_0 coincides with the Bohr radius of the hydrogen atom (see Sect. 12.2) Anyhow, the most significant result is that quantum mechanics predicts the stability of the atom.

[4] In this crude application of Eq. (4.43), we consider the minimum uncertainty and innore the factor 2 in the denominator.

4.6 Notes and References

WIG The condition Eq. (4.8) is equivalent to the equality of the amplitudes, i.e. $\langle \psi_{\vec{a}} | \phi_{\vec{a}} \rangle = \langle \psi | \phi \rangle$ apart from a phase factor. As proven by the Wigner theorem (Group Theory, Academic Press, 1959, Chap. 26), we can neglect the phase factor in most cases. An extended discussion of this point is in Chap. 7 of Ref. [GOT].

4.7 Problems

- Problem 1. The wave function of a particle is $\psi(x) = \int \mathrm{d}p\, \mathrm{e}^{\mathrm{i}px/\hbar} c_p$, where c_p is a constant in the momentum range $|p - p_0| \leq \Delta p$, and zero elsewhere. Determine the probability distribution of finding the particle at any position x.
- Problem 2. A free particle is confined on a segment $|q| \leq L$. Determine the standard deviation of its momentum and comment on the result in connection with the Heisenberg uncertainty relation.
- Problem 3. Prove that the state with minimum uncertainty $\Delta p \Delta q = \frac{\hbar}{2}$ is the ground state of harmonic oscillator.
- Problem 4. Consider two particles with the same mass m in the center-of-mass reference. Show that the relative position and the total momentum as well as the center-of-mass position and the relative momentum are simultaneously measurable observables.
- Problem 5. A particle is prepared in the state with wave function

$$\psi(x) = N \int_{-\infty}^{\infty} \mathrm{d}p\, \mathrm{e}^{-\frac{|p|}{p_0}} \mathrm{e}^{\mathrm{i}\frac{px}{\hbar}}.$$

  After traveling freely in space, it is detected. Calculate the probability of being detected in the state of momentum ranging in the interval $(-p', +p')$. (p_0 and N are constants).
- Problem 6. Consider a particle in a one-dimensional box with side L. The energy spectrum is

$$E_n = \frac{\hbar^2}{2m} \frac{\pi^2 n^2}{L^2}$$

  Compute the level density $N = \left(\frac{dE_n}{dn}\right)^{-1}$ and compare with the one-dimensional harmonic oscillator. Discuss its behaviour in the limit $\hbar \to 0$ and for small as well as large L.

Chapter 5
Elementary Phenomena

After having mastered the formalism of quantum mechanics, we address the application of the principle of superposition to simple phenomena, that involves either light or particles. Firstly, we shall consider the interference and diffraction of X-rays or particles from periodic structures such as crystal lattices, which are *natural laboratories* for such phenomena. Indeed, periodic structures with D spacing of a few Å are excellent substitutes for the lab-made slits. To observe these phenomena with particles, their wavelength must be comparable with that of light, e.g. X-rays of $\lambda \sim 10\,\text{Å}$. To convert λ to the beam kinetic energy, we apply the de Broglie relation $p = h/\lambda$,

$$K = \frac{p^2}{2m} = \frac{h^2}{2m\lambda^2}. \tag{5.1}$$

We find $K \approx 1\,\text{eV}$ for electrons, and $\approx 10^{-3}\,\text{eV}$ for neutrons, the ratio of neutron to electron mass being $\approx 2 \times 10^3$.

Second, we shall consider the collision between two identical particles. Since, in quantum mechanics, we cannot associate a trajectory to each of the two particles, we must assume that the two particles are indistinguishable. As a consequence, the state of each particle will be a superposition of the two possible classical trajectories with an interference effect similar to that of the two-slits experiment.

5.1 Double-Slit Interference

We re-examine the Young's experiment of the two slits (see Sect. 2.1) with photons, electrons, neutrons, or whatever kind of particles, making use of the formalism of quantum mechanics. Each one of the emitted particles travels with a constant

Supplementary Information The online version contains supplementary material available at https://doi.org/10.1007/978-981-97-1670-8_5.

U. Lombardo et al., *Quantum Mechanics*, UNITEXT for Physics,
https://doi.org/10.1007/978-981-97-1670-8_5

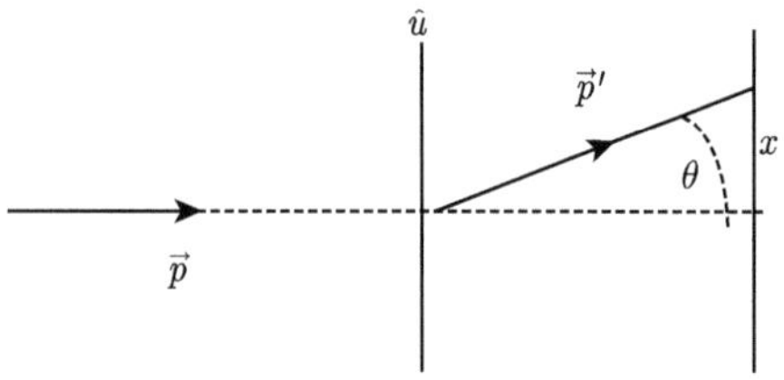

Fig. 5.1 Simplified plot of Young's experiment

momentum $\vec{p}$; hence, it is in a plane wave state $|\vec{p}\,\rangle$. Such a particle, when crossing one of the two slits, either F_1 or F_2, changes direction (see Fig. 2.1). At asymptotic distances, the final state can still be described as a plane wave state $|\vec{p}\,'\rangle$, in the θ direction, as shown in the schematic Fig. 5.1.

We call **A** the transition operator which changes the momentum state of the incident particle when it crosses the slit F_1.[1] Then the probability amplitude for the particle to reach the position x on the detector is given by

$$A_1(x) = \langle \vec{p}\,'|\mathbf{A}|\vec{p}\,\rangle. \tag{5.2}$$

The two slits are identical except that F_2 is translated by a vector $\vec{D}$ from F_1, $\vec{D}$ being a vector of modulus D and direction $\hat{u}$ orthogonal to the incident beam. Hence, the corresponding transition operator differs from **A** by a rigid translation, namely $\mathbf{A}_D = \mathbf{U}_D^\dagger \mathbf{A} \mathbf{U}_D$, where $\mathbf{U}_D = \mathbf{e}^{\frac{i}{\hbar}\vec{\mathbf{D}}\cdot\vec{\mathbf{p}}}$ is the unitary translation operator. Then, the probability amplitude for particle to reach the position x on the detector is given by

$$A_2(x) = \langle \vec{p}\,'|\mathbf{U}_D^\dagger \mathbf{A} \mathbf{U}_D|\vec{p}\,\rangle = \langle \vec{p}\,'|\mathbf{A}|\vec{p}\,\rangle e^{\frac{i}{\hbar}p' D \sin\theta}, \tag{5.3}$$

where the eigenvalue equation has been applied to the translation operator

$$\mathbf{U}_D|\vec{p}\,\rangle = |\vec{p}\,\rangle \tag{5.4}$$

$$\mathbf{U}_D|\vec{p}\,'\rangle = |\vec{p}\,'\rangle e^{-\frac{i}{\hbar}p' D \sin\theta}. \tag{5.5}$$

The superposition principle dictates that, when both slits are open, the probability of detecting the outgoing particle at the position x is

$$\begin{aligned} P(x) &= |\langle \vec{p}\,'|\mathbf{A}|\vec{p}\,\rangle + e^{\frac{i}{\hbar}pD\sin\theta}\langle \vec{p}\,'|\mathbf{A}|\vec{p}\,\rangle|^2 \\ &= P_0(x)|1 + e^{\frac{i}{\hbar}pD\sin\theta}|^2 \\ &= 4P_0(x)\cos^2\left(\frac{pD}{2\hbar}\sin\theta\right), \end{aligned} \tag{5.6}$$

where $P_0(x)$ is the probability associated to a single slit and $p' = p$. Thus, we find the same result as that of Eq. (2.1), but within a full quantum mechanical formalism.

[1] The transition operator will be defined in Sect. 11.2.

5.2 Diffraction Grating

Diffraction grating represents the one-dimensional idealization of the surface of a crystal lattice, which is made of equidistant atoms with D spacing comparable to the wavelength of an electron with a kinetic energy of a few hundred eV [DGE]. It can be viewed as an extension of the two-slit to N-slit screen, where eventually $N \to \infty$, and the diffraction from a grating as an extension of Young's experiment of interference from two slits. Under the weak source hypothesis (see Sect. 2.4), one particle at a time with momentum $\vec{p}$ elastically scatters from one of the grating atoms and is diffracted with momentum $\vec{p}\,'$ and at the θ angle, as depicted in Fig. 5.1. The probability amplitude (see Chap. 13) for scattering off the nth atom is $\langle \vec{p}\,'|\mathbf{A}_n|\vec{p}\,\rangle$. Thus, since the nth scattering amplitude is related to the $(n+1)^{\text{th}}$ by a translation, we can generalize Eq. (5.6) to N slits. The result is

$$P(x) = P_0(x)|1 + \mathrm{e}^{\frac{\mathrm{i}}{\hbar}p\,D\sin\theta} + \mathrm{e}^{\frac{\mathrm{i}}{\hbar}2pD\sin\theta} + \mathrm{e}^{\frac{\mathrm{i}}{\hbar}3pD\sin\theta} + \cdots|^2, \tag{5.7}$$

with the length of the grating $L = N \cdot D$ $(L \gg D)$. Equation (5.7) contains a succession whose argument is $\Delta\phi = \mathrm{e}^{\frac{\mathrm{i}}{\hbar}pD\sin\theta}$. By replacing it with its sum, we have

$$P(x) = P_0(x)\frac{\sin^2[NDp\sin\theta/2\hbar]}{\sin^2[Dp\sin\theta/2\hbar]}. \tag{5.8}$$

We highlight the maximum scattering probability which corresponds to $\theta = 0$ (or $x = 0$),

$$\begin{aligned} P(x) &= P(0)\lim_{N\to\infty}\frac{\sin^2(NDp\sin\theta/2\hbar)}{N^2\sin^2(Dp\sin\theta/2\hbar)} \\ &= P(0)\frac{\sin^2(Lp\sin\theta/2\hbar)}{(Lp\sin\theta/2\hbar)^2}. \end{aligned} \tag{5.9}$$

Equation (5.9) is a generalization of Eq. (5.6). The latter can be obtained from the former for $N = 2$. The multiplicity of the scattering centers enhances the interference phenomenon.

5.3 Double-Layer Reflection

A beam of electrons with momentum $\vec{p}$ impinges on the surface of a crystal, as in Fig. 5.2. The electrons might be reflected by the first layer, or by the second, if it has enough kinetic energy. For elastic collision with diffusion centers, the angle of incidence θ is equal to that of reflection, as depicted in Fig. 5.2. The emerging electrons have momentum $\vec{p}\,'$, regardless of whether they have been reflected by the

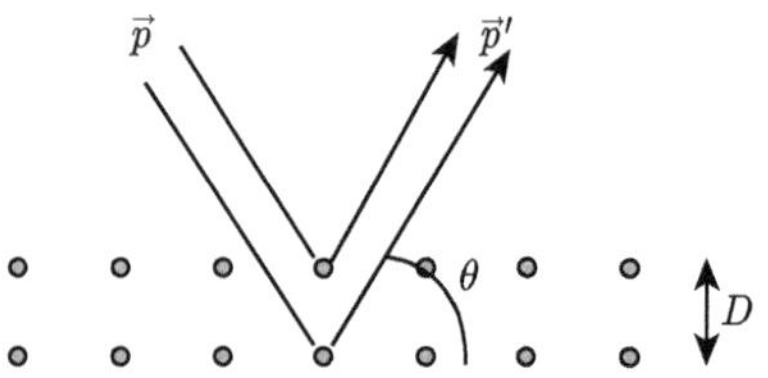

Fig. 5.2 Schematic representation of Bragg's diffraction

first or the second layer. In the case of a single layer, the probability of detecting an electron with momentum $\vec{p}\,'$, at the angle θ, is

$$P_0(\theta) = |\langle \vec{p}\,'|\mathbf{A}|\vec{p}\,\rangle|^2, \tag{5.10}$$

where $\mathbf{A}$ is the transition operator. In the case of two layers, the particle can be scattered either from the first layer or from the second one. The two transition operators differ by a translation of the vector, $\vec{D}$, where D is the spacing and $\hat{u}$ is a unitary vector orthogonal to the layers. Thus, resorting again to the superposition principle, the new probability is

$$P(\theta) = |\langle \vec{p}\,'|\mathbf{A}|\vec{p}\,\rangle + \langle \vec{p}\,'|\mathbf{U}_D^\dagger \mathbf{A}\mathbf{U}_D|\vec{p}\,\rangle|^2 \tag{5.11}$$

$$= P_0(\theta)|1 + \mathrm{e}^{\frac{2\mathrm{i}}{\hbar}pD\sin\theta}|^2, \tag{5.12}$$

where we have applied the eigenvalue equations of the translation operators

$$\mathbf{U}_D|\vec{p}\,'\rangle = |\vec{p}\,'\rangle \mathrm{e}^{\frac{\mathrm{i}}{\hbar}pD\sin\theta}, \tag{5.13}$$

$$\mathbf{U}_D|\vec{p}\,\rangle = |\vec{p}\,\rangle \mathrm{e}^{-\frac{\mathrm{i}}{\hbar}pD\sin\theta}. \tag{5.14}$$

Equation (5.12) is the quantum interpretation of the empirical Bragg law for the diffraction of X-rays by a crystal [BRA].

5.4 Scattering of Identical Particles

According to classical mechanics, identical particles are distinguishable, as we can follow them in their own trajectory at any time. Conversely, quantum mechanics rejects the concept of trajectory. Hence, we cannot distinguish identical particles participating together in a physical process. This has remarkable implications on the properties of many identical particle systems, such as electron clouds in atoms, nucleons in nuclei, and quantum fluids, as we will see in the next chapters.

The simplest physical process is the elastic scattering between two identical particles, which is depicted in Fig. 5.3 in the center-of-mass frame. A detector R is located in correspondence with the scattering angle θ. In the case of the upper side of Fig. 5.3,

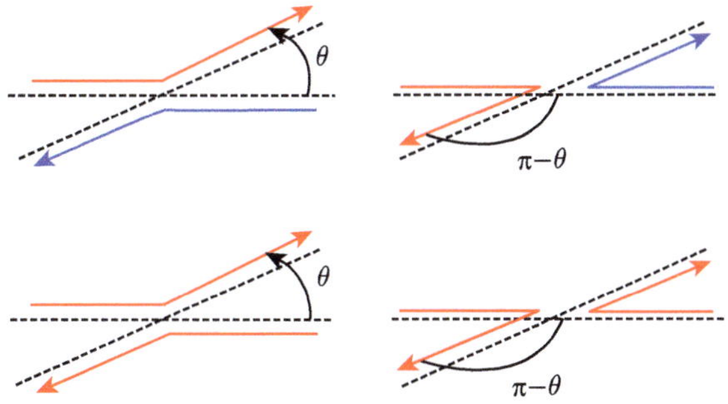

Fig. 5.3 Scattering between identical particles in the center-of-mass frame. Upper side: the two trajectories of two distinguishable particles detected at θ and $\pi - \theta$ angle, respectively. Lower side: the same for non-distinguishable particles

the two identical particles travel through two distinct trajectories, identified by different colours. The detector, supposed to distinguish the particles, will be counting the number of particles from the left trajectory, with probability amplitude $f_l(\theta)$ and number of particles from the right trajectory with probability amplitude $f_r(\theta)$, the latter being equal to the number of particles at the angle $\pi - \theta$. As a result, the total probability[2] will be

$$\mathcal{P}(\theta) = |f_l(\theta)|^2 + |f_r(\pi - \theta)|^2. \tag{5.15}$$

It is to be noticed that the amplitude of the right process is defined up to an arbitrary phase factor $e^{i\delta}$, which however is irrelevant when summing the probabilities.

Since the particles are identical, the detector cannot distinguish particles coming from the left or right trajectory. The situation is analogous to the double-slit experiment, where the detector placed at θ angle, cannot distinguish which slit the particle comes from. We must apply the superposition principle so that the total probability will be

$$\mathcal{P}(\theta) = |f(\theta) + e^{i\delta} f(\pi - \theta)|^2, \tag{5.16}$$

where we have consistently omitted the subscripts from the scattering amplitudes. The phase factor $e^{i\delta}$, instead, is no more irrelevant, because we have to sum up the amplitudes, exactly the same as in all interference processes. The phase factor can be fixed using the π rotation invariance: after the first π rotation, we have $f(\theta) = e^{i\delta} f(\pi - \theta)$, and after the second π rotation, $f(\theta) = e^{2i\delta} f(\theta)$. As a result $e^{2i\delta} = 1$, that is $e^{i\delta} = \pm 1$. The choice of the sign can be made after introducing the spin of particles. Particles with half-integer spin, known as with fermions, require the minus sign; particles with integer spin, known as bosons, require the plus sign. Therefore, the scattering probability of the identical particles is

[2] The total probability corresponds to the differential cross section and the probability amplitude to the scattering amplitude, to be discussed in Chap. 13.

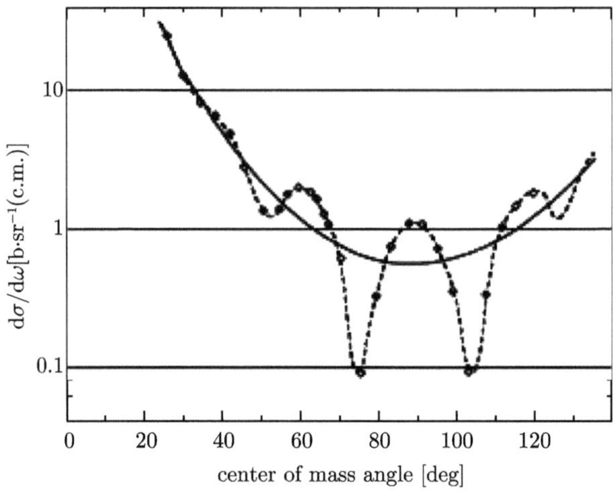

Fig. 5.4 Angular distribution for the elastic scattering channel of the collision Carbon on Carbon at 5 MeV. The solid line represents the theoretical prediction for identical particles. The open circles represent the experimental data [CCC]. Courtesy of American Physical Soc

$$P(\theta) = |f(\theta) \pm f(\pi - \theta)|^2. \tag{5.17}$$

with the *minus* sign for fermions and *plus* sign for bosons. The superposition of the amplitudes gives rise to interference effects which have been highlighted by experiments with elastic collisions between identical particles, as in the case of two carbon atoms at $E_{\rm cm} = 5\,\text{MeV}$ plotted in Fig. 5.4. The differential cross section exhibits oscillations around the value expected for distinguishable particles. The maximum effect of the experimental oscillations at 75° and 105° center-of-mass angles is estimated as large as six times the classical value.

5.5 Notes and References

DGE This property of the lattice had allowed Davisson and Germer to detect electron diffraction for the first time.

BRA W.L. Bragg, Proc. Cambridge Philos. Soc., **17** (1913) 43–57. Bragg derived his law

$$n\lambda = 2D \sin\theta$$

from the difference of the optical paths along the two layers.

CCC D.A. Bromley, J.A. Kuehner, and M. Alqvist, Physical Review, 123 (1961) 878.

5.6 Problems

- Problem 1. Calculate the beam intensity at the detector from a triple-slit layer diffraction.
- Problem 2. Determine the relation between the quantum probability of reflection from a double layer, Eq. (5.12), and empirical Bragg's law $n\lambda = 2D\sin\theta$, where n is an integer number.
- Problem 3. Calculate the cross section of the elastic scattering between two identical fermions at $\theta = \pi/2$ in the center of mass and discuss the result.
- Problem 4. Calculate the cross section of the elastic scattering between two identical fermions with spin $\frac{1}{2}\hbar$.
- Problem 5. A beam splitter (see figure) separates an incoming particle beam (A) into two components: one reflected (D) and one transmitted (C). The former is phase-shifted by π with respect to the incident beam (see Eq. (7.23)). Two beams of identical particles (one from A and one from B) cross the splitter as shown in the figure. Determine the state of two particles whether they are bosons or fermions.

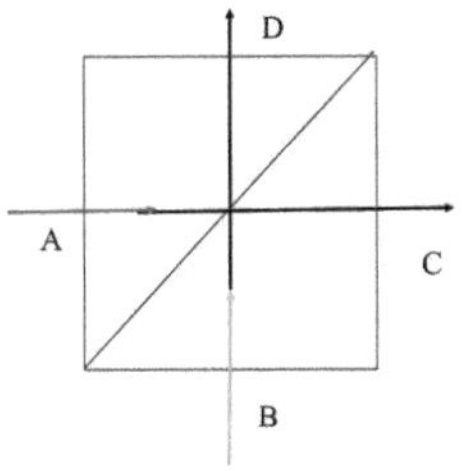

Chapter 6
Space Rotations and Angular Momentum

In this chapter the orbital angular momentum will be introduced as the generator of the space rotations. Since rotations are a non-commutating (non-Abelian) group, the components of the orbital angular momentum are not simultaneously measurable in quantum mechanics, at variance with classical mechanics. The spin will also be introduced as generator of the SU(2) transformations in intrinsic space, with the same algebraic properties as the angular momentum. The coupling of angular momentum with spin raises the issue of extending the symmetries of physical systems under rotations to more general transformations.

6.1 Space Rotations

In analogy with the space translations, we now move on, introducing the rotations as unitary transformations in Hilbert space. Let us consider a quantum system in the state $|\psi\rangle$ and perform a rigid rotation of an angle θ, around an arbitrary axis $\hat{e}$, denoted as $\mathbf{R} \equiv \mathbf{R}_{\hat{e}}(\theta)$. The rotation operator $\mathbf{U_R}$ establishes a one-to-one correspondence between any vector of Hilbert space and the rotated vector. This correspondence defines the operator $\mathbf{U_R}$ as

$$\begin{aligned} \mathbf{U_R}|\psi\rangle &= |\psi_{\mathbf{R}}\rangle, \\ \mathbf{U_R}|\vec{r}\,\rangle &= |\mathbf{R}\vec{r}\,\rangle, \end{aligned}$$

where $\mathbf{R}\vec{r}$ is the rotate $\vec{r}$ vector (a summary of the space rotations in the most used representations is given in the Complement to this section). The rotations form a

---

Supplementary Information The online version contains supplementary material available at https://doi.org/10.1007/978-981-97-1670-8_6.

U. Lombardo et al., *Quantum Mechanics*, UNITEXT for Physics,
https://doi.org/10.1007/978-981-97-1670-8_6

class of equivalent observers; in other words, the quantum probability is invariant under space rotations by virtue of the isotropy of the space.

$$|\langle \psi_R | \phi_R \rangle|^2 = |\langle \psi | \phi \rangle|^2,$$

in keeping with Eq. (4.8) for the translational invariance. As a consequence, the rotation operators are unitary operators

$$\mathbf{U}_{\mathbf{R}}^{\dagger} = \mathbf{U}_{\mathbf{R}}^{-1}.$$

The group of rotations is a non-Abelian group, because the total rotation, resulting from two successive rotations, depends on the order in which the two rotations are made. Correspondingly, the rotation operators do not commute, which has substantial effects on the properties of the quantum angular momentum.

Since the rotation operators do not commute, we need to determine the commutation rules. In full generality, these can be obtained from the isomorphism between the group of rotation operators and the group of rotations themselves [SAK]. In the representation of the orthogonal cartesian coordinates the rotations are represented by 3×3 matrices, which in the case of an infinitesimal rotation angle $\delta\phi$ fulfill the commutation relations Eq. (6.17), as shown in the Complement, Sect. 6.8. The infinitesimal rotation operators, by virtue of the isomorfism, also fulfill the same commutation rules

$$\mathbf{U}_x(\delta\phi)\mathbf{U}_y(\delta\phi) - \mathbf{U}_y(\delta\phi)\mathbf{U}_x(\delta\phi) = \mathbf{U}_z(\delta\phi^2) - \mathbf{1}, \tag{6.1}$$

$$\mathbf{U}_y(\delta\phi)\mathbf{U}_z(\delta\phi) - \mathbf{U}_z(\delta\phi)\mathbf{U}_y(\delta\phi) = \mathbf{U}_x(\delta\phi^2) - \mathbf{1}, \tag{6.2}$$

$$\mathbf{U}_z(\delta\phi)\mathbf{U}_x(\delta\phi) - \mathbf{U}_x(\delta\phi)\mathbf{U}_z(\delta\phi) = \mathbf{U}_y(\delta\phi^2) - \mathbf{1}, \tag{6.3}$$

where $\mathbf{U}_x(\delta\phi)$ ($\mathbf{U}_y(\delta\phi)$,$\mathbf{U}_z(\delta\phi)$) describes a rotation of the angle $\delta\phi$ around the axis x (y, z).

6.2 Orbital Angular Momentum as Generator of Infinitesimal Rotations

The rotation operator of an infinitesimal angle ϕ around the $\hat{e}$ axis can be written in full generality

$$\mathbf{U}_{\hat{e}}(\delta\phi) = \mathbf{1} - \frac{i}{\hbar}\delta\vec{\phi} \cdot \vec{\mathbf{l}}, \tag{6.4}$$

where $\delta\vec{\phi} = \delta\phi\hat{e}$. In consideration of the relationship between space isotropy and the conservation of angular momentum, we identify the generator $\vec{\mathbf{l}}$ of the rotation of angle $\delta\vec{\phi}$ with the projection of the orbital angular momentum $\vec{\mathbf{l}}$ onto the $\hat{e}$-axis, i.e. $\vec{\mathbf{l}} \cdot \hat{e}$. Since the rotation operator is unitary it follows that the orbital angular momentum is hermitian as requested by the quantum–mechanical formalism. From the commutation rules of the rotation operator, Eqs.(6.1) and (6.3), we obtain the commutation rules for the components of the angular momentum in orthogonal cartesian coordinates,

$$\left[\mathbf{l}_x, \mathbf{l}_y\right] = i\hbar\,\mathbf{l}_z, \qquad \left[\mathbf{l}_y, \mathbf{l}_z\right] = i\hbar\,\mathbf{l}_x, \qquad \left[\mathbf{l}_z, \mathbf{l}_x\right] = i\hbar\,\mathbf{l}_y. \tag{6.5}$$

In the formulation of the quantum mechanics based on the principle of correspondence between the classical and quantum dynamical variables, the orbital quantum angular momentum is defined as

$$\vec{\mathbf{l}} = \vec{\mathbf{r}} \times \vec{\mathbf{p}}. \tag{6.6}$$

The commutation rules among its components, derived from the commutation rules between position and momentum, coincide with Eq. (6.5), but the procedure previously shown is set in a more general context. In Sect. 6.6, we shall introduce the spin variable, which fulfills the same algebra as the orbital angular momentum, as the generator of rotations in a space other than the coordinate space. As a consequence, the foregoing commutation rules can be extended to the spin. The spin of a particle couples to its orbital angular momentum to give the total angular momentum $\vec{\mathbf{j}} = \vec{\mathbf{l}} + \vec{\mathbf{s}}$. In turn, the individual total angular momenta of each particle couple together, giving the total angular momentum of many-particle systems like electron clouds of atoms or the nucleons of atomic nuclei. In the case of atoms, the angular momenta of the Z electrons couple to give the total angular momentum of the cloud

$$\vec{\mathbf{J}} = \sum_{k=1,Z} \vec{\mathbf{j}}_k.$$

In general, the total angular momentum of the whole system comes into play rather than the individual angular momenta of the particles, because only this is a good quantum number.

6.3 Properties of the Angular Momentum

In this section, we shall study the general properties of angular momentum $\vec{\mathbf{j}}$, regardless of whether it is orbital angular momentum, spin, or their sum. The commutation rules, Eqs. (6.1), (6.2) and (6.3), reveal that the components of the angular momentum form a closed algebra. From such rules, we can infer the angular momentum properties by means of purely algebraic methods, rather than solving differential equations [MES].

Since the three components of $\vec{\mathbf{j}}$ do not commute, they do not have simultaneous eigenstates. In other words, it is not possible to assign to a quantum system neither three nor two components of the angular momentum. However, the angular momentum magnitude $\mathbf{j}^2 = \mathbf{j}_x^2 + \mathbf{j}_y^2 + \mathbf{j}_z^2$ is an invariant for rotations, thus it commutes with each of the three components. Hence, we can determine the simultaneous eigenstates of $\mathbf{j}^2$ and one of the components. Conventionally, we choose $\mathbf{j}_z$, and we solve the eigenvalue equations

$$\mathbf{j}^2|jm\rangle = |jm\rangle\hbar^2 j(j+1), \tag{6.7}$$
$$\mathbf{j}_z|jm\rangle = |jm\rangle\hbar m. \tag{6.8}$$

The eigenvalues are expressed in $\hbar$ units, because $\hbar$ has the dimensions of an action as well as angular momentum. Nevertheless, to simplify the formalism, in the following, we will put $\hbar = 1$. We have written the $\mathbf{j}^2$ eigenvalues in the form $j(j+1)$ (with $j > 0$) for our convenience. This does not lead to any restriction, because they have to be greater than or equal to zero. As eigenstates of physical observables, the $|jm\rangle$ states have to satisfy the completeness and orthonormality relations. Therefore, we write

$$\sum_{j,m}|jm\rangle\langle jm| = \mathbf{1}, \tag{6.9}$$
$$\langle jm|j'm'\rangle = \delta_{jj'}\delta_{mm'}. \tag{6.10}$$

We define two auxiliary operators $\mathbf{j}_\pm = \mathbf{j}_x \pm i\mathbf{j}_y$. They are the adjoint operators of each other. From Eqs. (6.1)–(6.3), it is easy to prove the following commutation rules,

$$[\mathbf{j}_z, \mathbf{j}_\pm] = \pm\mathbf{j}_\pm,$$

$$[\mathbf{j}_+, \mathbf{j}_-] = 2\mathbf{j}_z,$$

$$[\mathbf{j}^2, \mathbf{j}_+] = [\mathbf{j}^2, \mathbf{j}_-] = 0,$$

from which we obtain the identity

$$\mathbf{j}_\pm\mathbf{j}_\mp = \mathbf{j}^2 - \mathbf{j}_z(\mathbf{j}_z \pm 1).$$

From Eqs.(6.7) and (6.8), it follows that $\mathbf{j}_\pm|jm\rangle$ is the eigenstate of $\mathbf{j}^2$,

$$\mathbf{j}^2\mathbf{j}_\pm|jm\rangle = \mathbf{j}_\pm\mathbf{j}^2|jm\rangle = j(j+1)\mathbf{j}_\pm|jm\rangle,$$

and $\mathbf{j}_\pm|jm\rangle$ is eigenstate of $\mathbf{j}_z$ as well,

$$\begin{aligned}\mathbf{j}_z\mathbf{j}_\pm|jm\rangle &= [\mathbf{j}_z, \mathbf{j}_\pm]|jm\rangle + \mathbf{j}_\pm\mathbf{j}_z|jm\rangle \\ &= \pm\,\mathbf{j}_\pm|jm\rangle + m\mathbf{j}_\pm|jm\rangle = (m \pm 1)\mathbf{j}_\pm|jm\rangle.\end{aligned}$$

Hence, $\mathbf{j}_+$ rises and $\mathbf{j}_-$ lowers the value of m by one.

Then, we obtain the squared norm of the vectors $\mathbf{j}_\pm|jm\rangle$,

$$\langle jm|\mathbf{j}_\mp\mathbf{j}_\pm|jm\rangle = j(j+1) - m(m \pm 1) \geq 0. \tag{6.11}$$

This equation corresponds to a system of two inequalities whose solution is

$$-j \leq m \leq +j.$$

Applying p times the $\mathbf{j}_+$ operator to one eigenstate $|jm\rangle$, we get an eigenstate having the eigenvalue $m + p$[1]. Iterating such a procedure, since j represents the upper extreme of the $\mathbf{j}_z$ eigenvalues, we will reach a $p_{\max}$ number for which $m + p_{\max} = j$. Analogously, by applying for q times the operator $\mathbf{j}_-$ to an eigenstate $|jm\rangle$, we obtain an eigenstate with eigenvalue $m - q$. Since $-j$ represents the lower extreme of the $\mathbf{j}_z$ eigenvalues, there will be a $q_{\max}$ for which $m - q_{\max} = -j$. The member-to-member subtraction of the expression for j with the one for $-j$ gives $2j = q_{\max} + p_{\max}$. Thus, the possible j values are integers or half-integers. The eigenvalues of m are also assumed to be integer or half-integer values that vary by integer units within the interval $-j \leq m \leq j$. For a fixed j value, there are $2j + 1$ values of m. For instance, if $j = 2$, the corresponding m values are $-2,\ -1,\ 0,\ 1,\ 2$. The eigenvalues spectrum of $\mathbf{j}_z$ is observed in the Stern–Gerlach experiment: the number of components in which the atom beam decomposes after the interaction with the inhomogeneous magnetic field, is equal to the number of eigenvalues of $\mathbf{j}_z$. Therefore, according to the $2j + 1$ rule, such a number provides the j value. In particular, for an odd number of components j is an integer number, while if the number of components is even, j will be a half-integer number. We highlight that the observation of an even number of components, j half-integer, is evidence of the existence of the electron spin. We will further see that, if the angular momentum is purely orbital (which means zero spin), its eigenvalues will be integer numbers, and the eigenvalues of its projection $\mathbf{j}_z$ will be integer numbers as well.

We have determined the simultaneous eigenvalues spectrum of the $\mathbf{j}^2$ and $\mathbf{j}_z$ operators. We have seen that the related eigenvectors can be generated by iteratively applying the raising operator $\mathbf{j}_+$, or the lowering operator $\mathbf{j}_-$, to the $|jm\rangle$ state. The only issue of how to generate further eigenstates is that, we need to know at least one. We shall show how to determine the $|jj\rangle$ eigenstate, with $m = j$, in the next section.

Let us consider a composite system of two particles with angular momentum $\vec{j}_1$ and $\vec{j}_2$, respectively, and we have

$$[\mathbf{j}_{1i}, \mathbf{j}_{2j}] = 0, \quad i, j = x, y, z.$$

The eigenstates of $(\mathbf{j}_1^2, \mathbf{j}_{1z})$ is $|j_1 m_1\rangle$, where we have applied the eigenvalue equations

$$\begin{aligned} \mathbf{j}_1^2 |j_1 m_1\rangle &= j_1(j_1 + 1)\hbar^2 |j_1 m_1\rangle, \\ \mathbf{j}_{1z} |j_1 m_1\rangle &= m_1 \hbar |j_1 m_1\rangle, \end{aligned}$$

[1] We follow the algebraic demonstration of Ref.[MES].

and resumed the Planck constant $\hbar$. Similarly,

$$\begin{aligned}\mathbf{j}_2^2|j_2m_2\rangle &= j_2(j_2+1)\hbar^2|j_2m_2\rangle,\\ \mathbf{j}_{2z}|j_2m_2\rangle &= m_2\hbar|j_2m_2\rangle.\end{aligned}$$

For any state of the two-particle system, the states $|j_1m_1j_2m_2\rangle = |j_1m_1\rangle|j_2m_2\rangle$ are simultaneous eigenstates of $(\mathbf{j}_1^2, \mathbf{j}_{1z}, \mathbf{j}_2^2, \mathbf{j}_{2z})$. The two angular momenta can couple to a total angular momentum $\vec{J} = \vec{j}_1 + \vec{j}_2$, and the eigenstate of $(\mathbf{J}^2, \mathbf{J}_z)$ is denoted as $|JM\rangle$ where

$$\begin{aligned}\mathbf{J}^2|JM\rangle &= J(J+1)\hbar^2|JM\rangle,\\ \mathbf{J}_z|JM\rangle &= M\hbar|JM\rangle.\end{aligned}$$

The $|JM\rangle$ states can be obtained from states $|j_1m_1j_2m_2\rangle$ by a unitary transformation of dimension $(2j_1+1)(2j_2+1)$, whose the matrix elements are denoted as $\langle j_1m_1j_2m_2|JM\rangle$. Therefore,

$$|JM\rangle = \sum_{m_1m_2}\langle j_1m_1j_2m_2|JM\rangle|j_1m_1j_2m_2\rangle,$$

where $J = |j_1 - j_2|, |j_1 - j_2| + 1, \ldots, j_1 + j_2$, and $M = m_1 + m_2$.

The transformation matrix elements are known as Clebsch–Gordan coefficients (see Appendix C).

6.4 Orbital Angular Momentum in Polar Coordinates

In this section, we come back to the orbital angular momentum $\vec{\mathbf{l}}$ as a generator of space rotations. Let us examine how a wave function $\psi(\vec{r})$ transforms under infinitesimal space rotations. We first recall how the position vector $\vec{r}$ transforms under infinitesimal rotation of an angle $\delta\phi$ around the $\hat{u}$ direction

$$\vec{r} + \delta\vec{r} = \vec{r} + \delta\vec{\phi}\times\vec{r},$$

where $\vec{\delta\phi} = \delta\phi\hat{u}$ ($\hat{u}$ is a unit vector whose orientation is the one from which the rotation is seen take place counterclockwise). We consider now the effect of the infinitesimal rotation on the wave function

$$\begin{aligned}\langle\vec{r}\,|\mathbf{U}_{\hat{u}}(\delta\phi)|\psi\rangle &= \psi(\vec{r} - \delta\vec{r})\\ &= \psi(\vec{r}) - \delta\vec{\phi}\times\vec{r}\cdot\nabla\psi\\ &= \psi(\vec{r}) - \delta\vec{\phi}\cdot\vec{r}\times\nabla\psi, \qquad (6.12)\end{aligned}$$

where we exchanged the vector product with the scalar product. From the expression of the rotation operator, Eq. (6.4), it follows that

$$\frac{i}{\hbar}\langle \vec{r}\,|\delta\vec{\phi}\cdot \mathbf{l}|\psi\rangle = \delta\vec{\phi}\cdot(\vec{r}\times\nabla\psi).$$

Considering an infinitesimal rotation around the z axis, we have $\mathbf{l}_u = \frac{\hbar}{i}(x\partial_y - y\partial_x)$. Such a rotation corresponds to a variation of $\delta\phi$ of the azimuthal angle ϕ, while the radial angle θ is unchanged. It is easy to prove that

$$\frac{\hbar}{i}(x\partial_y - y\partial_x)\psi = \frac{\hbar}{i}\frac{\partial\psi}{\partial\phi}. \tag{6.13}$$

Expressing rotations around the x axis, or the y axis is a more tedious task, because in these cases, the variation of θ together with ϕ is involved (see Problem 10). We just write down the expressions for their combinations $\mathbf{l}_\pm = \mathbf{l}_x \pm i\mathbf{l}_y$

$$\langle \vec{r}\,|\mathbf{l}_\pm|\psi\rangle = e^{\pm i\phi}\left(\pm\frac{\partial}{\partial\theta} + i\cot\theta\frac{\partial}{\partial\phi}\right)\psi, \tag{6.14}$$

that will be used in the following. From Eq. (6.13) and (6.14), we infer the expression of $\mathbf{l}^2$ in polar coordinates,

$$\mathbf{l}^2 = \mathbf{l}_+\mathbf{l}_- + \mathbf{l}_z(\mathbf{l}_z + \mathbf{1}) = -\left[\frac{1}{\sin\theta}\frac{\partial}{\partial\theta}\left(\sin\theta\frac{\partial}{\partial\theta}\right) + \frac{1}{\sin^2\theta}\frac{\partial^2}{\partial\phi^2}\right].$$

Now, we have all the ingredients in hand to determine the eigenstates of the orbital angular momentum $|\,lm\rangle$ in the polar coordinates. We refer to $\psi_{lm}(\vec{r}\,)$ as the simultaneous eigenfunctions of $\mathbf{l}^2$ and $\mathbf{l}_z$. According to Eq. (6.13), the eigenvalues equation for the $\mathbf{l}_z$ operator in polar coordinates is the following:

$$\frac{\hbar}{i}\frac{\partial\psi(r,\theta,\phi)}{\partial\phi} = \hbar m\psi(r,\theta,\phi).$$

The integral of this equation is

$$Y_{lm}(\theta,\phi) = f_{lm}(\theta)e^{im\phi}.$$

We come now to an important step: to interpret such eigenfunctions as probability amplitudes, they have to be single-valued functions. In other words, the exponential has to be a periodic function, i.e.

$$e^{im\phi} = e^{im(\phi+2\pi)},$$

hence m has to be an integer number, and consequently l has to be an integer number as well. Therefore, the eigenvalues of the orbital angular momentum are integer numbers.

In closing, we consider the eigenstate $|ll\rangle$, which, since $m \leq l$, has the property $\mathbf{l}_+|ll\rangle = 0$. In polar coordinates, this means that

$$\begin{aligned} \langle \vec{r}\,|\mathbf{l}_+|ll\rangle &= \mathrm{e}^{\mathrm{i}\phi}\left[\frac{\partial}{\partial\theta} + \mathrm{i}\cot\theta\frac{\partial}{\partial\phi}\right] Y_{ll}(\theta,\phi) \\ &= \mathrm{e}^{\mathrm{i}(1+l)\phi}\left[\frac{\partial f_{ll}(\theta)}{\partial\theta} - l\cot\theta f_{ll}(\theta)\right] = 0. \end{aligned}$$

The solution of this differential equation is $\sin^l\theta$. Therefore, we conclude that, in the case $m = l$, the simultaneous eigenfunctions of $\mathbf{l}^2$ and $\mathbf{l}_z$ are

$$Y_{ll}(\theta,\phi) = \text{const}\cdot\sin^l(\theta)\mathrm{e}^{\mathrm{i}l\phi}$$

The constant is to be fixed by imposing the normalization of $Y_{ll}(\theta,\phi)$. Once we obtain the eigenfunction related to the eigenvalues l and $m = l$, by iteratively applying the $\mathbf{l}_-$ operator (in the form of Eq. (6.14) to the Y_{ll} eigenfunction, it is possible to determine the other eigenfunctions with eigenvalue l and $m < l$. The wave functions Y_{lm} are called spherical harmonics. It is proved that they form a complete set of orthonormal functions, as they should, to be eigenfunctions of physical observable. Hence, we write

$$\sum_{l=0}^{\infty}\sum_{m=-l}^{l} Y^*_{lm}(\Omega)Y_{lm}(\Omega') = \delta(\Omega,\Omega'), \tag{6.15}$$

$$\int \mathrm{d}\Omega Y^*_{lm}(\Omega)Y_{l'm'}(\Omega) = \delta_{ll'}\delta_{mm'}, \tag{6.16}$$

where $\Omega \equiv (\theta,\phi)$ and $\mathrm{d}\Omega \equiv \mathrm{d}\sin\theta d\phi$. Equations (6.15) and (6.16) correspond to the completeness and orthonormality properties of the state vectors $|lm\rangle$ given by Eqs. (6.9) and (6.10).

6.5 Reflection of Axes and Parity

Consider the space coordinates (x, y, z). The flip in the sign of two coordinates, e.g. (x, y, z) into $(-x, -y, z)$, is a rotation; instead, the flip of the three coordinates (x, y, z) into $(-x, -y, -z)$ is a reflection. The operator corresponding to the reflection is called *parity* Π. It has the property $\Pi^2 = \Pi$, so its eigenvalues are ± 1. A state such that $\psi(x, y, z) = \psi(-x, -y, -z)$ is said to be a state with *even* parity, and a state such that $\psi(x, y, z) = -\psi(-x, -y, -z)$ is said to be a state with *odd* parity. Since the orbital angular momentum $\mathbf{l}$ and the parity operator Π commute, the eigenvectors of $\mathbf{l}$ have a definite parity. The reflection in polar coordinates is $(r, \theta, \phi) \to (r, \pi - \theta, \phi + \pi)$. The eigenfunctions of orbital angular momentum transform under reflection as

$$Y(r, \theta, \phi) = (-1)^l Y(r, \pi - \theta, \phi + \pi).$$

Therefore, their parity is $(-1)^l$. As we will see later, the eigenfunctions of systems submitted to central potentials are spherical harmonics and hence their parity $(-1)^l$ is a good quantum number.

6.6 Spin

The Stern-Gerlach experiment is a clear and direct evidence of the existence of spin as discussed in Chap. 2. The silver atom in its ground state has orbital angular momentum equal to zero; nevertheless, after the silver beam has gone through the non-uniform magnetic field, it splits into two beams, and the atoms concentrate only in the two points A and B, as shown in Fig.2.6. This means that we are faced with a magnetic moment along the z-axis, which is quantized with only two possible values. Hence the dynamical variable $\vec{s}$, responsible for this magnetic moment, must have the properties of an angular momentum with half-integer modulus equal to $\frac{1}{2}$ (in $\hbar$ units), because its state multiplicity must be $2s + 1 = 2$. This variable, known as spin, has nothing to do with the orbital angular momentum, even if it shares with the latter the property of interacting with the magnetic field. Theoretically, the spin degree of freedom was predicted by Dirac in the framework of the relativistic theory of electrons (see Chap. 15). In analogy with space–time symmetry transformations, we introduce the unitary transformations $\mathbf{U}(\vec{\theta})$ as rotations in a two-dimensional spin space, leaving the probability amplitudes of spin states invariant[2]. The spin is identified as the Hermitian dynamical variable generating the SU(2) infinitesimal transformations

$$\mathbf{U}(\vec{\theta}) = \mathbf{1} - \frac{\mathrm{i}}{\hbar}\vec{\theta} \cdot \vec{s},$$

where $\vec{s}$ is a three-component vector, each component being a 2×2 matrix. In Chap. 15, it is shown that the simplest representation is that of the Pauli matrices

$$\sigma_x = \begin{pmatrix} 0 & 1 \\ 1 & 0 \end{pmatrix}, \quad \sigma_y = \begin{pmatrix} 0 & -\mathrm{i} \\ \mathrm{i} & 0 \end{pmatrix}, \quad \sigma_z = \begin{pmatrix} 1 & 0 \\ 0 & -1 \end{pmatrix}$$

[2] The operator **U**, belonging to SU(2) group, is unitary and with determinant equal to one. From these two properties it can be shown that its generator fulfills the same algebra as the angular momentum.

with $\vec{\sigma} = 2\vec{s}$. As well as the Pauli matrices, the spin components satisfy the same algebra as angular momenta

$$\begin{aligned}[s_i, s_j] &= i\hbar\epsilon_{ijk}s_k,\\ [s_i, s^2] &= 0,\end{aligned}$$

where ϵ_{ijk} is the Levi-Civita antisymmetric tensor [LC].

The analogy with orbital angular momentum is a pure coincidence, because, as we said, the spin and angular momentum are quite different physical variables. In fact, the algebras of angular momentum is not exactly the same as that of spin. If one rotates by $\theta = 2\pi$ once around an axis, e.g. y-axis, spin changes the sign, whereas angular momentum does not.

The two spin states are simultaneous eigenstates of $\mathbf{s}^2$ and $\mathbf{s}_z$,

$$\begin{aligned}\mathbf{s}^2|s, s_z\rangle &= \hbar^2 s(s+1)|s, s_z\rangle,\\ \mathbf{s}_z|s, s_z\rangle &= \hbar s_z|s, s_z\rangle,\end{aligned}$$

where $s = \frac{1}{2}$ and $s_z = \pm\frac{1}{2}$ (in $\hbar$ units). The two spin eigenstates form an orthonormal complete ensemble in the two-dimensional Hilbert's space.

The eigenvalue of $\mathbf{s}^2$ is equal to $\frac{3}{4}\hbar^2$. Therefore,

$$\mathbf{s}_x^2 = \mathbf{s}_y^2 = \mathbf{s}_z^2 = \frac{1}{4}\hbar^2.$$

We then introduce the $\mathbf{s}_\pm$ operators, which are analogous to the $\mathbf{j}_\pm$ operators. Since the $\mathbf{s}_z$ eigenvalue cannot vary by two units,

$$\mathbf{s}_+^2 = \mathbf{s}_-^2 = 0,$$

from which we get to the anti-commutation rules

$$\mathbf{s}_x\mathbf{s}_y + \mathbf{s}_y\mathbf{s}_x = \mathbf{s}_y\mathbf{s}_z + \mathbf{s}_z\mathbf{s}_y = \mathbf{s}_z\mathbf{s}_x + \mathbf{s}_x\mathbf{s}_z = 0.$$

For a particle with spin s in a state with orbital angular momentum l, e.g. electron in an atomic shell, the state of total angular momentum is obtained by coupling the two variables

$$|jm\rangle = \sum_{l_z s_z}\langle ll_z, ss_z|jm\rangle|ll_z\rangle|ss_z\rangle,$$

where $j = \left|l - \frac{1}{2}\right|, l + \frac{1}{2}$, and $m = l_z + s_z$. The symbols $\langle ll_z, ss_z|jm\rangle$ are the Clebsch–Gordan coefficients, see Appendix C.

6.7 The Rigid Rotor

Space rotations are commonly experienced by atoms and molecules. They give rise to the so-called rotational bands, due to the coupling of the electronic transitions with the rotation states of the system as a whole. Often, the centrifugal force stretches the shape of the system or, if this can vibrate, rotations couple with vibrations. The simplest rotating system is the rigid rotor. The Hamiltonian of a rigid rotational motion in three dimensions is

$$\mathbf{H} = \frac{\hbar^2 \mathbf{l}_x^2}{2I_x} + \frac{\hbar^2 \mathbf{l}_y^2}{2I_y} + \frac{\hbar^2 \mathbf{l}_z^2}{2I_z}.$$

$\vec{l}$ is the orbital angular momentum and I_i are the components of the moment of inertia, which are constant in the rigid rotor. If $I_x = I_y = I_z$ the rotor is isotropic. Let us suppose that the rotation occurs around a chosen axis, say the z axis. In this case, the Hamiltonian becomes $\mathbf{H} = \frac{\mathbf{l}_z^2}{2I}$, where $I = I_z$. The eigenvalue equation of the Hamiltonian of an isotropic rigid rotator is the same as the eigenvalue equation of the $\mathbf{l}^2$ operator, thus

$$\mathbf{H}|lm\rangle = \frac{\hbar^2 \mathbf{l}^2}{2I}|lm\rangle = \frac{\hbar^2 l(l+1)}{2I}|lm\rangle.$$

Those eigenstates having the same eigenvalue l, but different eigenvalue m have the same energy. The consequence is that the energy eigenvalue $E_l = \frac{\hbar^2 l(l+1)}{2I}$ is $(2l+1)$-fold degenerate. This is due to the existence of $(2l+1)$ distinct eigenstates, belonging to the same l. Similar considerations can be made for rotations associated to spin or total angular momentum.

6.8 Complement to Sect. 6.1: Infinitesimal Space Rotations

In Chap. 1, we learned that in classical mechanics the space rotations belong to a class of canonical transformations whose generator is the orbital angular momentum. As well as the space translations, any rotation can be decomposed into three independent rotations around, respectively, the x-axis, the y-axis and the z-axis. But, unlike translations, the order in which the three rotations are performed is essential, as illustrated in Fig. 6.1, Above, a vector, lying on the y-axis, is rotated first around the x-axis, then around the y-axis and, finally, rotating around the z-axis, it returns to the starting position. Below, the order of the three rotations is reversed, but the final position of the vector is not the same as the starting one. It stems from the fact that, as we shall see below, the three components of the orbital angular momentum do not commute, which has profound implications for the properties of the quantum systems.

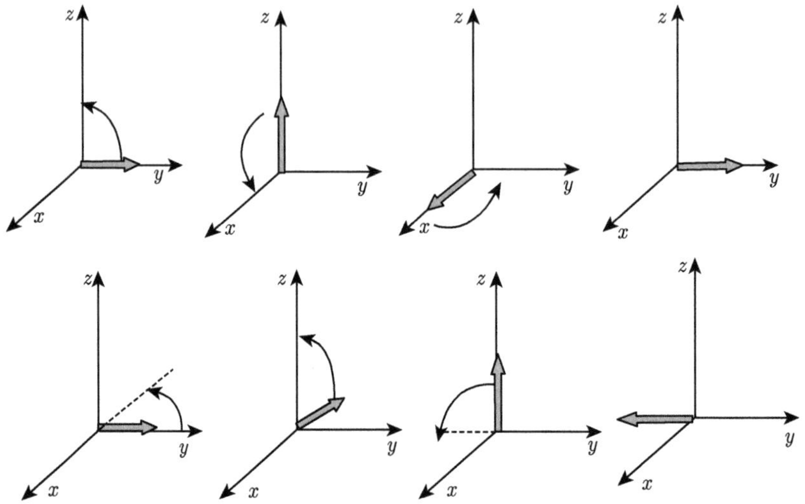

Fig. 6.1 Space rotations of a vector: above, three successive 90 degrees rotations of a vector around the x, y and z axes, below, the opposite order, i.e. around z, y, and x axes. The final position of the vector is different in the two cases

Let us develop the formalism of the infinitesimal space rotations in spherical coordinates [SAK]. Let $Rz(\delta\phi)$ be the rotation matrix that transforms vector $\vec{r} = (x, y, z)$ into $\vec{r}' = (x', y', z')$. The transition from Cartesian to spherical coordinates to the second order in $\delta\phi$ is

$$\begin{aligned} x' &= r\cos\theta\cos(\phi + \delta\phi) \\ &= x - y\delta\phi - \frac{x}{2}\delta\phi^2, \\ y' &= r\cos\theta\sin(\phi + \delta\phi) \\ &= y + x\delta\phi - \frac{y}{2}\delta\phi^2, \\ z' &= z. \end{aligned}$$

$$\mathbf{R}_z(\delta\phi) = \begin{pmatrix} 1 - \frac{1}{2}\delta\phi^2 & -\delta\phi & 0 \\ \delta\phi & 1 - \frac{1}{2}\delta\phi^2 & 0 \\ 0 & 0 & 1 \end{pmatrix}.$$

The $\mathbf{R}_y(\delta\phi)$ rotation matrix can be obtained from the $\mathbf{R}_z(\delta\phi)$ matrix by exchanging y with $-z$ (the minus sign ensures the left-handed triplet of axes),

$$\mathbf{R}_y(\delta\phi) = \begin{pmatrix} 1-\frac{1}{2}(\delta\phi)^2 & 0 & \delta\phi \\ 0 & 1 & 0 \\ -\delta\phi & 0 & 1-\frac{1}{2}(\delta\phi)^2 \end{pmatrix}.$$

The $\mathbf{R}_x(\delta\phi)$ rotation matrix is obtained by the exchange of z with x,

$$\mathbf{R}_x(\delta\phi) = \begin{pmatrix} 1 & 0 & 0 \\ 0 & 1-\frac{1}{2}(\delta\phi)^2 & -\delta\phi \\ 0 & \delta\phi & 1-\frac{1}{2}(\delta\phi)^2 \end{pmatrix}.$$

The dependence of the rotation on the order in which the three rotations are performed implies that the related rotation matrices, $\mathbf{R}_x(\delta\phi)$, $\mathbf{R}_y(\delta\phi)$, and $\mathbf{R}_z(\delta\phi)$, do not commute among themselves. Indeed, the calculation of the commutators up to the second order in $\delta\phi$ gives

$$\mathbf{R}_x(\delta\phi)\mathbf{R}_y(\delta\phi) - \mathbf{R}_y(\delta\phi)\mathbf{R}_x(\delta\phi) = \mathbf{R}_z(\delta\phi^2) - \mathbf{1}. \tag{6.17}$$

Analogous commutation relations can be obtained from the circular permutation of x, y, z.

6.9 Notes and References

MES A. Messiah, *Quantum Mechanics Vol.2*, North-Holland, 1961, Chap.XIII.

SAK J. J. Sakurai, *Modern Quantum Mechanics*, Addison-Wesley Publishing Company, 1994, Chap. 3.

LC Levi-Civita antisymmetric tensor: Defining $\varepsilon 123 \equiv 1$, by antisymmetry it follows that

$$\varepsilon 123 = \varepsilon 231 = \varepsilon 312 = +1$$
$$\varepsilon 132 = \varepsilon 213 = \varepsilon 321 = -1$$

If any two indices are the same, it is vanishing.

6.10 Problems

- Problem 1. Determine the uncertainty relation for a simultaneous measurement of the angular momentum components $\mathbf{l}_x$ and $\mathbf{l}_y$.
- Problem 2. Using the invariance under space rotations prove that the operator $\mathbf{p}^2$ commutes with the components of the orbital angular momentum $\vec{\mathbf{l}} = \vec{\mathbf{r}} \times \vec{\mathbf{p}}$.
- Problem 3. Derive the commutation rules of orbital angular momentum, $\vec{\mathbf{l}} = \vec{\mathbf{r}} \times \vec{\mathbf{p}}$, from the commutation rules between position and momentum.
- Problem 4. Let $|1, 0\rangle$ be simultaneous eigenstate of $\mathbf{l}^2$ and $\mathbf{l}_z$. Show that it is also an eigenstate of $\mathbf{l}_x^2$.
- Problem 5. Let us consider a system prepared in the state $|\psi\rangle = \sum_{m=-l}^{l} c_m |l\, m\rangle$. Calculate the minimal uncertainty in a measurement of $\mathbf{l}_x$ and $\mathbf{l}_y$.
- Problem 6. Let us consider a non-polarized atomic beam with $l = 1$. Determine from the average value $\langle l_z \rangle$ and the standard deviation $\sqrt{\langle(\mathbf{l}_z - \langle \mathbf{l}_z \rangle)^2\rangle}$ the probability of detecting an atom of the beam in each eigenstate of $\mathbf{l}_z$.
- Problem 7. Let us consider a spin rotation. Prove that if the spin is rotated once around an arbitrary axis, e.g. y-axis, by 2π, then the spin changes sign, whereas the angular momentum does not.
- Problem 8. Calculate the expressions of $\mathbf{l}_x$ and $\mathbf{l}_y$ in polar coordinates.

Chapter 7
Time Translations and Hamiltonian

In the latter of the three chapters devoted to the basic dynamical variables the Hamiltonian will be introduced as the generator of time translations. The time-dependent equation of motion will be derived as well as the Schrödinger equation for stationary states. By solving the stationary equation for simple one-dimensional systems, remarkable quantum effects will be encountered, including the striking barrier penetration, known as *tunneling*.

7.1 Time Evolution Operator

The time evolution of a physical system can be compared to a translation along the time dimension. Thus, in analogy to space translations, we postulate that there is a one-to-one correspondence between the states that the system goes through during its temporal evolution between any two instants of time t and t'. This correspondence is formally described by the *evolution operator* $U(t', t)$ as

$$|\psi(t')\rangle = \mathbf{U}(t', t)|\psi(t)\rangle.$$

Two noticeable properties of the evolution operator are the inverse operator $\mathbf{U}^{-1}(t'.t) = \mathbf{U}(t, t')$ and the composition law $\mathbf{U}(t', t) = \mathbf{U}(t', t'')\mathbf{U}(t'', t)$. Firstly, we consider a system, which is isolated or at most submitted to a perturbation independent of time. In this case the time uniformity dictates that the evolution operator can only be dependent on the time interval $(t' - t)$ rather than on the individual time instants, i.e. $\mathbf{U}(t', t) \equiv \mathbf{U}(t' - t)$, and any probability amplitudes must be independent of time, i.e.

$$\langle\phi(t' + t_0)|\psi(t + t_0)\rangle = \langle\phi(t')|\psi(t)\rangle. \tag{7.1}$$

Supplementary Information The online version contains supplementary material available at https://doi.org/10.1007/978-981-97-1670-8_7.

U. Lombardo et al., *Quantum Mechanics*, UNITEXT for Physics,
https://doi.org/10.1007/978-981-97-1670-8_7

By plugging the evolution operator into Eq. (7.1), we obtain

$$\langle \phi(t')|\mathbf{U}^\dagger(t_0)\mathbf{U}(t_0)|\psi(t)\rangle = \langle \phi(t')|\psi(t)\rangle,$$

and $\mathbf{U}^\dagger\mathbf{U} = \mathbf{1}$. By multiplying this relation times the inverse $\mathbf{U}^{-1}$, we find the condition for the evolution operator to be unitary, namely $\mathbf{U}^\dagger = \mathbf{U}^{-1}$. In the case of systems submitted to time dependent perturbations, the time uniformity breaks down, nonetheless the unitarity of the evolution operator can be justified from the general property of the probability conservation. Indeed, suppose we submit the system to the measurement of an observable $\boldsymbol{O}$ with eigenvalues O_k and eigenstates $|k\rangle (k = 1, D)$. The result is the set of probabilities $P_k(t) = |\langle k|\mathbf{U}(t - t_0)|\psi(t_0)\rangle|^2$ at a time t associated to the values O_k, when the system is prepared in the state $|\psi(t_0)\rangle$ at time t_0. Since no probability loss is in principle allowed, we must impose the condition $\sum_k P_k(t) = \sum_k P_k(t_0) = 1$, i.e.

$$\begin{aligned}&\sum_k |\langle k|\mathbf{U}(t - t_0)|\psi(t_0)\rangle|^2 \\ &= \langle \psi(t_0)|\mathbf{U}^\dagger(t - t_0)\left(\sum_k |k\rangle\langle k|\right)\mathbf{U}(t - t_0)|\psi(t_0)\rangle = 1,\end{aligned}$$

so that the evolution operator must once again be unitary.

In analogy to the space translations and rotations, we define an infinitesimal time evolution operator, generating a time translation of a small time interval δt, as follows:

$$\mathbf{U}(t + \delta t, t) = \mathbf{1} - \frac{\mathrm{i}}{\hbar}\delta t\, \mathbf{H}(t). \tag{7.2}$$

We refer to $\mathbf{H}(t)$ as the generator of the time translation to be interpreted as the quantum analog of the classical Hamiltonian of the system. From the duration composition law, a finite interval of time, $t' - t$, can be broken down into N infinitesimal intervals with $N\delta t = t' - t$; then, by exploiting the composition law, we have

$$\begin{aligned}\mathbf{U}(t', t) &= \mathbf{U}(t', t_1)\mathbf{U}(t_1, t_2)...\mathbf{U}(t_{N-1}, t_N)\mathbf{U}(t_N, t) \\ &= \mathrm{e}^{-\frac{\mathrm{i}}{\hbar}\delta t\mathbf{H}(t')}\mathrm{e}^{-\frac{\mathrm{i}}{\hbar}\delta t\mathbf{H}(t_1)}\cdots\mathrm{e}^{-\frac{\mathrm{i}}{\hbar}\delta t\mathbf{H}(t_N)},\end{aligned} \tag{7.3}$$

where we applied Eq. (7.2), converted into exponential form. We would like to combine the exponential factors into a single exponential, with exponent equal to the sum of the single exponents, but this can only be done if the Hamiltonian is independent of time, or otherwise if

$$\left[\mathbf{H}(t_i), \mathbf{H}(t_j)\right] = 0, \qquad t_i \neq t_j.$$

In the first case Eq. (7.3) is written as

$$\mathbf{U}(t',t) = \mathrm{e}^{-\frac{\mathrm{i}}{\hbar}\mathbf{H}(t'-t)}, \tag{7.4}$$

and, in the second case,

$$\begin{aligned}\mathbf{U}(t',t) &= \mathrm{e}^{-\frac{\mathrm{i}}{\hbar}\delta t[\mathbf{H}(t')+\mathbf{H}(t_1)+\ldots+\mathbf{H}(t_N)]} \\ &= \mathrm{e}^{-\frac{\mathrm{i}}{\hbar}\int_t^{t'}\mathrm{d}\tau\mathbf{H}(\tau)}. \end{aligned}\tag{7.5}$$

In all other cases it is necessary to adopt suitable approximation methods, which will be presented in Chap. 11. In the case of Eq. (7.4) we observe that the time evolution operator depends on the duration $t'-t$, which means that the Hamiltonian does not violate the uniformity of time, and hence its properties do not depend on the time chosen as origin of the time axis. In other words, such kind of Hamiltonian describes either isolated systems or those submitted to time independent external perturbation.

7.2 Equations of Motion

In this section, we shall derive the equations of motion in terms of the time evolution of the state of a physical system. We start with the general relation between the states at different times

$$|\psi(t')\rangle = \mathbf{U}(t',t)|\psi(t)\rangle. \tag{7.6}$$

Once the Hamiltonian in the time interval $t'-t$ is determined, the state of the system at the time instant t' will be univocally determined by the state at the time instant t. Such a point highlights an important feature of quantum mechanics. The time evolution of quantum states is deterministic, meaning that they obey rigorous motion laws in the same way as classical mechanics, whereas the states have an inherent probabilistic nature. In other words, in quantum mechanics, assigning to the particle a wave function $\psi(\vec{r},t)$ at a time instant t, the laws of motion uniquely determine the wave function $\psi(\vec{r},t')$ at t', whereas in classical mechanics, assigning to particle a defined position and momentum at a time instant t, the laws of motion uniquely determine position and momentum at any later time instant t'.

In the two special cases discussed above the time evolution of the state is simply obtained from Eq. (7.4) and Eq. (7.5)

$$|\psi(t')\rangle = \mathrm{e}^{-\frac{\mathrm{i}}{\hbar}\mathbf{H}(t'-t)}|\psi(t)\rangle,$$

$$|\psi(t')\rangle = \mathrm{e}^{-\frac{\mathrm{i}}{\hbar}\int_t^{t'}\mathrm{d}\tau\mathbf{H}(\tau)}|\psi(t)\rangle,$$

respectively. In order to establish the equations of motion in a general form, we submit the system at time t to a time translation of infinitesimal duration δt. The state translated of δt, is then

$$
\begin{aligned}
|\psi(t+\delta t)\rangle = \mathbf{U}(t+\delta t, t)|\psi(t)\rangle &= \left(\mathbf{1} - \frac{\mathrm{i}}{\hbar}\delta t \mathbf{H}(t)\right)|\psi(t)\rangle \\
&= |\psi(t)\rangle - \frac{\mathrm{i}}{\hbar}\delta t \mathbf{H}(t)|\psi(t)\rangle,
\end{aligned}
$$

which can be rewritten as

$$
\frac{|\psi(t+\delta t)\rangle - |\psi(t)\rangle}{\delta t} = -\frac{\mathrm{i}}{\hbar}\mathbf{H}(t)|\psi(t)\rangle.
$$

Finally, for $\delta t \rightarrow 0$, we get

$$
\mathrm{i}\hbar\frac{\mathrm{d}}{\mathrm{d}t}|\psi(t)\rangle = \mathbf{H}(\mathbf{t})|\psi(t)\rangle. \tag{7.7}
$$

This equation is called *time-dependent Schrödinger equation*. This equation can be recast as an equation of motion of the time evolution operator. Indeed, taking the time derivative of Eq. (7.6)

$$
\mathrm{i}\hbar\frac{\mathrm{d}}{\mathrm{d}t'}\mathbf{U}(t', t)|\psi(t)\rangle = \mathrm{i}\hbar\frac{\mathrm{d}}{\mathrm{d}t'}|\psi(t')\rangle,
$$

and applying Eq. (7.7) we get

$$
\mathrm{i}\hbar\frac{\mathrm{d}}{\mathrm{d}t'}\mathbf{U}(t', t) = \mathbf{H}(t')\mathbf{U}(t', t),
$$

for any $|\psi(t)\rangle$.

The time evolution of the probability density $\rho(\vec{r}, t) = |\psi(\vec{r}, t)|^2$ can alternatively replace the time evolution of the wave function $\psi(\vec{r}, t)$. This offers two advantages not only for the interpretation of the wave function but also for the constraints to be imposed to make it physically acceptable. For simplicity we consider the motion of a particle in a potential field $V(\vec{r})$. From the Schrödinger equation, Eq. (7.7), we easily get

$$
\frac{\partial\rho(\vec{r}, t)}{\partial t} + \vec{\nabla}\cdot\vec{J}(\vec{r}, t) = 0, \tag{7.8}
$$

where

$$
\vec{J}(\vec{r}, t) = \frac{\hbar}{2im}[\psi^*(\vec{r}, t)\vec{\nabla}\psi(\vec{r}, t) - \psi(\vec{r}, t)\vec{\nabla}\psi^*(\vec{r}, t)] \tag{7.9}
$$

is the probability current. Equation (7.8) is the quantum analog of the fluid dynamical *continuity equation*, i.e. the conservation law of fluid during its motion, but here the conserved quantity is the probability.[1] Of interest are the processes, in which the probability is not conserved, e.g. collisions with absorption of particles, described by complex potentials. In such cases the imaginary part of potential replaces the zero in the righthand side of Eq. (7.8).

[1] The continuity equation will be taken up again in Chap. 17.

In the present context, the interest for the continuity equation is that the probability conservation requires the wave function to be a continuous function everywhere in space along with its derivative. This imposes on the wave function strict restrictions in the points, in which the potential shows a discontinuity, the same thing that happens with the potential well.

7.3 Stationary Schrödinger Equation

A class of solutions of the Schrödinger equation, Eq. (7.7), is that of an isolated system, whose Hamiltonian is independent of time. These solutions have the form

$$|\psi(t)\rangle = \mathrm{e}^{-\frac{\mathrm{i}}{\hbar}Et}|\psi\rangle, \tag{7.10}$$

where the time dependence is only in the phase factor. By calculating the average value of any physical observable with respect to such states, one easily finds

$$\langle\psi(t)|\mathbf{O}|\psi(t)\rangle = \langle\psi|\mathbf{O}|\psi\rangle,$$

which means that, when the system occupies one of these states, its properties are preserved as time passes. In other words, the system is in a *stationary state*. The parameter E in Eq. (7.10) plays the special role of energy associated to a stationary state. By plugging Eq. (7.10) into the Schrödinger equation, it follows the *time-independent Schrödinger equation*:

$$\mathbf{H}|\psi\rangle = E|\psi\rangle.$$

This is an eigenvalue equation for the Hamiltonian H, and its eigenvalues are the energy values of the system. In conclusion, the search for the stationary states of a system is equivalent to solving the eigenvalue problem of its Hamiltonian, the eigenvalues being the energy values corresponding to the eigenstates. The solution (or diagonalization) of the eigenvalue equation for the Hamiltonian of various physical systems is one of the main issues addressed in the next sections as well as in the following chapters.

7.3.1 Schrödinger Equation for Potential Wells

In this section we shall solve the stationary Schrödinger equation for elementary systems. The free particle is the simplest one. Its Hamiltonian is the kinetic energy of the particle,

$$\mathbf{H} = \frac{\mathbf{p}^2}{2m}.$$

The operators $\mathbf{H}$ and $\vec{p}$, being $\mathbf{H}$ a function of $\vec{\mathbf{p}}$, have a common set of eigenstates, the plane waves, as we learned in Chap. 3. The first non-trivial system is made of two interacting particles with mass m whose Hamiltonian has the form

$$\mathbf{H} = \frac{\mathbf{p}_1^2}{2m} + \frac{\mathbf{p}_2^2}{2m} + V(\vec{\mathbf{r}}_1 - \vec{\mathbf{r}}_2). \tag{7.11}$$

Since this Hamiltonian is invariant under space translations, the two-particle motion can be split into the center-of-mass and relative motion

$$\mathbf{H} = \frac{\mathbf{P}^2}{2M} + \frac{\mathbf{p}^2}{m} + V(\vec{\mathbf{r}}),$$

where $\vec{\mathbf{P}} = \vec{\mathbf{p}}_1 + \vec{\mathbf{p}}_2$ is the total momentum and $\vec{\mathbf{r}} = \vec{\mathbf{r}}_1 - \vec{\mathbf{r}}_2$ is the relative coordinate. The two motions are independent of each other. As a result, the eigenfunctions of the Hamiltonian Eq. (7.11) are the product between the eigenfunctions of the free center-of-mass motion and the eigenfunctions of the relative motion

$$\psi(\vec{r}_1, \vec{r}_2) = \mathrm{e}^{\frac{i}{\hbar}\vec{P}\cdot\vec{R}}\psi_{\mathrm{rel}}(\vec{r}),$$

$\vec{R}$ being the center-of-mass position. We conclude that the study of the relative motion of two interacting particles amounts to that of one particle with mass $m/2$, named reduced mass (see Eq. (7.12)), subjected to a central potential. In the case of two particles with different mass, the reduced mass is

$$m = \frac{m_1 m_2}{m_1 + m_2} \tag{7.12}$$

Postponing to Chap. 12 the study of the full two-body problem, we discuss the case of a particle in a one-dimensional potential, starting with the general properties of the solution of the Schrödinger in the simplest case of a potential well

$$V(q) = \begin{cases} V_0 & \text{for } |q| < l \\ 0 & \text{elsewhere.} \end{cases}$$

Such a function shows two discontinuity points, $q = -l$ and $q = l$, for which the potential changes abruptly from 0 to a constant value $V_0 \neq 0$. short-range potentials, like the nuclear and ionic ones, can be approximated by potential wells; conversely, this is not possible for long-range potentials like, for instance, the Coulomb potential. The sign of V_0 establishes whether the well is attractive, $V_0 < 0$, or repulsive, $V_0 > 0$. It is never too much to remember that only the difference of the potentials between two points, one placed at $q = \infty$, has physical significance, hence $V_0 < 0$ means $V_0(P) - V_0(\infty) < 0$ with the usual choice $V_0(\infty) = 0$. Despite these simplifications, much of the properties of these simple systems can be extended to general cases. Let us write the Schrödinger equation,

$$\left[-\frac{\hbar^2}{2m}\frac{d^2}{dq^2}+V(q)\right]\psi_E(q)=E\psi_E(q). \tag{7.13}$$

We will discuss the two cases: attractive and repulsive well. For $|q|>l$, the potential has the value $V(q)=0$, and the wave function satisfies the equation,

$$-\frac{\hbar^2}{2m}\frac{d^2}{dq^2}\psi_E(q)=E\psi_E(q).$$

The solutions of Eq. (7.13) depend on the sign of E:

(1) $E>0$: the wave function is a plane wave.
(2) $V_0<E<0$: the wave function has an exponential behaviour, which can be either increasing or decreasing.
(3) $E<V_0$: there are no solutions of physical interest because the velocity would have an imaginary value.

The $E>0$ case represents a state in which the particle is not confined inside the well. In reality, this situation describes a particle that approaches the well, interacts with it and then moves away. In classical mechanics this corresponds to the motion over a limitless orbit, that is, an elastic collision of the particle with a scattering center. The Schrödinger equation has solutions for arbitrary values of $E>0$; the eigenvalue range is known as *continuum spectrum* of the Hamiltonian.

The $V_0<E<0$ case, instead, corresponds to a bound state of the particle. It is confined inside the well, but with a probability exponentially decreasing from inside to outside. The exponentially increasing solution has to be rejected, because it would give, for $q\to\infty$, an infinitely increasing probability. Classical Mechanics depicts this situation as a motion over a closed orbit. At variance with classical mechanics such bound states only exist for some special values of $E<0$, forming the so-called *discrete spectrum* of the Hamiltonian. We point out that such a property is called *energy quantization*.

For $E<V_0$ the particle would move inside the well with an imaginary momentum in classical mechanics or negative kinetic energy in quantum mechanics. Thus, such a solution is rejected by the both of them.

7.3.2 Attractive Well: $V_0<0$

Let us examine the solutions of the Schrödinger equation for the case $V_0<E<0$, where V_0 is non-vanishing in the interval $-l<q<l$ and vanishing outside (see Fig. 7.1). Since the potential is invariant under the inversion of the q-axis, we expect to find two kinds of solution, i.e. symmetric and antisymmetric solutions. The Schrödinger equation, Eq, (7.13), can be written

$$\psi''(q)+k^2\psi(q)=0, \tag{7.14}$$

where $k = \tilde{k} = \sqrt{2m(E - V_0)/\hbar^2}$ inside the well and $k = k_0 = \sqrt{2mE/\hbar^2}$ outside. Since the energy E is negative, k_0 is a pure imaginary number, but $\tilde{k}$ is real and positive. The general solution in the three space regions is

$$(I) \quad \psi_I(q) = a\, e^{-ik_0 q} q < -l$$

$$(II) \quad \psi_{II}(q) = b\, \sin(\tilde{k}q + \phi) \quad |q| < l \tag{7.15}$$

$$(III) \quad \psi_{III}(q) = c\, e^{ik_0 q} q > l, \tag{7.16}$$

taking into account that for bound states the solutions must be asymptotically decreasing outside the well. The continuity conditions on the wave function and on its first derivative, at the points of separation $q = l$ and $q = -l$ of the three intervals allow us to determine the three coefficients as well as the energy eigenvalues. These conditions are written

$$\psi_I(-l) = \psi_{II}(-l); \qquad \psi_I'(-l) = \psi_{II}'(-l) \tag{7.17}$$

$$\psi_{II}(l) \ = \psi_{III}(l); \qquad \psi_{II}'(l) \ = \psi_{III}'(l). \tag{7.18}$$

The parity conservation significantly simplifies the problem. It is enough to consider only the solutions with even parity, that is $\psi(q) = \psi(-q)$. In this case, this symmetry requires that $a = c$ ($a = -c$ in the odd parity case) and

$$\sin(\tilde{k}q + \phi) \ = \ \sin(-\tilde{k}q + \phi). \tag{7.19}$$

This condition is satisfied when $\phi = (2n + 1)\pi/2$, with integer n, so that $\psi_{II}(q) = \cos(\tilde{k}q)$ ($\phi = n\pi$ and $\psi_{II} = \sin(\tilde{k}q)$ in the case of odd parity). Thus, only two of the four boundary conditions, Eq. (7.18), are independent. By employing Eq. (7.16), the two conditions at the discontinuity $q = l$ are written

$$b\, \cos(\tilde{k}l) = a\, e^{ik_0 l}, \qquad -\tilde{k}b \sin(\tilde{k}l) = ik_0 a e^{ik_0 l}. \tag{7.20}$$

By squaring and adding them, we get the ratio of the coefficients

$$\frac{b^2}{a^2} = \left(1 - \frac{k_0^2}{\tilde{k}^2}\right) e^{2ik_0 l}, \tag{7.21}$$

with which the determination of the coefficients is completed, being $c = a$. The missing coefficient is arbitrary, because the eigenfunctions are determined less than an arbitrary factor. The coefficients depend on $\tilde{k}$, and hence from the energy eigenvalues E. These must first be calculated, which can be done considering the ratio of the two conditions in Eq. (7.20), i.e.

$$\tan(\tilde{k}l) \ = \ -\frac{ik_0}{\tilde{k}}\,. \tag{7.22}$$

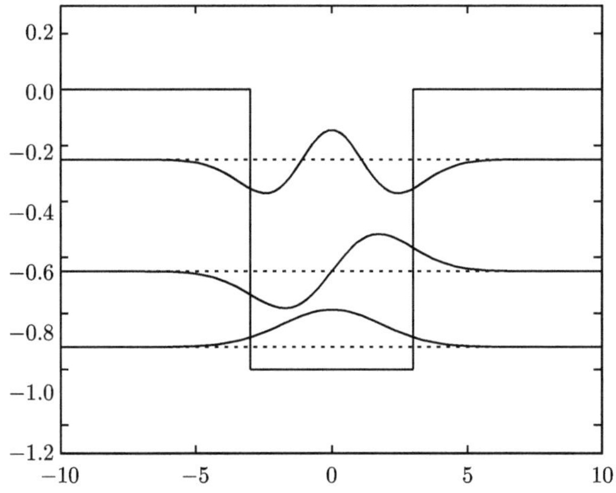

Fig. 7.1 First three eigenfunctions of the attractive well

This equation can be solved only graphically: the intersections of the two functions at first and second member give the quantized eigenvalues of $\tilde{k}$, and hence of the energy E. Figure 7.1 shows the first three eigenfunctions. Notice that the lowest energy curve is the ground-state eigenfunction, symmetric around $q = 0$ (even parity), the next one is the first excited state eigenfunction, antisymmetric (odd parity), and the third one the second excited state eigenfunction symmetric again.

7.3.3 *Repulsive Well:* $V_0 > 0$

We refer to the repulsive well as a potential barrier. In such a case there are only unbound states with positive energy. They describe two physical phenomena occurring when a particle hits the barrier, as illustrated in Fig. 7.2. They are known as barrier reflection when the barrier bounces the particle back, and transmission, when the particle crosses the barrier. We know that, in a classical physics scenario, the two processes are alternative, meaning that for $E < V_0$ the barrier bounces the particle back and for $E > V_0$ the particle crosses the barrier. In the quantum mechanics

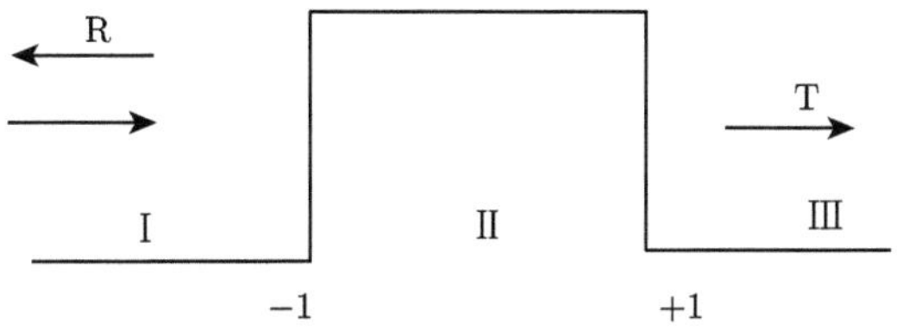

Fig. 7.2 Sketch of the potential barrier

scenario, on the other hand, regardless of the E magnitude with respect to V_0, both situations can occur simultaneously with different probabilities. We examine the $E < V_0$ and $E > V_0$ case, separately.

7.3.4 Potential Barrier: $0 < E < V_0$

According to the previous considerations, we search for solutions of the following form

$$\psi(q) = \begin{cases} e^{ikq} + Re^{-ikq} & q < -l \\ ae^{-\tilde{k}q} + be^{\tilde{k}q} & |q| < l \\ Se^{ikq} & q > l, \end{cases} \tag{7.23}$$

where $k = \frac{\sqrt{2mE}}{\hbar}$ and $\tilde{k} = \frac{\sqrt{2m(V_0 - E)}}{\hbar}$. The coefficients R and S regard, respectively, the reflection and the transmission of the incoming wave. We then calculate the density of the probability currents concerning reflection and transmission by means of Eq. (7.8)

$$j_{\text{trans}} = \frac{\hbar k}{m}|S|^2,$$
$$j_{\text{refl}} = \frac{\hbar k}{m}(1 - |R|^2).$$

In view of the continuity equation, Eq. (7.8), we observe that for stationary states $\vec{\nabla} \cdot \vec{j} = 0$, this means that the current is constant at each point of space and it follows that,

$$|S|^2 + |R|^2 = 1. \tag{7.24}$$

The physical meaning of Eq. (7.24) is straightforward: the incoming particle can be transmitted to or reflected by the barrier and there are no other possibilities. The reflection and transmission probabilities, respectively, correspond to the numbers $|R|^2$ and $|S|^2$. The boundary conditions for the wave function and its first space derivative make it easy to calculate the four coefficients a, b, R and S appearing in Eq. (7.23),

$$a = \frac{(1 + i\eta)e^{(\tilde{k}-ik)l}}{(1 - \eta^2)\sinh(2\tilde{k}l) + 2i\eta\cosh(2\tilde{k}l)},$$
$$b = \frac{(1 - i\eta)e^{-(\tilde{k}+ik)l}}{(\eta^2 - 1)\sinh(2\tilde{k}l) - 2i\eta\cosh(2\tilde{k}l)},$$

where $\eta = \frac{\tilde{k}}{k} = \sqrt{\frac{V_0 - E}{E}}$ and the transmission coefficient T is given by

$$T = |S|^2 = \frac{4E(V_0 - E)}{V_0^2 \sinh^2(2\tilde{k}l) + 4E(V_0 - E)}.$$

The reflection coefficient can be obtained from Eq. (7.24). The relevant result is that the probability for the particle to go through the barrier, i.e. the transmission coefficient T, is different from zero. This is a purely quantum effect called *tunneling*.

7.3.5 Potential Barrier: $E > V_0 > 0$

We search for solutions of the form

$$\psi(q) = \begin{cases} e^{ikq} + Re^{-ikq} & q < -l \\ \sin(\tilde{k}q) + b\cos(\tilde{k}q) & |q| < l \\ Se^{ikq} & q > l, \end{cases}$$

where $k = \frac{\sqrt{2mE}}{\hbar}$ and $\tilde{k} = \frac{\sqrt{2m(E - V_0)}}{\hbar}$, in this case, the wave function inside the well has an oscillating behaviour due to $E > V_0$. We proceed in the same way as the previous case to obtain the integration coefficients,

$$a = \frac{-e^{-ikl}}{\sin(\tilde{k}l) + i\eta\cos(\tilde{k}l)},$$

$$b = \frac{e^{-ikl}}{\cos(\tilde{k}l) - i\eta\sin(\tilde{k}l)},$$

where $\eta = \frac{\tilde{k}}{k}$. Conversely, to the previous case, we write the expression for the reflection coefficient,

$$|R|^2 = 1 - |S|^2 = \frac{V_0^2 \sin^2(2\tilde{k}l)}{4E(E - V_0) + V_0^2 \sin^2(2\tilde{k}l)}. \tag{7.25}$$

Equation (7.25) highlights another relevant quantum effect: despite the fact that the energy of the particle E is greater than the barrier's height V_0, there exists a finite probability $|R|^2$ for the particle to be reflected.

7.4 Problems

- Problem 1. Derive the continuity equation, Eq. (7.8), when the potential contains an imaginary part, i.e. $\mathbf{U} = \mathbf{U_R} + \mathbf{iU_I}$.
- Problem 2. Solve the eigenvalue equation for a particle with mass m freely moving on a ring with radius R. (Hint: Express the free Hamiltonian in terms of angular coordinates and solve the differential equation with periodic boundary conditions $\psi(\phi) = \psi(\phi + 2\pi)$ and $\frac{\mathrm{d}\psi(\phi)}{\mathrm{d}\phi} = \frac{\mathrm{d}\psi(\phi + 2\pi)}{\mathrm{d}\phi}$.
- Problem 3. A particle with wave function $\psi(x) = x \cdot (|L| - x)$ is inside a potential well

$$V(x) = \begin{cases} 0 & \text{for } |x| \le L \\ \infty & \text{elsewhere} \end{cases}$$

  Calculate the probability of finding the particle in any stationary state of the well.
- Problem 4. A beam of particles of mass m and energy E, running along the x-axis, meets a potential step $V = V_0 > 0$ for $x > 0$. Calculate the reflected current and the probability to find a particle in the interval $0 < x < L$ in the case $E < V_0$.
- Problem 5. Consider the attractive square potential $V(q) = -V_0$ in the interval $|q| < L$ and equal to zero elsewhere. From the solutions of the Schrödinger equation with energy $E > 0$, determine the transmission coefficient and its properties.

Chapter 8
Harmonic Oscillations

Harmonic oscillations are frequently encountered in several fields of physics. Remarkable examples are the electromagnetic waves, lattice vibrations in solid state physics, giant resonances in nuclear physics, neutrino oscillations in elementary particle physics. The reason for this is that physical systems are commonly in a stable state, that is, guaranteed by the existence of a minimum of the potential. But when a small perturbation displaces the system from equilibrium, the latter makes small harmonic oscillations. Let us consider, for simplicity purposes, a one-dimensional system subjected to an attractive potential $V(q)$ with a minimum in q_0, corresponding to the equilibrium position. If the system departs from q_0, it carries out small shifts $\delta q = q - q_0$. Its motion will be determined after the potential $V(q)$ has been developed in power series of δq truncated to the first non-vanishing order

$$V(q) = \frac{1}{2} V''(q_0)(q - q_0)^2,$$

being $V'(q_0)$=0. The energy scale has been set such that $V(q_0)$=0. By solving the classical equations of motion, it is easily seen that $V''(q_0)$, known as string constant, determines the oscillations with angular frequency ω, i.e. $V''(q_0) = m\omega^2$, where m is the mass of the system. Thus, the potential can be written in the well-known form

$$V(q) = \frac{1}{2} m\omega^2 (q - q_0)^2 \tag{8.1}$$

called harmonic oscillator potential. So far this has been a general discussion.

In classical mechanics, the equation of motion of a particle with mass m in harmonic oscillator potential yields harmonic oscillations $q(t) - q_0 = A\cos(\omega t)$ with amplitude A, angular frequency ω, and energy $E = m\omega^2 A^2$. Position $q(t)$ and

---

Supplementary Information The online version contains supplementary material available at https://doi.org/10.1007/978-981-97-1670-8_8.

U. Lombardo et al., *Quantum Mechanics*, UNITEXT for Physics,
https://doi.org/10.1007/978-981-97-1670-8_8

momentum $p(t) = m\dot{q}(t)$ are simultaneously determined at any time, and the energy E is a continuous function of the amplitude. The quantum mechanical picture is quite different: the values of $q(t)$ and $p(t)$ at the same time are submitted to the uncertainty principle and the energy is quantized.

8.1 Quantum Harmonic Oscillator

Moving to quantum mechanics we must replace the classical commuting numerical variables q and p with the quantum non-commuting operators $\mathbf{q}$ and $\mathbf{p}$. These latter obey the commutation rule $[\mathbf{q}, \mathbf{p}] = \mathrm{i}\hbar$. The one-dimensional harmonic oscillator Hamiltonian is

$$\mathbf{H} = \frac{\mathbf{p}^2}{2m} + \frac{1}{2}m\omega^2\mathbf{q}^2.$$

The Schrödinger equation

$$\left(\frac{\mathbf{p}^2}{2m} + \frac{1}{2}m\omega^2\mathbf{q}^2\right)|\psi_n\rangle = E_n|\psi_n\rangle$$

is a differential equation, but it can be solved algebraically using the Dirac method. Let us introduce the operators

$$\mathbf{a} \equiv \sqrt{\frac{m\omega}{2\hbar}}\left(\mathbf{q} + \frac{\mathrm{i}\mathbf{p}}{m\omega}\right), \quad \mathbf{a}^\dagger \equiv \sqrt{\frac{m\omega}{2\hbar}}\left(\mathbf{q} - \frac{\mathrm{i}\mathbf{p}}{m\omega}\right), \quad \mathbf{N} \equiv \mathbf{a}^\dagger\mathbf{a}, \tag{8.2}$$

which fulfill the commutation rules

$$[\mathbf{a}, \mathbf{a}^\dagger] = 1, \qquad [\mathbf{N}, \mathbf{a}] = -\mathbf{a}, \qquad [\mathbf{N}, \mathbf{a}^\dagger] = \mathbf{a}^\dagger. \tag{8.3}$$

In terms of the new operators the Schrödinger equation takes the form

$$\hbar\omega\left(\mathbf{N} + \frac{1}{2}\right)|\psi_n\rangle = E_n|\psi_n\rangle.$$

The eigenstates of $\mathbf{H}$ with eigenvalues E_n are also eigenstates of $\mathbf{N}$ with eigenvalues n, that is $|\psi_n\rangle \equiv |n\rangle$ and $E_n = \hbar\omega\left(n + \frac{1}{2}\right)$.

The following properties of $\mathbf{N}$ are easily deduced

- $\mathbf{a}^\dagger|n\rangle$ is eigenstate of $\mathbf{N}$ belonging to eigenvalue $n + 1$,
- $\mathbf{a}|n\rangle$ is eigenstate of $\mathbf{N}$ belonging to eigenvalue $n - 1$,
- the eigenvalues n are all integer numbers starting from 0.

The first two properties are a consequence of the commutation rules, Eq. (8.3); as for the third one, starting from the eigenstate $|n\rangle$ with $n \geqslant 0$[1] and by repeatedly applying the operator **a**, one arrives at the eigenstate $\mathbf{a}^m|n\rangle \sim |n-m\rangle$, with $m = n$. Since m is integer, n is also integer. Starting with the eigenstate $|0\rangle$ belonging to $n = 0$, one can build up whatever eigenstate of **N** repeatedly applying the operator $\mathbf{a}^\dagger$ as follows,

$$
\begin{aligned}
|1\rangle &= \mathbf{a}^\dagger|0\rangle, \\
|2\rangle &= \frac{\mathbf{a}^\dagger}{\sqrt{2}}|1\rangle = \frac{(\mathbf{a}^\dagger)^2}{\sqrt{2!}}|0\rangle, \\
|3\rangle &= \frac{\mathbf{a}^\dagger}{\sqrt{3}}|2\rangle = \frac{(\mathbf{a}^\dagger)^3}{\sqrt{3!}}|0\rangle, \\
\cdots\cdots & \;\cdot\; \cdots\cdots\cdots\cdots \\
|n\rangle &= \frac{(\mathbf{a}^\dagger)^n}{\sqrt{n!}}|0\rangle.
\end{aligned}
$$

These are all eigenstates of $\boldsymbol{H}$ with their respective eigenvalues

$$
E_n = \hbar\omega\left(n + \frac{1}{2}\right) \qquad (n = 0, 1, 2, 3, \cdots).
$$

We see that the energy levels are equally spaced. This property makes it easy to identify the collective oscillations in nuclear, atomic or molecular spectra. The gap between the two nearest neighbor levels $\hbar\omega$ is called excitation quantum. The eigenstate $|n\rangle$ is interpreted as a state containing n quanta, and the operator $\mathbf{a}^\dagger$ (**a**) is called creation (annihilation) operator because, when applied to $|n\rangle$, it adds (removes) a quantum $\hbar\omega$. This excitation quantum is called phonon when dealing with ion oscillations of a lattice and photon when dealing with electromagnetic oscillations of a black body. The ground state $|0\rangle$ is known as the vacuum, even though it has, in fact, a non vanishing energy equal to $\frac{1}{2}\hbar\omega$ (zero-point energy). This is consistent with the uncertainty principle $\Delta q \Delta p \geqslant \hbar/2$. Indeed it is possible to prove (Problem 1) that the uncertainty is the smallest with the harmonic oscillator, namely $\Delta q \Delta p = \hbar/2$.

8.1.1 Eigenfunctions of the Harmonic Oscillator

As was the case with the energy spectrum, the eigenfunctions of the harmonic oscillator can be obtained algebraically instead of solving the Schödinger equation in the form of a second-order differential equation. We follow an iterative procedure starting with the definition of vacuum state

[1] Notice that $n = \langle n|\mathbf{N}|n\rangle = \langle n|\mathbf{a}^\dagger\mathbf{a}|n\rangle = \|\mathbf{a}|n\rangle\|^2 \geqslant 0$.

$$\mathbf{a}|0\rangle = 0.$$

By using the transformation in Eq. (8.2), the vacuum wave function written is obtained from $\langle q|\mathbf{a}|0\rangle = 0$ integrating the first differential equation

$$\sqrt{\frac{m\omega}{2\hbar}}\langle q|\mathbf{q} + \frac{\mathrm{i}\mathbf{p}}{m\omega}|0\rangle = \sqrt{\frac{m\omega}{2\hbar}}\left(q + \frac{\hbar}{m\omega}\frac{d}{dq}\right)\psi_0(q) = 0,$$

Normalized to unity the result is

$$\psi_0(q) = \left(\frac{m\omega}{\pi\hbar}\right)^{\frac{1}{4}} \mathrm{e}^{-\frac{1}{2}\frac{m\omega}{\hbar}q^2}.$$

The wave functions of the excited states are obtained repeatedly applying the creation operator $\mathbf{a}^\dagger$ transformed according with Eq. (8.2). For example, the first excited state is obtained as follows,

$$\begin{aligned}\psi_1(q) = \langle q|1\rangle = \langle q|\mathbf{a}^\dagger|0\rangle &= \sqrt{\frac{m\omega}{2\hbar}}\left(q - \frac{\hbar}{m\omega}\frac{d}{dq}\right)\psi_0(q)\\ &= \left[\frac{4}{\pi}\left(\frac{m\omega}{\hbar}\right)^3\right]^{1/4} q\,\mathrm{e}^{-\frac{1}{2}\frac{m\omega}{\hbar}q^2}.\end{aligned}$$

The eigenfunctions are the Hermite polynomials, whose properties are illustrated in Appendix C.

8.2 Vibrations of a Crystal Lattice

In this section small oscillations in quantum systems will be studied in view of important applications in various fields. Due to the strong similarity between the quantum and classical treatments, the small oscillations will be developed in both approaches.

8.2.1 Small Oscillations in Classical Approach

Consider a system of N interacting particles with mass m, e.g. nucleons in a nucleus, atoms in a molecule or ions in a lattice. For simplicity purposes we consider each particle of the system in one-dimensional motion. In this case the classical Hamiltonian is

$$H = \sum_{k=1,N} \frac{p_k^2}{2m} + V(q_1, q_2, \cdots, q_N).$$

Suppose the system has a stable equilibrium configuration $q_0 \equiv [q_0^{(1)}, q_0^{(2)}, \cdots, q_0^{(N)}]$ corresponding to the minimum of the interaction potential. This condition permits the onset of small collective excitations around q_0, In the aforementioned physical cases, these could be giant resonances in nuclei, inter-atomic vibrations in molecular gases, or ion vibrations in crystal lattices. With the same procedure leading to the harmonic oscillator potential, Eq. (8.1), we develop the interaction in series up to the second-order of the displacement of the coordinates $\delta q_i = q_i - q_0^{(i)}$ from the minimum

$$V(\delta q_1, \delta q_2, \cdots, \delta q_N) = \frac{1}{2} \sum_{j,k=1,N} M_{jk}(q_0)\delta q_j \delta q_k, \tag{8.4}$$

where M_{jk} are the second derivatives of the interaction with respect to the displacements calculated in q_0. The sum restriction $j < k$ is replaced by double counting, which is compensated by the introduction of the factor $\frac{1}{2}$ in front of the sum. In this way, the potential is real and symmetric, that is $M_{jk} = M_{kj}$. Given the Hamiltonian, the small oscillations are obtained solving the equations of motion in the Hamilton formalism

$$\begin{aligned} \delta\dot{q}_k &= \{\delta q_k, H\} = \frac{p_k}{m}, \\ \dot{p}_k &= \{p, H\} = -\sum_i M_{ki}\delta q_i. \end{aligned}$$

The two coupled equations can be combined taking the time derivative of the first equation and applying the latter equation to the result. We get

$$m\, \delta\ddot{q}_k = -\sum_i M_{ki}\delta q_i.$$

Since the Hamiltonian has no explicit time dependence, we can search for a solution of the form $\delta q_k = a_k \mathrm{e}^{\mathrm{i}\omega t}$, so that

$$\sum_k \left(-m\omega^2 \delta_{ik} + M_{ik}\right) a_k = 0.$$

This is none other than an eigenvalue equation in N-dimensional space. In general there are N different eigenvalues ω_α^2 (non-degenerate case), which are obtained by solving the secular equation

$$\det\left[M - m\omega^2 \mathbf{1}\right] = 0.$$

Since the matrix M is real and symmetric, the eigenvalues are real and positive. In addition, the eigenvectors with components δ_i^α are orthogonal and can be normalized to unity so that $\sum_i \delta q_i^\alpha \delta q_i^\beta = \delta_{\alpha\beta}$. These modes with angular frequency ω_α are called normal excitations modes. The general solution to the equations of motion is a superposition of these modes

$$\delta q_i = \sum_\alpha \delta q_i^\alpha e^{i\omega t}.$$

Lastly, we want to express the Hamiltonian in terms of normal coordinates. We have to perform the transformation to the new coordinates δQ_α by means of the eigenvector components

$$\delta q_i = \sum_\alpha \delta q_i^\alpha \delta Q_\alpha.$$

By imposing this transformation to the Hamiltonian we easily achieve

$$H = \sum_\alpha \frac{1}{2m}\left[P_\alpha^2 + \omega_\alpha^2 \delta Q_\alpha^2\right], \tag{8.5}$$

after having applied the eigenvalue equation and the orthogonality of the eigenvectors. In conclusion, we have proven that a system of coupled harmonic oscillators can be broken down into a sum of independent harmonic oscillators corresponding to the normal modes of vibration.

An important example of a many-body system making small oscillations is the crystal lattice. In this case the particles are arranged in periodic configurations. Consistent with the above discussed model, we can consider a one-dimensional lattice of infinite length, where identical atoms of mass m are equally spaced at a distance a apart (lattice constant) as shown in Fig. 8.1.

Each atom n makes small oscillations $\delta q_n \ll a$ around the equilibrium positions $q^0 \equiv [q_1^0, q_2^0, \cdots]$. Expanded then at the second order around q^0, the inter-atomic

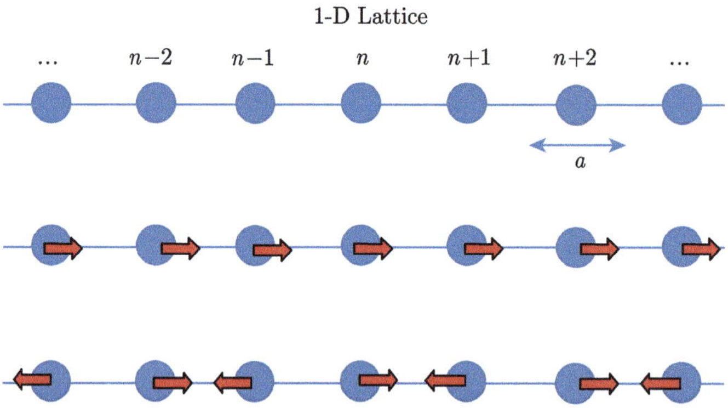

Fig. 8.1 Three configurations of a one-dimensional lattice at the instant t. Upper chain: equilibrium configuration. Middle chain: vibrations with $k < \pi/a$. Lower chain: vibrations with $k = \pi/a$

interaction takes the form of a multi-dimensional harmonic oscillator potential, as shown in Eq. (8.5). To simplify the formal developments we assume that the inter-atomic potential is short-range so that each atom interacts only with the two nearest atoms. As a result, the Hamiltonian takes the form,

$$H = \sum_n \left[\frac{p_n^2}{2m} + \frac{1}{2} M_n (\delta q_{n+1} - \delta q_n)^2 \right],$$

and the equations of motion for the displacements δq_n are

$$m\,\delta \ddot{q}_n = -M(\delta q_{n+1} - \delta q_n) - M(\delta q_n - \delta q_{n-1}), \tag{8.6}$$

where $M_n \equiv M$ is constant due to the translational symmetry of the lattice. The same symmetry suggests looking for solutions in the form of plane waves

$$\delta q_n = A e^{i(k q_n^0 - \omega t)},$$

where $q_n^0 = na$. In Fig. 8.1, two collective vibration modes are depicted along with the equilibrium configuration. By imposing the plane wave solution to the equation of motion, we can easily obtain the *dispersion relation* connecting the wave number k with the angular frequency $\omega(k)$.

$$\omega(k) = 2\omega_0 |\sin(ka/2)|,$$

where $\omega_0 = \sqrt{\dfrac{M}{m}}$. This dispersion relation is plotted in Fig. 8.2. In the most popular case of light, the dispersion relation between wave number and angular frequency is $\omega = ck$, where c is the velocity of light. Similarly, the lattice vibration propagates with phase velocity ω/k. In general, the group velocity $\partial\omega/\partial k$ does not coincide with

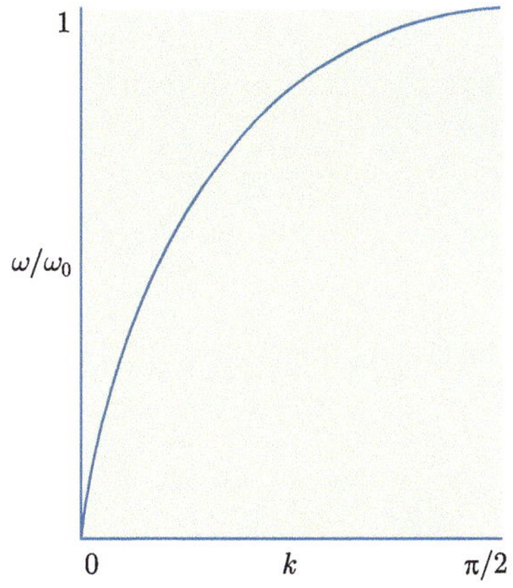

Fig. 8.2 Dispersion relation $\omega = \omega(k)$ for 1-D lattice vibrations ($k > 0$)

the phase velocity. In the limiting case $k \to 0$, the angular frequency is $\omega = \omega_0 ka$ and the propagation velocity $v = \sqrt{a^2 \frac{M}{m}}$ turns out to be constant. The collective propagation results from in-phase oscillations of atoms as shown in Fig. 8.1. This behaviour is similar to what happens with the wave propagation in a continuous elastic medium, because the propagation of waves with wavelength $\lambda \gg a$, in particular $\lambda = \frac{1}{k} \to \infty$, is insensitive to the discontinuous lattice structure. However, as soon as the wavelength becomes comparable to the lattice constant the structure effects appear and at $k = \pi/a$ the atom vibrations abruptly change from phase accordance to phase opposition, exhibiting the maximum restoring force.

8.2.2 Small Oscillations in Quantum Approach

The classical description of the collective vibrations in terms of normal modes, so-called phonons, provides for a satisfactory understanding of the phonon band structure in crystals, but it cannot explain the experimental thermal properties. As a result, we need to extend this study to quantum mechanics.

We again consider the one-dimensional lattice and extend the formalism developed in the last subsection in accordance with quantum mechanics. The lattice is made of N equally spaced atoms with lattice constant a. The Hamiltonian is formally identical

$$\mathbf{H} = \sum_n \left[\frac{\mathbf{p}_n^2}{2m} + \frac{1}{2} M_n (\delta \mathbf{q}_{n+1} - \delta \mathbf{q}_n)^2 \right],$$

but now the displacement coordinates and conjugate momenta are Hermi- tian operators following the commutation rules

$$\left[\delta \mathbf{q}_n, \mathbf{p}_{n'} \right] = \mathrm{i}\hbar \delta_{n,n'}.$$

Expressed in terms of their Fourier transforms with suitable boundary conditions they are

$$\delta \mathbf{q}_n = \frac{1}{\sqrt{N}} \sum_k \mathrm{e}^{ikna} \delta \mathbf{q}_k, \qquad \mathbf{p}_n = \frac{1}{\sqrt{N}} \sum_k \mathrm{e}^{ikna} \mathbf{p}_k.$$

The Fourier transforms fulfill the commutation rules

$$\left[\delta \mathbf{q}_k, \mathbf{p}_{k'}^{\dagger} \right] = \mathrm{i}\hbar \delta_{k,k'},$$

where $\delta\mathbf{q}_k^\dagger = \delta\mathbf{q}_k$ and $\mathbf{p}_k^\dagger = \mathbf{p}_k$. Owing to the boundary conditions, the wave numbers take discrete values $2\pi m/(Na)$. In terms of the new parameters the Hamiltonian takes the form

$$\mathbf{H} = \sum_k \left[\frac{\mathbf{p}_k^\dagger \cdot \mathbf{p}_k}{2m} + \frac{m\omega_k^2}{2} \delta\mathbf{q}^\dagger{}_k \cdot \delta\mathbf{q}_k \right].$$

This is the quantum Hamiltonian of N independent harmonic oscillators. The ground state of the lattice is the state where each oscillator is at the lowest level (no phonons). The energy states of the overall system are the result of the collection of excitation levels n_k of each individual oscillator k. The total energy will be $E = \sum_k n_k \hbar\omega_k$. The occupation numbers n_k can be interpreted as the number of phonons of energy $\hbar\omega_k$ occupying the state k. We postpone in Chap. 16 the thermodynamical implications of such a picture.

8.3 Three-Dimensional Harmonic Oscillator

The three-dimensional harmonic oscillator is a combination of three independent one-dimensional oscillations. The Hamiltonian, expressed in orthogonal cartesian coordinates is

$$\mathbf{H} = \left(\frac{\mathbf{p}_x^2}{2m} + \frac{m}{2}\omega_x^2 \mathbf{x}^2 \right) + \left(\frac{\mathbf{p}_y^2}{2m} + \frac{m}{2}\omega_y^2 \mathbf{y}^2 \right) + \left(\frac{\mathbf{p}_z^2}{2m} + \frac{m}{2}\omega_z^2 \mathbf{z}^2 \right), \tag{8.7}$$

where, in general, the three angular frequencies ω_x, ω_y and ω_z can be different. Its diagonalization is simplified by the fact that the three modes are independent of each other, and hence we can apply the results of Sect. 8.1. In the isotropic case, i.e. $\omega_x = \omega_y = \omega_z = \omega$, the Hamiltonian

$$\mathbf{H} = \frac{\mathbf{p}^2}{2m} + \frac{m}{2}\omega^2 \mathbf{r}^2 \tag{8.8}$$

is rotationally invariant and hence it commutes with the angular momen- tum. The simultaneous eigenstates of $\mathbf{H}$, $\mathbf{l}^2$, and $\mathbf{l}_z$ can be determined in the representation of spherical coordinates. In fact, in this representation, these eigenstates are more suitable for application to systems with the same symmetry e.g. spherical nuclei (see Sect. 14.4). We refer to Sect. 14.6, the treatment in spherical coordinates. With the choice of Cartesian cooorditates, indeed, the Hamiltonian takes the simple form

$$\begin{aligned}\mathbf{H} &= \left(\frac{\mathbf{p}_x^2}{2m} + \frac{m}{2}\omega^2 \mathbf{x}^2 \right) + \left(\frac{\mathbf{p}_y^2}{2m} + \frac{m}{2}\omega^2 \mathbf{y}^2 \right) + \left(\frac{\mathbf{p}_z^2}{2m} + \frac{m}{2}\omega^2 \mathbf{z}^2 \right) \\ &= \mathbf{H}_x + \mathbf{H}_y + \mathbf{H}_z,\end{aligned}$$

Table 8.1 Energy eigenvalues of the isotropic three-dimensional harmonic oscillator. First row: ground state energy. From second to fourth row: first excited state

n_x	n_y	n_z	E_n
0	0	0	$\frac{3}{2}\hbar\omega$
0	0	1	$\frac{5}{2}\hbar\omega$
0	1	0	$\frac{5}{2}\hbar\omega$
1	0	0	$\frac{5}{2}\hbar\omega$

corresponding to three oscillations with the same angular frequency ω.

Thus, by applying the results of Sect. 8.1, we are faced with three independent one-dimensional eigenvalue equations

$$\mathbf{H}_x|n_x\rangle = \left(n_x + \frac{1}{2}\right)\hbar\omega|n_x\rangle,$$
$$\mathbf{H}_y|n_y\rangle = \left(n_y + \frac{1}{2}\right)\hbar\omega|n_y\rangle,$$
$$\mathbf{H}_z|n_z\rangle = \left(n_z + \frac{1}{2}\right)\hbar\omega|n_z\rangle.$$

The eigenvalue equation for $\mathbf{H}$, resulting from their combination, is written as

$$\mathbf{H}|n_x, n_y, n_z\rangle \equiv (\mathbf{H}_x + \mathbf{H}_y + \mathbf{H}_z)|n_x, n_y, n_z\rangle = \left(n_x + n_y + n_z + \frac{3}{2}\right)\hbar\omega|n_x, n_y, n_z\rangle.$$

where $|n_x, n_y, n_z\rangle = |n_x\rangle|n_y\rangle|n_z\rangle$ are simultaneous eigenstates of $\mathbf{H}_x$, $\mathbf{H}_y$ and $\mathbf{H}_z$ and the eigenfunctions are the product of their respective eigenfunctions, that is

$$\langle x, y, z|n_x, n_y, n_z\rangle = \langle x|n_x\rangle\langle y|n_y\rangle\langle z|n_z\rangle = \varphi_{n_x}(x)\varphi_{n_y}(y)\varphi_{n_z}(z).$$

Let us focus on the eigenvalues spectrum. Looking at Table 8.1, we notice that for $n = 0$ the zero-point energy is equal to $\frac{3}{2}\hbar\omega$ and, in general, for a given positive integer $n = n_x + n_y + n_z$, this is

$$E_n = (n_x + n_y + n_z + \frac{3}{2})\hbar\omega = \left(n + \frac{3}{2}\right)\hbar\omega \tag{8.9}$$

and the corresponding eigenstate is $\frac{(n+1)(n+2)}{2}$ times degenerate.

The extension of these results to the anisotropic harmonic oscillator spectrum is quite straightforward. The eigenvalue spectrum is reported in Table 8.2 It is worth-

Table 8.2 Energy eigenvalues of the anisotropic three-dimensional harmonic oscillator

n_x	n_y	n_z	E_n
0	0	0	$\frac{\hbar}{2}(\omega_x + \omega_y + \omega_z)$
0	0	1	$\frac{\hbar}{2}\omega_x + \frac{\hbar}{2}\omega_y + \frac{3}{2}\hbar\omega_z$
0	1	0	$\frac{\hbar}{2}\omega_x + \frac{3}{2}\hbar\omega_y + \frac{\hbar}{2}\omega_z$
1	0	0	$\frac{3}{2}\hbar\omega_x + \frac{\hbar}{2}\omega_y + \frac{\hbar}{2}\omega_z$

while noticing that, in this case, the eigenvalues are no more degenerate. As a consequence of the spherical symmetry breaking.

8.4 Notes and References

[MSH] A. Messiah, *Quantum Mechanics* Vols. 1 & 2, North-Holland (1966).

8.5 Problems

- Problem 1. Derive the uncertainty relation between position and momentum of a particle in the ground-state of a one-dimensional harmonic oscillator.
- Problem 2. Given a particle in the ground-state of an one-dimensional harmonic oscillator, calculate the probability of finding the particle in the classically forbidden region.
- Problem 3. A particle is subject to a harmonic oscillator potential of angular frequency ω and period $T = \frac{2\pi}{\omega}$. Let us define $dP = 2\frac{dt}{T}$ as the "classical" probability density of finding it in the interval of its trajectory dq around q_0, dt being the time it takes the particle to cross dq. Calculate $dP(q)$ when the classical energy of the particle coincides with that of the quantum ground-state. Compare classical and quantum probabilities.
- Problem 4. A particle with mass m and charge q is submitted to an elastic force with elasticity constant k and also to an electric field, both directed along the x-axis. Calculate the energy spectrum.
- Problem 5. Solve the eigenvalue equation for the Hamiltonian $\mathbf{H} = \frac{\mathbf{p}^2}{2m} + \mathbf{V}(\mathbf{q})$, where $V(\mathbf{q}) = \infty$ for $q < 0$ and $V(\mathbf{q}) = \frac{1}{2}\mathbf{m}\omega^2\mathbf{q}^2$ for $q > 0$.
- Problem 6. For a system in the ground-state of the harmonic oscillator, demonstrate the Virial theorem $\langle K \rangle = \langle V \rangle$, i.e. the average values of the kinetic and potential energy are equal to each other.

- Problem 7. *Coherent state* is the state which satisfies the eigenvalue equation

$$\mathbf{a}|\alpha\rangle = |\alpha\rangle\alpha,$$

where the operator **a** is the annihilation operator of the harmonic oscillator and the eigenvalue α is a complex number. Determine the eigenvector $|\alpha\rangle$ and calculate the position-momentum uncertainty for that state. (Hint: Use the representation of the eigenstates of the harmonic oscillator, expressed in terms of the creation and annihilation operators).
- Problem 8. Demonstrate that for the harmonic oscillator the following sum-rule is true

$$\sum_m (E_m - E_n)|\langle m|\mathbf{x}|n\rangle|^2 = \frac{\hbar^2}{2m}.$$

Chapter 9
Approximations to the Stationary Schrödinger Equation

The Schrödinger equation can be solved without approximations only for a few simple systems. As a matter of fact, each physical system requests a specific method of approximation, nevertheless, some general methods have been devised to be employable for particular classes of systems. One of these classes includes systems with Hamiltonian of the form $\mathbf{H_0} + \mathbf{V}$, where $\mathbf{H_0}$ can be treated exactly and $\mathbf{V}$ as a small perturbation. Its effects can be seen as corrections to those of the unperturbed Hamiltonian $\mathbf{H_0}$. Another one includes systems with Hamiltonian, from which we cannot identify the dominant part. This approximation is directed towards the ground state, with the intent of seeking among the states consistent with some expected general features of the system, the one with the lowest energy.

9.1 Perturbation Theory

We consider a system with Hamiltonian of the form $\mathbf{H} = \mathbf{H_0} + \mathbf{V}$, where $\mathbf{H_0}$ is the unperturbed term and $\mathbf{V}$ is an interaction with a small strength. Examples, that interest us, are the electromagnetic field inducing electron transitions between atomic orbitals or the spin-orbit interaction splitting orbitals. Indeed both are weak interactions that can be treated with perturbation methods.

9.1.1 Non-degenerate Case

Let us first consider the eigenstate $|\psi_k\rangle$ with eigenvalue E_k fulfilling the full Schrödinger equation

---

Supplementary Information The online version contains supplementary material available at https://doi.org/10.1007/978-981-97-1670-8_9.

U. Lombardo et al., *Quantum Mechanics*, UNITEXT for Physics,
https://doi.org/10.1007/978-981-97-1670-8_9

$$(\mathbf{H_0} + \mathbf{V})|\psi_k\rangle = E_k|\psi_k\rangle, \tag{9.1}$$

and develop both sides in series around the corresponding unperturbed state $|\psi_k^0\rangle$ whose eigenvalue E_k^0 is assumed to be non-degenerate[1]

$$\begin{aligned}|\psi_k\rangle =& |\psi_k^0\rangle + |\psi_k^1\rangle + |\psi_k^2\rangle + \cdots,\\ E_k =& E_k^0 + E_k^1 + E_k^2 + \cdots.\end{aligned}$$

Fixing the normalization of $|\psi_k\rangle$ such that $\langle\psi_k^0|\psi_k\rangle = 1$ and assuming $\langle\psi_k^0|\psi_k^0\rangle = 1$, one has

$$\langle\psi_k^0|\psi_k\rangle = \langle\psi_k^0|\psi_k^0\rangle + \langle\psi_k^0|\psi_k^1\rangle + \langle\psi_k^0|\psi_k^2\rangle + \cdots = 1. \tag{9.2}$$

This implies that, except for the first one, all other terms in the right-hand side of Eq.7 (9.2), are zero,

$$\langle\psi_k^0|\psi_k^1\rangle = \langle\psi_k^0|\psi_k^2\rangle = \cdots = 0. \tag{9.3}$$

Including the above expansions into Eq. (9.1), we get the equation

$$\begin{aligned}&(\mathbf{H}_0 + \mathbf{V})(|\psi_k^0\rangle + |\psi_k^1\rangle + |\psi_k^2\rangle + \cdots)\\ =&(E_k^0 + E_k^1 + E_k^2 + \cdots)(|\psi_k^0\rangle + |\psi_k^1\rangle + |\psi_k^2\rangle + \cdots),\end{aligned}$$

which splits into a set of equations. The first equation is just the unperturbed eigenvalue equation ($\mathbf{V} = \mathbf{0}$). The second one is the first-order perturbation

$$\mathbf{H_0}|\psi_k^1\rangle + \mathbf{V}|\psi_k^0\rangle = E_k^0|\psi_k^1\rangle + E_k^1|\psi_k^0\rangle. \tag{9.4}$$

Projecting onto the state $|\psi_k^0\rangle$ and applying Eq. (9.3), we obtain

$$E_k^1 = \langle\psi_k^0|\mathbf{V}|\psi_k^0\rangle, \tag{9.5}$$

which is the first-order approximation to the energy $E_k \approx E_k^0 + E_k^1 = E_k^0$. The corresponding correction to the state $|\psi_k\rangle$ is obtained by projecting Eq. (9.4) onto the eigenstates of the unperturbed Hamiltonian except for $|\psi_k^0\rangle$. In this way, we obtain its j-components ($j \neq k$)

$$\langle\psi_j^0|\psi_k^1\rangle = \frac{\langle\psi_j^0|\mathbf{V}|\psi_k^0\rangle}{E_j^0 - E_k^0}.$$

Higher order approximations are obtained using the same procedure.

---

[1] This condition implies that $E_k \to E_k^0$ as $\mathbf{V} \to \mathbf{0}$.

9.1.2 Degenerate Case

If the unperturbed level under consideration is degenerate, i.e. several eigenstates $|\psi^0_{k\alpha}\rangle$, with $\alpha = 1, \cdots, G$, belong to the same energy eigenvalue E^0_k, the limit $V \to 0$ does not lead to a definite eigenstate. In this case, the state $|\psi^0_k\rangle$ must be understood as a superposition of the degenerate states

$$|\psi^0_k\rangle = \mathcal{P}_k|\psi^0_k\rangle = \sum_\alpha c_\alpha|\psi^0_{k\alpha}\rangle,$$

where the operator $\mathcal{P}_k$ projects onto the subspace of the states $|\psi^0_{k\alpha}\rangle$, $\alpha = 1, \cdots, G$. In Eq. (9.4), we must replace $|\psi^0_k\rangle$ with the preceding superposition

$$\sum_\alpha c_\alpha\langle\psi^0_{k\beta}|\mathbf{V} - E^1_k|\psi^0_{k\alpha}\rangle = 0. \tag{9.6}$$

This latter is the eigenvalue equation for perturbing potential $\mathbf{V}$, where c_α are the components of the corresponding eigenvectors. In general, the number of energy eigenvalues is equal to the degeneracy number G, so the effect of the perturbation is a splitting of a single level into G levels. Let us illustrate this procedure with an example: the hydrogen atom embedded in a constant electric field E directed along the z-axis, whose potential is

$$V = -e\mathcal{E}z = -e\mathcal{E}r\cos\theta,$$

z being the z-projection of the electron position $\vec{r}$. If the electron is in the ground state ($l = m = 0$), the first order perturbation does not affect the energy as

$$E^1_0 = \langle 100|V|100\rangle = -e\mathcal{E}\langle 100|r\cos\theta|100\rangle \sim \langle 1 \mid r \mid 1\rangle \int \mathrm{d}\Omega Y^*_{00}Y_{10}Y_{00} = 0,$$

which amounts to saying that the electric potential carries an angular momentum $l = 1$ ($\cos\theta \sim Y_{10}$), whereas the electron ground state is a $l = 0$ state. If we consider the first excited state with energy E^0_2, the perturbation is expected to be non-vanishing because its angular momentum is not zero. But this state has degeneracy $G = 4$: one state with $l = m = 0$, and three states with $l = 1, m = 0$ or $m = \pm 1$. In order to apply Eq. (9.6), we must calculate the matrix elements of the perturbing potential, most of which are vanishing. The only non-vanishing matrix elements are $V_{01} = \langle 00|V|10\rangle$ and $V_{10} = \langle 10|V|00\rangle = V^*_{01}$, hence the relevant matrix is

$$V = \begin{pmatrix} 0 & V_{01} \\ V_{10} & 0 \end{pmatrix}.$$

Fig. 9.1 Stark effect

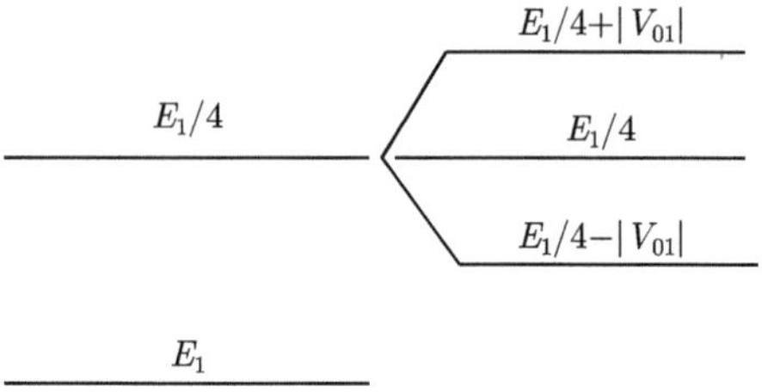

By solving the secular equation, we easily find the two energy eigenvalues $E_2^1 = \pm\sqrt{V_{01}V_{10}}$. As a result, the effect of the perturbation is to remove the degeneracy of the first excited state: a single level splits into three states, as shown in Fig. 9.1 (Stark effect). The corresponding eigenstates with the two new levels are

$$|\psi\rangle \sim |00\rangle \pm |10\rangle. \tag{9.7}$$

9.2 Variational Approach

In quantum mechanics, the Schrödinger equation can be derived from a variational principle as well as the equations of motion in classical mechanics. For this purpose we consider the energy expectation value $\left(\frac{\langle\psi|\mathbf{H}|\psi\rangle}{\langle\psi|\psi\rangle}\right)$, which is a functional of the wave function ψ defined in the Hilbert space of a given system. The Ritz principle states that a $|\psi_0\rangle$ state minimizing the energy functional in the whole Hilbert space i.e.

$$\delta E = \delta\frac{\langle\psi|\mathbf{H}|\psi\rangle}{\langle\psi|\psi\rangle} = 0.$$

amounts to solve the Schrödinger equation.

After moving the normalization factor to the left hand side, we perform the variation

$$\delta E\langle\psi|\psi\rangle + E\langle\delta\psi|\psi\rangle + E\langle\psi|\delta\psi\rangle = \langle\delta\psi|\mathbf{H}|\psi\rangle + \langle\psi|\mathbf{H}|\delta\psi\rangle.$$

By imposing the minimization condition δE=0, we obtain

$$E\langle\delta\psi|\psi\rangle + E\langle\psi|\delta\psi\rangle = \langle\delta\psi|\mathbf{H}|\psi\rangle + \langle\psi|\mathbf{H}|\delta\psi\rangle. \tag{9.8}$$

Since the variation is arbitrary, we can replace $|\delta\psi\rangle$ with $\mathrm{i}|\delta\psi\rangle$ and obtain

$$(-\langle\delta\psi|\psi\rangle + \langle\psi|\delta\psi\rangle)E = -\langle\delta\psi|\mathbf{H}|\psi\rangle + \langle\psi|\mathbf{H}|\delta\psi\rangle. \tag{9.9}$$

Subtracting Eq. (9.9) from Eq. (9.8), we get

$$\langle \delta\psi|\psi\rangle E = \langle \delta\psi|\mathbf{H}|\psi\rangle,$$

and since the variation $\langle \delta\psi|$ is arbitrary, we end up with the Schrödinger equation

$$|\psi\rangle E = \mathbf{H}|\psi\rangle.$$

As a result, the stationary states, i.e. those minimizing the energy functional in Hilbert space, are eigenstates of the Hamiltonian. As already pointed out, the Ritz principle is the quantum counterpart of the classical principle of least action. The classical trajectory is the one minimizing the action and the eigenstates of the Hamiltonian are those minimizing the average energy value. The lowest energy corresponds to the ground state. An important outcome of the Ritz principle is that the energy of the ground-state is less or equal to the average value $\langle\psi|\mathbf{H}|\ \rangle$ for any state ψ of Hilbert space of the considered system. It is easily obtained with the following steps

$$\begin{aligned} E =& \frac{\langle\psi|\mathbf{H}|\psi\rangle}{\langle\psi|\psi\rangle} = \sum_k \frac{\langle\psi|\mathbf{H}|\psi_k\rangle\langle\psi_k|\psi\rangle}{\langle\psi|\psi\rangle} \\ =& \sum_k E_k \frac{|\langle\psi_k|\psi\rangle|^2}{\langle\psi|\psi\rangle} \geqslant E_0 \sum_k \frac{|\langle\psi_k|\psi\rangle|^2}{\langle\psi|\psi\rangle} = E_0, \end{aligned} \tag{9.10}$$

where a representation of the identity operator has been introduced in terms of the eigenstates $|\psi_k\rangle$ of the Hamiltonian. This implies that any approximate estimate of the ground-state energy is always bigger than the exact value. The variational principle provides a powerful approximation when the interaction of the considered system is too strong to apply the perturbation theory, and we have a physical insight into the ground state of the wave function. In this case, we first guess a trial wave function $\phi(p_1, p_2, \cdots)$, which scans a subspace of Hilbert space varying the parameters p_i; afterwards, we search for the minimum of the energy functional $\langle\phi|\mathbf{H}|\phi\rangle$ as a function of the parameters p_i. This estimate in that subspace is an upper bound of the true energy, according to Eq. (9.10). The closer the trial subspace is to the true ground state the more accurate = the energy estimate will be. If the ground state belongs to the subspace, the estimate will coincide with the exact value. In practice, this event never happens because normally the ground-state energy is unknown. How to choose the trial wave function is the main point of the variational approximation. For instance, if the Hamiltonian exhibits some symmetries, one should choose a subspace, whose states have the same symmetry.

9.3 Perturbation Versus Variational Approximations For ^{4}He

For the purpose of comparing the two approximation methods, we estimate the effect of the Coulomb interaction on the ^{4}He binding energy. Its Hamiltonian is

$$\mathbf{H} = \mathbf{H}_0 + \mathbf{V}_{12} = \mathbf{H}_0 + \frac{e^2}{|\vec{\mathbf{r}}_1 - \vec{\mathbf{r}}_2|}, \tag{9.11}$$

where $\mathbf{H_0}$ is the Hamiltonian of the two electrons interacting with the nucleus and $\mathbf{V_{12}}$ is the interaction potential between the two electrons. The nucleus with its two protons exerts on each electron a force twice that on the hydrogen electron and hence the exact diagonalization of $\mathbf{H_0}$ is straightforward. It is useful to resume the energy and the wave function of the hydrogen electron (see Chap. 12)

$$E_{\text{H}} = -\frac{e^2}{2r_0} = -13.5\,\text{eV} \qquad \phi_0(r) = \frac{1}{\sqrt{\pi r_0^3}} \exp(-r/r_0),$$

where r_0 is the *Bohr radius*.

9.3.1 Perturbation Method

Neglecting $\mathbf{V_{12}}$, the ground-state energy of each electron would be the same as in the hydrogen atom, except for the double charge of the nucleus containing two protons. Hence

$$\psi_0^0(\vec{r}_1, \vec{r}_2) = \frac{1}{\pi r^3} \text{e}^{-(r_1+r_2)/r},$$

where the radius r is half the Bohr radius r_0, as

$$r = \frac{\hbar^2}{m(2e)(e)} = \frac{r_0}{2}.$$

So, with double the nuclear charge and half the Bohr radius, the unperturbed binding energy of each electron turns out to be four times that of the hydrogen electron. For two electrons, the binding energy is

$$E_0^0 = -2\frac{(2e)(e)}{2(r_0/2)} = 8E_{\text{H}}.$$

The repulsive effect of $\mathbf{V_{12}}$ is calculated from the perturbation theory by applying Eq. (9.5)

$$E_0^1 = \langle\psi_0^0|\frac{e^2}{|\mathbf{\vec{r}}_1 - \mathbf{\vec{r}}_2|}|\psi_0^0\rangle = \int \mathrm{d}^3\vec{r}_1 \int \mathrm{d}^3\vec{r}_2 \left(\frac{\pi}{r^3}\right)^2 \frac{\mathrm{e}^{-2(r_1+r_2)/r}}{|\vec{r}_1 - \vec{r}_2|}. \tag{9.12}$$

The change of the variable $\vec{r}_{1,2} = \vec{s}_{1,2}\, r$, makes the integral dimensionless. Bringing together the physical quantities into a single factor proportional to E_H

$$E_0^1 = 2E_\mathrm{H}\frac{1}{\pi^2}\int \mathrm{d}^3\vec{s}_1 \int \mathrm{d}^3\vec{s}_2 \mathrm{e}^{-2(s_1+s_2)} = -2\frac{5}{4}E_\mathrm{H}.$$

As $E_\mathrm{H} = -13.5\,\mathrm{eV}$, the helium ground-state is $E_0 \approx E_0^0 + E_0^1 = (-108 + 34)$ eV$= -74\,\mathrm{eV}$, to the first perturbation order. From the comparison with the experimental value $-78.6\,\mathrm{eV}$, the first-order contribution of (34 eV) is not small with respect to the unperturbed value, hence we should go to the second-order E_0^2.

9.3.2 Variational Estimate

The main problem is to set the subspace where the minimum of the average value of the Hamiltonian $\mathbf{H}$ of the ^{4}He atom has to be found. Since the unperturbed wave function works with the perturbation method, we can still use it as a trial wave function in the variational method, with the proviso that we also include the effect of the perturbing potential. The Coulomb repulsion on a single electron is to weaken the attraction of the nucleus with the other electron. We can incorporate this effect on the nuclear interaction by replacing its charge $2e$ with an effective charge λe, λ being a parameter to be determined by a minimization procedure. The effective charge must be introduced only in the wave function (in particular in the radius r), but not in the full Hamiltonian. As a result, the trial wave function is

$$\psi_0^0(\lambda) = \frac{\pi}{\left(\frac{r_0}{\lambda}\right)^3}\mathrm{e}^{-(r_1+r_2)/r},$$

with $r = r_0/\lambda$. The minimum of the energy functional fulfills the condition

$$\frac{\delta E(\lambda)}{\delta\lambda} = \frac{\delta}{\delta\lambda}\langle\psi_0^0(\lambda)|\mathbf{H}|\psi_0^0(\lambda)\rangle = 0.$$

The average kinetic energy of each electron in the state $\psi_0^0(\lambda)$ is obtained from the hydrogen kinetic energy $K = E_\mathrm{H} - V = e^2/2r_0$ by replacing the hydrogen nuclear charge e with λe and the Bohr radius r_0 with r_0/λ. Thus one has

$$K_1 + K_2 = -2\lambda^2 E_\mathrm{H}.$$

The average potential energy in the state $\psi_0^0(\lambda)$ is obtained from that of hydrogen $V = -e^2/2r$ with the hydrogen nuclear charge e replaced by the double charge $2e$ (warning: the λ effect, as we said, is not to be included in the Hamiltonian!). Therefore we have

$$V_1 + V_2 = 8\lambda E_{\rm H}.$$

Finally, the average value of $\mathbf{V_{12}}$ with respect to $\psi_0^0(\lambda)$ is obtained by replacing in Eq. (9.12) r with λr (warning: but not in the $\mathbf{V_{12}}$ strength e^2). The sum of the kinetic and potential energies yields

$$E(\lambda) = -2\lambda^2 E_{\rm H} + 8\lambda E_{\rm H} - \frac{5}{4}\lambda E_{\rm H}.$$

The minimum is reached with $\lambda_0 = 2(1 - 5/32)$, the charge correction being $5/32$. The ground-state energy is

$$E(\lambda_0) = -2\lambda_0^2 E_{\rm H} + 8\lambda_0 E_{\rm H} - \frac{5}{4}\lambda_0 E_{\rm H} = -76.6\,\text{eV}.$$

The deviation from the experimental value is 2 eV to be compared with the value of 4.6 eV obtained from the perturbation method. Therefore, the variational estimate, despite the extremely simple trial wave function, turns out to be much better than the perturbation one, at least in this case. At last, it is worthwhile noticing that the variational estimate of -76.6 eV is an upper bound of the experimental value of -78.6 eV, in agreement with Eq. (9.10).

9.4 Problems

- Problem 1. Consider a diatomic molecule with a dipole moment $\vec{d}$ submitted to an electric field $\vec{E} = const$ directed along the x axis. The effect of this field on the energy spectrum can be tackled with the Hamiltonian model of a charged planar rotor $\mathbf{H} = \mathbf{L}_z^2/2I$, where I is the moment of inertia. Calculate the effect of the perturbation $\mathbf{V} = -\vec{\mathbf{d}} \cdot \vec{\mathbf{E}} = -\mathbf{dE}\,\mathbf{cos}\phi$ on the energies of the ground state and first excited state.
- Problem 2. Two identical fermions with spin $1/2$ are enclosed into the lowest energy state of an infinite square well of length l. Determine the first-order energy correction, when they are submitted to a contact interaction $V = V_0\delta(q_1 - q_2)$. Assume $c >> \hbar_2/I$.
- Problem 3. Estimate the ground-state energy of a perturbed rigid rotor with Hamiltonian $\mathbf{H} = \dfrac{\mathbf{L^2}}{2I} + c\mathbf{L}_x$, using the trial state $|\psi\rangle = |10\rangle + \beta|11\rangle$. The factor c is a positive constant and β is the variational parameter.

- Problem 4. A particle of mass m is constrained to move on a segment of length L. Assuming it is submitted to an external potential

$$\mathbf{V}(x) = \mathbf{V}_0 \cos^2\left(\frac{\pi \mathbf{x}}{L}\right),$$

calculate the first order correction to the ground-state energy.
- Problem 5. A particle of mass m is constrained to move on a ring of radius R on the xy-plane. Calculate the energy eigenfunctions and eigenvalues. Suppose it is submitted to an external potential

$$\mathbf{V}(q) = \mathbf{V}_0 \left(\frac{x}{R}\right)^2$$

Calculate the first-order correction to the ground-state energy.
- Problem 6. Two one-dimensional harmonic oscillators of mass m and angular frequency ω interact via the potential $\mathbf{V} = \mathbf{V}_0(\mathbf{x}_1 - \mathbf{x}_2)^2$. Calculate the first order correction to the energy of ground state.
- Problem 7. Two one-dimensional harmonic oscillators of mass m and angular frequency ω oscillate one along the x-axis and the other along the y-axis. They interact via the potential $\mathbf{V} = k\mathbf{xy}$. Calculate the first and the second order corrections to the ground-state energy. (Hint: use the representation of the occupation numbers)
- Problem 8. A particle of mass m is subjected to a potential $\mathbf{V} = \lambda|\mathbf{x}|$. Using the trial wave function $\psi(x) = Ae^{-\beta x^2}$, with β as variational parameter, determine an upper bound to the ground-state energy. The parameter λ is a constant.
- Problem 9. Determine the binding energy of the deuteron by using the trial wave function $\psi(r) = Ae^{-\beta r^2}/r$, with β as variational parameter. The interaction potential between the neutron and the proton is $V(r) = V_0\delta(r)$, $\delta(r)$ being the Dirac delta function.

Chapter 10
Time-Dependent Equations of Motion

In Chap. 7, we derived the time-dependent Schrödinger equation from the properties of time translations applied to the system states. With this approach, the time evolution of the average values of physical variables in non-stationary conditions can be determined from the time evolution of states. An alternative approach is the so-called Heisenberg representation: the time evolution can be studied by applying the corresponding unitary operator $\mathbf{U}(t, t')$ to the observables, rather than the states. This approach makes it easier to compare quantum and classical dynamics, and indeed the semi-classical limit is better investigated in such representation, as shown in Sect. 10.4. One more advantage of this representation regards the study of the relationship between space–time symmetries of systems and conservation laws of physical observables.

10.1 Heisenberg Representation

Consider the average value of a dynamical variable $\mathbf{O}$ in the state $|\psi(t)\rangle$ of a system at the time t, being in the state $|\psi_0\rangle$ at t_0. We can move the time dependence from the state to the operator with the transformation

$$\langle\psi(t)|\mathbf{O}|\psi_t\rangle = \langle\psi_0|\mathbf{U}^\dagger(t, t_0)\mathbf{O}\mathbf{U}(t, t_0)|\psi_0\rangle = \langle\psi_0|\mathbf{O}_H(t)|\psi_0\rangle. \tag{10.1}$$

where we have defined the time-dependent operator

$$\mathbf{O}_H(t) = \mathbf{U}^\dagger(t, t_0)\mathbf{O}\mathbf{U}(t, t_0). \tag{10.2}$$

Supplementary Information The online version contains supplementary material available at https://doi.org/10.1007/978-981-97-1670-8_10.

U. Lombardo et al., *Quantum Mechanics*, UNITEXT for Physics,
https://doi.org/10.1007/978-981-97-1670-8_10

Hence, the introduction of the unitary operator, responsible for time translations, provides a procedure to describe the dynamical evolution of a quantum system alternative to the time-dependent Schrödinger equation. It transfers the time dependence from the states to the observables while keeping the probability amplitudes and the average values of observables unchanged. However, the two representations refer to two different classes of quantum processes: the time-dependent Schrödinger wave functions conveniently describe processes evolving continuously over time, in the spirit of wave mechanics; conversely, the Heisenberg operators, by means of the transition amplitudes, are appointed to describe the discontinuous evolution, like the electron transitions between atomic orbitals induced by the electromagnetic (e.m.) field.

After Eq. (10.2) has been differentiated, applying the time-dependent Schrödinger equation (see Eq. (7.7), we obtain the equation of motion for the operator $\mathbf{O}_H(t)$

$$i\hbar\frac{\mathrm{d}}{\mathrm{d}t}\mathbf{O}_H(t) = [\mathbf{O}_H(t), \mathbf{H}_H(t)] + i\hbar\frac{\partial}{\partial t}\mathbf{O}_H(t). \tag{10.3}$$

The last term on the right-hand side of Eq. (10.3) is a compact form of the expression

$$\frac{\partial}{\partial t}\mathbf{O}_H(t) = \mathbf{U}^{\dagger}(t, t_0)\left(\frac{\partial}{\partial t}\mathbf{O}(t)\right)\mathbf{U}(t, t_0).$$

The partial derivative $\frac{\partial}{\partial t}$, different from the total time derivative $\frac{\mathrm{d}}{\mathrm{d}t}$, only operates on the physical variables explicitly depending on time, e.g. e.m. fields. Otherwise, such a derivative vanishes for observables explicitly independent time.

10.2 Two-Level Quantum System

The solution of the Heisenberg equations of motion requires approximation methods which will be developed in the next chapter. There are, however, simple cases in which the exact solution is possible. One of them is the two-level system, which is suitable for studying physical systems of considerable interest, despite of its simplicity. Such systems are spin $\frac{1}{2}$ particles in a magnetic field, pairing correlations responsible for superfluidity and superconductivity as well as quantum computing.

10.2.1 Unperturbed Hamiltonian

The Hamiltonian of a particle with $\frac{1}{2}$ spin in a magnetic field $\vec{B}_0 \equiv (0, 0, B_0)$ directed along the z-axis is $\mathbf{H}_0 = \gamma \boldsymbol{s}_z B_0$, where $\gamma \boldsymbol{s}_z$ is the magnetic moment. In the spin space, the two states $\left|s = \frac{1}{2}, s_z \pm \frac{1}{2}\right\rangle \equiv |\pm\rangle$, which are eigenstates of $\mathbf{s}_z$, can describe any spin state. In the spin space the matrix elements of H_0 between two possible spin states are $\langle\sigma|\mathbf{H}_0|\sigma'\rangle$ and are also its eigenvalues. The eigenvalue equation is

$$\begin{pmatrix} \epsilon_+ & 0 \\ 0 & \epsilon_- \end{pmatrix} \begin{pmatrix} 1 \\ 0 \end{pmatrix} = \epsilon_+ \begin{pmatrix} 1 \\ 0 \end{pmatrix}, \quad \begin{pmatrix} \epsilon_+ & 0 \\ 0 & \epsilon_- \end{pmatrix} \begin{pmatrix} 0 \\ 1 \end{pmatrix} = \epsilon_- \begin{pmatrix} 0 \\ 1 \end{pmatrix},$$

where $\epsilon_\pm = \pm\frac{1}{2}\gamma\hbar B_0$.

10.2.2 Perturbation Potential

By adding a further magnetic field B_1, directed along the x-axis, to the previous magnetic field B_0, the new Hamiltonian $\mathbf{H} = \mathbf{H}_0 + \mathbf{V} = \mathbf{H}_0 + B_1\gamma\mathbf{s}_x$ is no longer diagonal. The matrix elements of the new field are now

$$\begin{aligned} \langle\pm|\mathbf{V}|\pm\rangle &= \gamma B_1\langle\pm|\mathbf{s}_x|\pm\rangle = 0, \\ \langle+|\mathbf{V}|-\rangle &= \gamma B_1\langle+|\mathbf{s}_x|-\rangle = \frac{1}{2}\gamma B_1\langle+|\mathbf{s}_+|-\rangle = \frac{1}{2}\hbar\gamma B_1 = V, \\ \langle-|\mathbf{V}|+\rangle &= \gamma B_1\langle-|\mathbf{s}_x|+\rangle = \frac{1}{2}\gamma B_1\langle-|\mathbf{s}_-|+\rangle = \frac{1}{2}\hbar\gamma B_1 = V. \end{aligned}$$

The off-diagonal matrix elements are responsible for the transition between the two levels $|\pm\rangle$. The eigenvalue equation for the Hamiltonian is

$$\mathbf{H}|\phi_\pm\rangle = \begin{pmatrix} \epsilon_+ & V \\ V & \epsilon_- \end{pmatrix} |\phi_\pm\rangle = E_\pm|\phi_\pm\rangle. \tag{10.4}$$

From now on, we denote $\epsilon_- = -\epsilon$ and $\epsilon_+ = +\epsilon$. The secular equation of the eigenvalue problem (10.4) is

$$\begin{vmatrix} \epsilon - E & V \\ V & -\epsilon - E \end{vmatrix} = -(\epsilon - E)(\epsilon + E) - V^2 = 0.$$

From this, we obtain the eigenvalues $E_\pm = \pm\sqrt{\epsilon^2 + V^2}$. The eigenvectors are solutions of the homogeneous equation

$$\begin{pmatrix} \epsilon & V \\ V & \epsilon \end{pmatrix} \begin{pmatrix} \cos\theta_\pm \\ \sin\theta_\pm \end{pmatrix} = E_\pm \begin{pmatrix} \cos\theta_\pm \\ \sin\theta_\pm \end{pmatrix},$$

where the components of the eigenvectors are written so as to be normalized to unity. We find that the ratios of their components are

$$\cot\theta_\pm = \frac{V}{E_\pm - \epsilon}.$$

At $V = 0$, we obtain the eigenvalues and eigenvectors of the unperturbed Hamiltonian, and for $V \ll 1$, we can write a power expansion in terms of V of the energy of the ground state. To the lowest order

$$E_- \approx -\epsilon + \Delta E_2 = -\frac{V^2}{2\epsilon},$$

$$\Delta E_2 = \langle -|\mathbf{V}\frac{\mathbf{1}-\mathbf{P}}{\mathbf{H}_0 - \epsilon_-}\mathbf{V}|-\rangle = -\frac{V^2}{2\epsilon}.$$

To the first order in V, we have $\Delta E_1 = \langle -|\mathbf{V}|-\rangle = 0$. The term $\mathbf{1} - \mathbf{P}$ is the space complementary to $|+\rangle$. It is worth noting that this result is the same as the one obtained within the stationary perturbation theory.

10.2.3 Time-Dependent Hamiltonian

Suppose the perturbation considered above is turned on at the time $t = 0$ and off at $t = T$. As in the case of atoms irradiated by electromagnetic waves, the main concern is the transition induced by the perturbation between the two unperturbed eigenstates. The system is very simple, thus the equation of motion can be solved exactly. At times $t > T$, the system will be in the state $|\psi_T\rangle = \mathbf{U}(T, 0)|-\rangle$, being $|-\rangle$ the ground state at $t = 0$. The transition probability to the excited state $|+\rangle$ is $P_{+-} = |\langle +|\mathbf{U}(T, 0)|-\rangle|^2$. Since there are only two states, the probability of no transition is $1 - P_{+-}$. Moreover, the eigenstates $\phi_\pm$ of the Hamiltonian $\mathbf{H}$ form a complete set of states at any time, in general, we can write

$$\begin{aligned}|\psi(t)\rangle &= |\phi_+(t)\rangle\langle\phi_+(t)|\psi(t)\rangle + |\phi_-(t)\rangle\langle\phi_-(t)|\psi(t)\rangle \\ &= \mathrm{e}^{\frac{\mathrm{i}}{\hbar}E_+t}|\phi_+(0)\rangle\langle\phi_+(0)|\psi(0)\rangle + \mathrm{e}^{\frac{\mathrm{i}}{\hbar}E_-t}|\phi_-(0)\rangle\langle\phi_-(0)|\psi(0)\rangle,\end{aligned}$$

where we have used the property of the amplitudes to be invariant under time translations. Thus, the transition amplitude from the initial state, assumed to be $|\psi(0)\rangle = |-\rangle$, to the state $|+\rangle$ is

$$\begin{aligned}\langle +|\psi(t)\rangle &= \cos\theta_+ \sin\theta_+ \mathrm{e}^{\frac{\mathrm{i}}{\hbar}E_+t} + \cos\theta_- \sin\theta_- \mathrm{e}^{\frac{\mathrm{i}}{\hbar}E_-t} \\ &= \frac{1}{2}[\sin(2\theta_+)\mathrm{e}^{\frac{\mathrm{i}}{\hbar}E_+t} + \sin(2\theta_-)\mathrm{e}^{\frac{\mathrm{i}}{\hbar}E_-t}].\end{aligned}$$

Taking into account that $\sin 2\theta_- = -\sin 2\theta_+$, we end up with

$$P_{+-}(t) = \sin^2(2\theta_+)\sin^2[(E_+ - E_-)t/\hbar] = \frac{V^2}{\epsilon^2 + V^2}\sin^2[2(\epsilon^2 + V^2)^{\frac{1}{2}}t/\hbar]. \quad (10.5)$$

Equation (10.5) is known as Rabi's formula. The system oscillates between the two states, but the oscillation amplitude is never vanishing, because the excited state is never fully occupied, $P_{+-}(t) < 1$.

The classical analogue of this system corresponds to a system of two coupled pendula. Lastly, we consider the case in which the magnetic field perpendicular to the z-axis is rotating with angular frequency ω, i.e. $\mathbf{B}_1 = B_1[\cos(\omega t)\mathbf{x} + \sin(\omega t)\mathbf{y}]$, where **x** and **y** are the unitary vectors along the x- and y-axes. The matrix elements of the potential are

$$\langle \pm | \mathbf{s}_x \cos(\omega t) + \mathbf{s}_y \sin(\omega t) | \mp \rangle = \frac{1}{2} \langle \pm | \mathbf{s}_\pm | \mp \rangle \mathrm{e}^{\mp \mathrm{i}\omega t} = \mathbf{V} \mathrm{e}^{\mp \omega t},$$

and the new Hamiltonian is written as

$$\mathbf{H}(t) = \begin{pmatrix} \epsilon & V\mathrm{e}^{-\mathrm{i}\omega t} \\ V\mathrm{e}^{\mathrm{i}\omega t} & -\epsilon \end{pmatrix}.$$

Now, the new potential is explicitly time-dependent and we cannot apply the previous procedure to determine the transition probabilities. However, in the reference frame rotating with the same frequency the system does not move, hence we can restore the previous procedure. The unitary transformation corresponding to a rotation is

$$\hat{R}(t) = \begin{pmatrix} \mathrm{e}^{-\mathrm{i}\omega t} & 0 \\ 0 & \mathrm{e}^{\mathrm{i}\omega t} \end{pmatrix}.$$

In the rotating frame the time-dependent Schrödinger equation is

$$\mathrm{i}\hbar \frac{\mathrm{d}}{\mathrm{d}t} |\psi_R(t)\rangle = \mathbf{H}_R |\psi_R(t)\rangle + \frac{\hbar\omega}{2} \begin{pmatrix} 1 & 0 \\ 0 & -1 \end{pmatrix} |\psi_R(t)\rangle,$$

where $|\psi_R\rangle = \mathbf{R}|\psi\rangle$ and $\mathbf{H}_R = \mathbf{R}\mathbf{H}\mathbf{R}^\dagger$, and, in addition

$$\hat{H}_R + \frac{\hbar\omega}{2} \begin{pmatrix} 1 & 0 \\ 0 & -1 \end{pmatrix} = \begin{pmatrix} \epsilon - \dfrac{\hbar\omega}{2} & V \\ V & -\epsilon + \dfrac{\hbar\omega}{2} \end{pmatrix}.$$

As a result, in the rotating frame, the Hamiltonian is independent of time, but the energy gap between the two levels widens by $\hbar\omega$. Rabi's formula can be applied directly, replacing the gap 2ϵ with $2\epsilon - \hbar\omega$. For $2\epsilon = \hbar\omega$ a resonance appears. This means that, in the resonance conditions, the system oscillates from the spin *down* state to the spin *up* state (full spin flip), whereas, in a non-resonant state, the amplitude is less than unity and the spin–flip is only partial.

10.3 Relationship Between Symmetries and Conservation Theorems

The relations between space–time symmetries and conservation theorems of physical observables were established in the context of the infinitesimal canonical transformations. These observables are the generators of the symmetry transformations themselves. After introducing the quantum dynamical variables as generators of the symmetry transformations, described by unitary operators, we ask ourselves whether these relations are still true. With the equations of motion of physical observables, we can now investigate such an issue. From Eq. (10.3), we have the condition for the observable $\mathbf{O}_H$ to be a constant of motion:

- $\mathbf{O}$ must be explicitly independent of time,
- $\mathbf{O}_H$ must commutate with the Hamiltonian $\mathbf{H}_H$.

Since the two representations are related by a unitary transformation, the commutation relations remain unchanged

$$[\mathbf{O}_H(t), \mathbf{H}_H(t)] = [\mathbf{O}, \mathbf{H}] = 0.$$

The equation of motion of the Hamiltonian operator is

$$\mathrm{i}\hbar \frac{\mathrm{d}}{\mathrm{d}t}\mathbf{H}_H(t) = \mathrm{i}\hbar \frac{\partial}{\partial t}\mathbf{H}_H(t),$$

so that the Hamiltonian is a constant of motion only if it is explicitly independent of time.

Let us consider an observable $\mathbf{O}$ which is the generator of an infinitesimal symmetry transformation similar to space–time translations and space rotations. By applying this transformation to the Hamiltonian $\mathbf{H}$, we get the new Hamiltonian $\mathbf{H}'$

$$\mathbf{H}' = \mathbf{U}^{\dagger}_{\delta s}\mathbf{H}\mathbf{U}_{\delta s} = \left(\mathbf{1} + \frac{\mathrm{i}}{\hbar}\delta s\, \mathbf{O}\right)\mathbf{H}\left(\mathbf{1} - \frac{\mathrm{i}}{\hbar}\delta s\, \mathbf{O}\right) = \mathbf{H} + \frac{\mathrm{i}}{\hbar}[\mathbf{O}, \mathbf{H}]\delta s,$$

where δs is the smallness parameter and terms of the order $\delta s^2 \ll \delta s$ have been neglected.

To have $\mathbf{H}' = \mathbf{H}$, the Hamiltonian $\mathbf{H}$ must commute with the generator of the symmetry transformation. We conclude that the commutation of the Hamiltonian with the generator of a given symmetry transformation is the condition for the Hamiltonian to be invariant under that symmetry transformation as well as for the generator to be a constant of motion. We recover the relation between symmetries and conservation laws already established in classical mechanics. This property is a further justification for extending the identification of the classical dynamical variables with the generators of the symmetry transformations to quantum mechanics.

10.4 Classical Limit: Ehrenfest Theorem

Consider a particle in one-dimensional motion inside a potential $\mathbf{V}(q)$. We assume $\mathbf{q}(t)$ and $\mathbf{p}(t)$ to be, respectively, coordinate and momentum of the particle within the Heisenberg representation (we omit the lower case H for simplicity). They fulfill the equations of motion

$$\mathrm{i}\hbar \frac{\mathrm{d}\mathbf{q}(t)}{\mathrm{d}t} = [\mathbf{q}(t), \mathbf{H}] = \frac{\mathbf{p}(t)}{m}, \tag{10.6}$$

$$i\hbar\frac{\mathrm{d}\mathbf{p}(t)}{\mathrm{d}t} = [\mathbf{p}(t), \mathbf{H}] = [\mathbf{p}(t), \mathbf{V}(q)]. \tag{10.7}$$

Let us write the equations of motion for their average values. From Eq. (10.6), we get

$$\frac{\mathrm{d}\langle\mathbf{q}(t)\rangle}{\mathrm{d}t} = \frac{\langle\mathbf{p}(t)\rangle}{m}. \tag{10.8}$$

Equation (10.8) is identical to the first Hamilton equation of motion for the classical variables $q(t)$ and $p(t)$, which are replaced by the average values of their quantum equivalents. In other words, the average value of the position of the particle follows a classical trajectory. This, in no way, means that the particle follows a classical trajectory as we will see by studying its root mean square. From Eq. (10.7), we have

$$\begin{aligned} i\hbar\frac{\mathrm{d}\langle\mathbf{p}(t)\rangle}{\mathrm{d}t} =& \langle[\mathbf{p}(t), \mathbf{V}(\mathbf{q}(t))]\rangle = \int \mathrm{d}q\psi^*(q)(\mathbf{pV} - \mathbf{Vp})\psi(q) \\ =& \int \mathrm{d}q\psi^*(q)\frac{\hbar}{\mathrm{i}}\left(\frac{\mathrm{d}V}{\mathrm{d}q}\psi(q) + V(q)\frac{\mathrm{d}\psi}{\mathrm{d}q}\right) - \int \mathrm{d}q\psi^*(q)\frac{\hbar}{\mathrm{i}}V(q)\frac{\mathrm{d}\psi}{dq} \\ =& \frac{\hbar}{\mathrm{i}}\int \mathrm{d}q\psi^*(q)\frac{\mathrm{d}V(q)}{\mathrm{d}q}\psi(q) = \frac{\hbar}{\mathrm{i}}\left\langle\frac{\mathrm{d}\mathbf{V}(q)}{\mathrm{d}q}\right\rangle. \end{aligned} \tag{10.9}$$

Equation (10.9) looks like the classical equation of motion $\dot{p} = -\nabla V$. We notice that $\langle\mathbf{p}(t)\rangle$ does not behave like $\langle\mathbf{q}(t)\rangle$ unless we replace the average value of the gradient, $\langle\nabla V\rangle$, with the gradient of the average value, $\nabla\langle V\rangle$. This is true for some special cases: free particle (for which the potential is zero), a charged particle in a constant electric field $\mathbf{V}(q) = \mathcal{E}\mathbf{q}$, or a particle subject to a harmonic oscillator potential. It is generally true with polynomial potentials $\mathbf{V}(\mathbf{q}) \sim \mathbf{q}^n$, with $n \leqslant 2$. This correspondence between classical and quantum behaviour is the content of the Ehrenfest theorem. With the exception of the few above-mentioned cases, the behaviour of the quantum average values of dynamical variables drastically departs from the classical ones. In any case, this raises the issue of the conditions for which physical systems behave quasi-classically, specifically classically with quantum corrections. For this purpose, we develop the potential of Eq.(10.9) in Taylor series around a $\bar{q} \equiv \langle\, \mathrm{q}\rangle$, that is: $V(q) = V(\bar{q}) + V'(\bar{q})(q - \bar{q}) + \frac{1}{2}V''(\bar{q})(q - \bar{q})^2 + \frac{1}{6}V'''(\bar{q})(q - \bar{q})^3$, and we rewrite the average value of the gradient of the potential in terms of this expansion

$$\begin{aligned} \left\langle\frac{\mathrm{d}V}{\mathrm{d}q}\right\rangle =& V'(\bar{q})\int \mathrm{d}q|\psi(q)|^2 \\ &+ V''(\bar{q})\int \mathrm{d}q\psi^*(q)(q - \bar{q})\psi(q) \\ &+ \frac{1}{2}V'''(\bar{q})\int \mathrm{d}q\psi^*(q)(q - \bar{q})^2\psi(q) \\ =& V'(\bar{q}) + \frac{1}{2}V'''(q)\int \mathrm{d}q|\psi(q)|^2(q - \bar{q})^2, \end{aligned}$$

where the derivatives are calculated at $q = \bar{q}$ and $\int dq|\psi(q)|^2 = 1$. The first line represents the classical limit and provides the correspondence between the quantum mechanical average values and the classical values of the dynamical variables; the term in the second derivative of the potential (second line) disappears because $\langle q - \langle q \rangle \rangle = 0$; the term in the third derivative of the potential (last line) represents the first quantum correction. As a result, the accuracy of this truncated expansion not only depends on the properties of the potential but also depends on the root mean square of the position. This latter is a measure of the uncertainty in the location of the system, and hence its importance is related to the root mean square of the momentum by the Heisenberg uncertainty principle.

10.5 Particle Detection in Scattering Processes

We consider a laboratory experiment in which a particle is scattered by a finite potential. Let us suppose the detector is placed at a distance D from the scattering center and that the particle reaches the detector within a time interval t. When distance D is much wider than the range of the potential, we can assume that the scattered particle travels towards the detector as free particle. This condition corresponds to the maximum spread of the wave-packet. In this case, the probability amplitude of detecting the particle in a certain position is constant in the whole space. Hence, we would be unable to find out whether or not the detected particle comes from the potential region. In this section, we address this problem by studying the time evolution of the position dispersion of the wave-packet to evaluate for which distance D such a quantity has a minimum value. Let us write the dispersion of the position operator

$$(\Delta q(t))^2 = \langle \mathbf{q}^2(t) \rangle - \langle \mathbf{q}(t) \rangle^2 = \langle (\mathbf{q}(t) - q_0(t))^2 \rangle,$$

where $q_0 = \langle \mathbf{q} \rangle$. We can write the Heisenberg equation for the function $(\Delta \mathbf{q}(t))^2$ as follows::

$$i\hbar \frac{d}{dt} \langle (\mathbf{q}(t) - q_0(t))^2 \rangle = \langle [(\mathbf{q}(t) - q_0(t))^2, \mathbf{H}(t)] \rangle, \tag{10.10}$$

with $\mathbf{H}(t) = \dfrac{\mathbf{p}^2(t)}{2m}$ being the Hamiltonian of a free particle in the Heisenberg Representation. By developing the commutator of Eq.(10.10), we have

$$[(\mathbf{q}(t) - q_0(t))^2, \mathbf{H}(t)] = [\mathbf{q}^2, \mathbf{H}] - 2q_0[\mathbf{q}, \mathbf{H}]. \tag{10.11}$$

From now on we omit, for the sake of simplicity, the time dependence in the operators and related quantities. To calculate the first commutator on the right-hand side of Eq.(10.11), we make use of the relation: $[\mathbf{A}^2, \mathbf{B}] = [\mathbf{A}, \mathbf{B}]\mathbf{A} + \mathbf{A}[\mathbf{A}, \mathbf{B}]$, hence:

$$[\mathbf{q}^2, \mathbf{H}] = [\mathbf{q}, \mathbf{H}]\mathbf{q} + \mathbf{q}[\mathbf{q}, \mathbf{H}],$$

and the commutator (10.11) takes the form

$$[(\mathbf{q} - q_0)^2, \mathbf{H}] = \mathbf{q}[\mathbf{q}, \mathbf{H}] + [\mathbf{q}, \mathbf{H}]\mathbf{q} - 2q_0[\mathbf{q}, \mathbf{H}].$$

The commutator $[\mathbf{q}, \mathbf{H}]$ can be calculated by applying the following equation valid for any function $F(\mathbf{p})$ of the momentum operator: $[\mathbf{q}, F(\mathbf{p})] = \mathrm{i}\hbar\frac{\partial F}{\partial p}$. It results that

$$[\mathbf{q}, \mathbf{H}] = \left[\mathbf{q}, \frac{\mathbf{p}^2}{2m}\right] = \mathrm{i}\hbar\frac{\mathbf{p}}{m}, \tag{10.12}$$

we then rewrite Eq.(10.11)

$$[(\mathbf{q} - q_0)^2, \mathbf{H}] = \frac{2\mathrm{i}\hbar}{m}\left(\frac{\mathbf{qp}}{2} + \frac{\mathbf{pq}}{2} - q_0\mathbf{p}\right).$$

Finally, the Heisenberg equation for the dispersion of the position of the wave-packet has the following form:

$$\frac{\mathrm{d}}{\mathrm{d}t}(\Delta q(t))^2 = \frac{2}{m}\left(\frac{\langle\mathbf{qp}\rangle}{2} + \frac{\langle\mathbf{pq}\rangle}{2} - q_0 p_0\right). \tag{10.13}$$

Equation (10.13) includes the average value of the momentum operator $p_0(t) = \langle\mathbf{p}(t)\rangle$. The momentum of a free particle is a constant of motion, thus its average value does not change with time, that is $p_0(t) = const$. We now calculate a further time derivative of the position dispersion:

$$\frac{\mathrm{d}^2}{\mathrm{d}t^2}(\Delta q(t))^2 = \frac{2}{m}\left(\frac{1}{2\mathrm{i}\hbar}\langle[\mathbf{qp}, \mathbf{H}]\rangle + \frac{1}{2\mathrm{i}\hbar}\langle[\mathbf{pq}, \mathbf{H}]\rangle - p_0\frac{\mathrm{d}q_0}{\mathrm{d}t}\right). \tag{10.14}$$

We notice that $\frac{\mathrm{d}q_0}{dt} = \frac{p_0}{m}$ and that the development of the commutator on the right-hand side Eq. (10.14) gives:

$$[\mathbf{qp}, \mathbf{H}] = -[\mathbf{H}, \mathbf{q}]\mathbf{p} - \mathbf{q}[\mathbf{H}, \mathbf{p}],$$

$$[\mathbf{pq}, \mathbf{H}] = -[\mathbf{H}, \mathbf{p}]\mathbf{q} - \mathbf{p}[\mathbf{H}, \mathbf{q}].$$

In order to calculate these commutators, we make use of Eq.(10.12) and of the relation: $[\mathbf{p}, G(\mathbf{q})] = -\mathrm{i}\hbar\frac{\partial G}{\partial q}$, $G(\mathbf{q})$ being a general function of the position operator. It follows that

$$[\mathbf{qp}, \mathbf{H}] = [\mathbf{pq}, \mathbf{H}] = \mathrm{i}\hbar\frac{\mathbf{p}^2}{m}.$$

Equation (10.14) establishes a relation between the second time derivative of the dispersion of the position operator and the dispersion of the momentum operator

$$\frac{\mathrm{d}^2}{\mathrm{d}t^2}(\Delta q(t))^2 = \frac{2}{m^2}\langle(\mathbf{p} - p_0)^2\rangle = \frac{2}{m^2}(\Delta p(t))^2. \tag{10.15}$$

By integrating Eq. (10.15), we find out that the standard deviation of the position has a parabolic dependence on time

$$(\Delta q(t))^2 = \frac{(\Delta p(t))^2}{m^2}t^2 + ((\Delta q(0))^2)' + (\Delta q(0))^2, \tag{10.16}$$

where $((\Delta q(0))^2)'$ and $(\Delta q(0))^2$ are integration constants calculated for $t = 0$. We assume that, at this instant, we have the minimum uncertainty:

$$(\Delta q(0))^2(\Delta p(0))^2 = \frac{\hbar^2}{4}. \tag{10.17}$$

For a free particle, $(\Delta p(t))^2$ is also a constant of motion, that is $(\Delta p(t))^2 = (\Delta p(0))^2 = \text{const}$ and

$$((\Delta q(0))^2)'(\Delta p(0))^2 = 0,$$

for which $((\Delta q(0))^2)' = 0$. Furthermore, from Eq.(10.17), we have $(\Delta p(0))^2 = \dfrac{\hbar^2}{4(\Delta q(0))^2}$ and Eq.(10.16) turns into:

$$(\Delta q(t))^2 = (\Delta q(0))^2\left[1 + \frac{\hbar^2}{4m^2(\Delta q(0))^4}t^2\right].$$

We now introduce the quantities related to the experimental conditions. If $\vec{v} = \dfrac{\mathbf{p}}{m}$ is the velocity of the particle, from the de Broglie relation, we have

$$v = \frac{h}{m\lambda}.$$

Such a quantity is related to the distance D and to the time interval t during which the particle travels from the scattering center to the detector

$$t = \frac{D}{v} = \frac{Dm\lambda}{h}. \tag{10.18}$$

By replacing the variable t with expression (10.18) into Eq.(10.15), we finally obtain the relation between the spreading of the wave-packet and the distance at which the detector is placed:

$$(\Delta q(t))^2 = (\Delta q(0))^2 \left[1 + \frac{D^2 \lambda^2}{16\pi^2 (\Delta q(0))^4} \right], \qquad (10.19)$$

From Eq.(10.19), we observe that, in order to have the minimum spread of the wave-packet, i.e. $(\Delta q(t))^2 \approx (\Delta q(0))^2$, the following condition is required

$$D\lambda \ll 4\pi(\Delta q(0))^2.$$

This theoretical investigation on the wave-packet spreading has to be taken into account in the arrangements of the experimental set up for scattering processes. The distance D between the target and the detectors but also the detector geometry, have to match the wave-packet size.

10.6 Problems

- Problem 1. A box with volume V is divided into two equal parts, l and r, by a potential barrier. A particle, inside the box, can cross the barrier by tunnel effect. The crossing process can be described by the Hamiltonian $\mathbf{H} = c(|l\rangle\langle r| + |r\rangle\langle l|)$, where $|l\rangle$ ($|r\rangle$) is the state describing the particle in the l (r) side and c is a real constant. Calculate the probability of finding the particle in the right state at time T, if it was in the left state at $t = 0$.
- Problem 2. A system is in the ground-state of the Hamiltonian

$$\mathbf{H} = \begin{pmatrix} \varepsilon & V \\ V & 2\varepsilon \end{pmatrix},$$

  where $V = \sqrt{2}\varepsilon$ ($\varepsilon < 0$). Calculate the probability $P(t)$ of finding the system in the ground-state of the unperturbed Hamiltonian $\mathbf{H}_0$ with eigenvalue ε after removing the potential V at $t = 0$.
- Problem 3. Let us consider an electron moving in a uniform and constant electric field E. Calculate the time evolution of its position average, position, and momentum standard deviations. Compare the results with their classical limits.
- Problem 4. Calculate the two commutators $[\mathbf{p(t)}, \mathbf{x(0)}]$ and $[\mathbf{x(t)}, \mathbf{x(0)}]$ for a free particle in one-dimensional motion.
- Problem 5. At $t = 0$, a particle is in one of the two states

$$|\psi_1(0)\rangle = \frac{1}{\sqrt{2}}(|1\rangle - |2\rangle),$$

$$|\psi_2(0)\rangle = \frac{1}{\sqrt{2}}(|1\rangle + \mathrm{i}|2\rangle),$$

where $|1\rangle$ is the ground state of a one-dimensional infinite potential well, and $|2\rangle$ the first excited state. Calculate the momentum average in each of the two states at $t > 0$.

- Problem 6. A particle is subject to the one-dimensional harmonic oscillator potential $\mathbf{V} = \frac{1}{2}m\omega^2\mathbf{q}^2$. Demonstrate that the position and momentum average values are sinusoidal functions of time with angular frequency ω.
- Problem 7. Let $\mathbf{L}$ be the angular momentum operator. If $\mathbf{L}_x$ and $\mathbf{L}_y$ are constants of motion, prove that also $\mathbf{L}_z$ is a constant of motion.

Chapter 11
Time-Dependent Perturbation Theory

In this chapter, we deal with the quantum transitions between stationary states of a system subjected to a perturbation for a time lapse. The stationary states are described by the time-independent Hamiltonian $\mathbf{H_0}$ and the perturbation by the time-dependent interaction $\mathbf{V}(t)$. Except for a few cases, the Schrödinger equation with the full Hamiltonian $\mathbf{H}(t) = \mathbf{H}_0 + \mathbf{V}(t)$ cannot be solved exactly. Approximate solutions are based on the expansion of the time evolution operator $\mathbf{U}(t, t_0)$ in series of $\mathbf{V}(t)$ powers. A concern of the perturbation methods is to make sure of the convergence of the perturbative expansion, that is mainly entrusted to the strength of the interaction. The lowest perturbation order provides accurate predictions for sufficiently weak interactions such as the Coulomb force, whereas more elaborated methods are requested by strong interactions like the nuclear force.

Assume the system is in the state $|\psi_0\rangle$, eigenstate of $\mathbf{H_0}$ with energy eigenvalue E_0, at time $t = 0$, and then it is subjected to the perturbation $\mathbf{V}(t)$ for a time T. The final state at time T is obtained applying the evolution operator to the state at the initial time $t = 0$, i.e.

$$|\psi(T)\rangle = \mathbf{U}(T, 0)|\psi_0\rangle = |\psi_0\rangle - \frac{\mathrm{i}}{\hbar}\int_0^T \mathrm{d}t'\mathbf{H}(t')\mathbf{U}(t', 0)|\psi_0\rangle, \qquad (11.1)$$

where we have used Eq. (7.2). The integral equation for the evolution operator can be converted into a series of $\mathbf{H}(t)$ powers, but the unpleasant feature of this series is that also H_0 is involved in the expansion. We will get rid of this drawback by introducing the interaction (or Dirac) representation in the next section.

Supplementary Information The online version contains supplementary material available at https://doi.org/10.1007/978-981-97-1670-8_11.

U. Lombardo et al., *Quantum Mechanics*, UNITEXT for Physics,
https://doi.org/10.1007/978-981-97-1670-8_11

11.1 Interaction Representation

Assuming a Hamiltonian with the form $\mathbf{H}(t) = \mathbf{H}_0 + \mathbf{V}(t)$, we introduce a new evolution operator $\mathbf{U}_I(t, t_0)$, in addition to $\mathbf{U}_0(t, t_0) = \mathrm{e}^{-\mathrm{i}\mathbf{H}_0(t-t_0)/\hbar}$ whose generator is $\mathbf{H}_0$ and $\mathbf{U}(t, t_0) = \mathrm{e}^{-\mathrm{i}\mathbf{H}(t)(t-t_0)/\hbar}$ whose generator is $\mathbf{H}(t)$. It is defined as

$$\mathbf{U}(t, t_0) = \mathbf{U_0}(t, t_0)\mathbf{U}_I(t, t_0). \tag{11.2}$$

It is easily shown that $\mathbf{U}_I$ is also a unitary operator. Starting with the Heisenberg equations of motion for $\mathbf{U}$ and $\mathbf{U_0}$

$$\mathrm{i}\hbar\frac{\mathrm{d}}{\mathrm{d}t}\mathbf{U}(t, t_0) = \mathbf{H}(t)\mathbf{U}(t, t_0),$$
$$\mathrm{i}\hbar\frac{\mathrm{d}}{\mathrm{d}t}\mathbf{U_0}(t, t_0) = \mathbf{H_0}\mathbf{U_0}(t, t_0),$$

the equations of motion for $\mathbf{U_I}$ are derived

$$\mathrm{i}\hbar\frac{\mathrm{d}}{\mathrm{d}t}\mathbf{U_I}(t, t_0) = \mathbf{V_I}(t)\mathbf{U_I}(t, t_0), \tag{11.3}$$
$$\mathbf{V_I}(t) = \mathbf{U}_0^{-1}\mathbf{V}(t)\mathbf{U_0}.$$

The advantage of the interaction representation is that the new generator $\mathbf{V_I}(t)$ of the time translations incorporates the dynamical effects due to the unperturbed Hamiltonian $\mathbf{H}_0$. Hence, in this form, the equations of motion are more suitable for developing perturbation methods. The time integration transforms the differential equation Eq. (11.3) into an integral equation

$$\mathbf{U_I}(t, t_0) = \mathbf{1} - \frac{\mathrm{i}}{\hbar}\int_{t_0}^{t} \mathbf{V_I}(t')\mathbf{U_I}(t', t_0)\mathrm{d}t',$$

that, in turn, is converted into a series in powers of $\mathbf{V_I}$

$$\mathbf{U_I}(t, t_0) = \mathbf{1} - \frac{\mathrm{i}}{\hbar}\int_{t_0}^{t} \mathrm{d}t'\mathbf{V_I}(t') + \left(-\frac{\mathrm{i}}{\hbar}\right)^2 \int_{t_0}^{t} \mathrm{d}t'\mathbf{V_I}(t') \int_{t_0}^{t'} \mathrm{d}t''\mathbf{V_I}(t'') + \cdots, \tag{11.4}$$

through an iterative procedure. In general, this equation of motion can be integrated only in a few cases, e.g. the two-level quantum systems treated in Sect. 10.5, whereas we resort to the truncation of the series, whenever the interaction strength is small enough. An important example is that of a charged particle interacting with the electromagnetic field, whose strength, known as *fine structure constant*, $\alpha = \dfrac{1}{137}$ is small. In this case, a good approximation is the truncation of Eq. (11.4) already at the first order, the α^2 term (and higher powers) being expected to be negligible owing to $\alpha \gg \alpha^2$. Another significant class is that of interactions lasting a shorter time than the relaxation time of the system. In this case, approximations can be applied regardless of the strength of the interaction, as discussed in Sect. 11.4.

11.2 Electron Transitions in Atoms

Frequently, the perturbation is due to interactions active during a finite period T, large on a microscopic scale but small on a macroscopic scale. This is the case of atoms exposed to electromagnetic radiation. The radiation induces electron transitions (excitations or de-excitations) of electrons among energy levels, through the absorption or emission of photons. Let us start by considering a target made of hydrogen atoms. For all the time T, a photon beam irradiates the target, interacting with the electrons. The emitted photons are detected by a photomultiplier array. Theoretically, the transition probability from the ground state of an atom at $t = 0$ to a given excited state $|\psi_k\rangle$ at time t is determined projecting the state $|\psi(t)\rangle$, given by Eq. (11.1), into the excited state

$$P_k = |\langle\psi_k|\psi(t)\rangle|^2 = |\langle\psi_k|\mathbf{U}(t,0)|\psi_0\rangle|^2 = |\langle\psi_k|\mathbf{U}_I(t,0)|\psi_0\rangle|^2.$$

In the last step, $\mathbf{U}$ has been replaced by $\mathbf{U_I} = \mathbf{U_0^{-1}U}$, applying Eq. (11.2). Then $\mathbf{U_0^{-1}}$ disappears because the excited state is one eigenvector with eigenvalue $\mathrm{e}^{-\mathrm{i}E_k t}$ irrelevant for the square modulus. Since we are dealing with a weak interaction, we can apply Eq. (11.4) truncated to the first order of the interaction $\mathbf{V_I}$

$$\begin{aligned} P_k &= \hbar^{-2}\left|\int_0^T \mathrm{d}t'\langle\psi_k|\mathbf{V}_I(t')|\psi_0\rangle\right|^2 \\ &= \hbar^{-2}\left|\int_0^T \mathrm{d}t'\langle\psi_k|\mathbf{V}(t')|\psi_0\rangle \mathrm{e}^{-\frac{\mathrm{i}}{\hbar}(E_k-E_0)t'}\right|^2, \end{aligned}$$

where $\mathbf{V}(t) = \mathbf{V_0}\cos\omega t$ is the potential due to the electromagnetic field (see Sect. 11.5).

Performing the time integration, we get

$$P_k = \hbar^{-2}|\langle\psi_k|\mathbf{V_0}|\psi_0\rangle|^2 \cdot \left|\frac{\mathrm{e}^{\mathrm{i}(\omega-\omega_{k0})T}-1}{\mathrm{i}(\omega-\omega_{k0})} - \frac{\mathrm{e}^{-\mathrm{i}(\omega+\omega_{k0})T}-1}{\mathrm{i}(\omega+\omega_{k0})}\right|^2,$$

where $\omega_{k0} = \omega_k - \omega_0 = (E_k - E_0)/\hbar$. The resonant structure of this expression for $\omega \pm \omega_{k0}$ encompasses two different processes. If the hydrogen atom is in the ground state at $t = 0$, then $\omega_k > \omega_0$, hence the first term is resonant and the second is negligible. In this case, P_k is the transition probability from the ground state to the excited state $0 \to k$ (left side of Fig. 11.1)

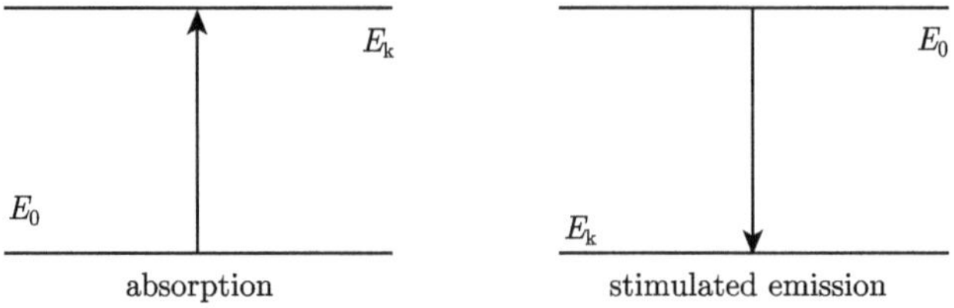

Fig. 11.1 Electronic transitions in atoms

$$P_k = \hbar^{-2}|\langle\psi_k|\mathbf{V_0}|\psi_0\rangle|^2 \cdot 4 \left|\frac{\sin[(\omega-\omega_{k0})T/2]}{(\omega-\omega_{k0})}\right|^2. \tag{11.5}$$

If the hydrogen atom at $t = 0$ is in an excited state, $\omega_0 > \omega_k$, the second term is resonant and the first is negligible. In this case, P_k is the transition probability from the excited state to the ground state $k \to 0$ (right side of Fig. 11.1)

$$P_k = \hbar^{-2}|\langle\psi_k|\mathbf{V_0}|\psi_0\rangle|^2 \cdot 4 \left|\frac{\sin[(\omega+\omega_{k0})T/2]}{(\omega+\omega_{k0})}\right|^2.$$

Let us focus on the first case. Assuming the exposure time to radiation $T \to \infty$, P_k tends to zero except for $\omega = \omega_k - \omega_0$, that is the transition becomes possible only if the photon energy is the difference between the two energy levels

$$\hbar\omega = E_k - E_0,$$

This condition was conjectured by Niels Bohr for the first time to interpret the spectral lines detected in the experiment in support of the Rutherford hypothesis on the atomic structure.

By applying the following realization of the Dirac δ-function (see Appendix A)

$$\lim_{t\to\infty} \frac{\sin^2 \alpha x t}{x^2 t} = \pi\alpha\delta(x),$$

the probability becomes

$$P_k = \frac{2\pi}{\hbar^2} T\, \delta(\hbar\omega - E_{k0})|\langle\psi_k|\mathbf{V_0}|\psi_0\rangle|^2.$$

We see that the probability linearly increases with time, as it should, because the number of transitions is proportional to the exposure time. In order to get rid of this trivial time dependence, we can define the transition probability *per* time unit. In experiments, the e.m. radiation is not exactly equal to a single frequency ω, but is distributed in a narrow region around ω that one can mimic with a continuous function $\rho(\omega)$ centered on ω. Hence, the probability will be

$$P_k/T = \int d\omega P_k(\omega)\rho(\omega) = \frac{2\pi}{\hbar}|\langle\psi_k|\mathbf{V_0}|\psi_0\rangle|^2\rho(\omega_{k0}).$$

Before closing this section, it is instructive to consider a constant interaction. It is enough to assume $\omega = 0$ in the transition probability and Eq. (11.5) simplifies:

$$P_k = \hbar^{-2}|\langle\psi_k|\mathbf{V_0}|\psi_0\rangle|^2 \cdot 4\left|\frac{\sin(\omega_{k0}T/2)}{\omega_{k0}}\right|^2.$$

The probability P_k displays damped oscillations with an infinite number of minima, which are solutions of the equation $\sin(\omega_{k0}T/2) = 0$. For large T, the probability before the first minimum is dominant, in other words, the probability beyond $\omega_{k0}T/2 \leq \pi$ is negligible. In terms of energy, this interval is

$$(E_k - E_0)T \leq 2\pi\hbar. \tag{11.6}$$

If we imagine the experiment with a perturbation $V \neq 0$ in the interval T, as a way of measuring the energy E_k, then tuning the duration T to larger and larger values, the uncertainty on E_k becomes less and less. It looks like the uncertainty principle between space and momentum [MES]. We should however remember that time is not an operator in non-relativistic quantum mechanics.

11.3 Dipole Approximation

The space component $\mathbf{V}_0$ of the e.m. field $\mathbf{V}$ has so far been neglected, but it entails another selection rule for the electron transitions we are going to infer in this section. From Eq. (11.8) of the Complement to this section, this is given by

$$\mathbf{V}_0 = -\frac{e}{mc}\vec{\mathbf{A}} \cdot \vec{\mathbf{p}},$$

where $\vec{p}$ is the electron momentum. Firstly, it should be noticed that the atomic dimension is much smaller than the radiation wavelength. Indeed, the energy of photons has the order of magnitude of the energy gap between the levels involved in the transition, e.g. $E_{\rm el} = \hbar\omega \approx e^2/r_0$ for hydrogen. Thus, one has

$$\lambda = \frac{2\pi c}{\omega} = \frac{2\pi c\hbar}{E_{\rm el}} = \frac{2\pi c\hbar}{e^2/r_0}$$

and hence

$$\frac{r_0}{\lambda} = \frac{e^2}{2\pi c\hbar} \approx \frac{1}{2\pi \cdot 137} \ll 1,$$

This implies that, in practice, the atom experiences a constant field on the atomic scale, at least for atoms that are not too big (since $r \sim r_0/Z$, where Z is the atomic number). As a consequence, we are allowed to only keep the first order in the multipole expansion of $\vec{\mathbf{A}}$ (dipole approximation). We easily get $\mathbf{V}_0 \approx \frac{e}{mc}\vec{\mathbf{p}} \cdot \vec{\mathbf{A}}_0$. Assuming the polarization of the field is parallel to the z-axis, the matrix element of the potential between two energy levels is

$$\langle \psi_k | \mathbf{V}_0 | \psi_0 \rangle = -\frac{e}{mc} \langle \psi_k | \mathbf{p_z} | \psi_0 \rangle \cdot \vec{\mathbf{A}}_0,$$

where

$$\langle \psi_k | \mathbf{p_z} | \psi_0 \rangle = -\frac{\mathrm{i}}{2m\hbar} \langle \psi_k | [\mathbf{H_0}, \mathbf{z}] | \psi_0 \rangle = -\frac{\mathrm{i}}{2m\hbar} (E_k - E_0) \langle \psi_k | \mathbf{z} | \psi_0 \rangle,$$

The eigenvalue equation has been applied for the states ψ_k and ψ_0, which are eigenstates of the atomic Hamiltonian $\mathbf{H_0}$. The operator $\mathbf{z}$ in polar coordinates is $\mathbf{r} \cos\theta \sim r Y_{10}$, so that

$$\langle \psi_k | \mathbf{p_z} | \psi_0 \rangle \sim \int r^2 \mathrm{d}r \psi_k(r) r \psi_0(r) \int \mathrm{d}\Omega Y^*_{l_k m_k} Y_{10} Y_{00} \sim \delta_{l_k,1} \delta_{m_k,0},$$

A new selection rule for the electron transitions comes out from the previous expression: a transition from the ground state is possible only if the angular quantum numbers of the excited state are $l_k = 1$ and $m_k = 0$. For transitions from the state $\psi_{k'}$ to the state ψ_k it is generalized to $|l_k - l_{k'}| = 1$. In addition, if the field is not polarized, the selection rule for the projection of angular momentum changes, because $\mathbf{p_x}$, $\mathbf{p_y}$ and spherical harmonics $Y_{1\pm1}$ also come into play. It is found that $m_k - m_{k'} = 0, \pm 1$. We can summarize the selection rules for the electron transition between two levels in atomic systems within the first-order time-dependent perturbation theory as follows:

- the photon energy must be equal to the energy gap between the two levels;
- the difference between the orbital angular momenta of the two levels must be equal to unity;
- the difference between their projections must be equal to 0 or ± 1.

These selection rules permit us to distinguish between allowed and prohibited transitions. If these rules are not satisfied, the photons impinging onto the atoms, are not absorbed, but they are elastically scattered.

11.4 Slow Versus Fast Processes

In the last section, we learned that the quantum transitions induced by e.m. perturbations on the atomic electrons are sensitive not only to the strength but also to the duration of the interaction. In general, there are processes where slow and fast variables coexist and their evolution depends on the interplay between the perturbation and response times. In extreme situations, some approximation methods have been devised to significantly simplify their study.

On the atomic scale, the β-decay of an atomic nucleus is frequently taken as an example. In such a process, a nucleus with Z protons and N neutrons spontaneously emits an electron, due to the decay of a neutron into a proton.[1] As a consequence, the atomic number Z grows by one unit and the electron cloud around the nucleus must rearrange itself according to the change of Coulomb potential. The kinetic energy of the electron from β-decay is quite large when compared to the transition energy gap ΔE of electrons from the cloud. Thus, the variation of the Coulomb field is much faster than the electronic rearrangement. The latter is given by Eq. (11.6) so that

$$\tau \ll \frac{\hbar}{\Delta E}.$$

When the electron from β-decay leaves the atom, the electron cloud of the atom must still start to rearrange itself. Therefore, the electronic transition probability between two states before and after the β-decay can be approximated as

$$P_{ab} = |\langle \psi_a^{Z+1} | \mathbf{U}(t, t_0) | \psi_b^Z \rangle|^2 \sim |\langle \psi_a^{Z+1} | \psi_b^Z \rangle|^2$$

being $\mathbf{U}(t, t_0) \sim 1$ for $t - t_0 \geq \tau$. This approximation is known as the sudden approximation.

On a nuclear scale, the coexistence of slow and fast variables is met in high-energy reactions, e.g. the Coulomb breakup of the deuteron (d, pn). In this process, the proton belonging to the deuteron experiences a fast Coulomb interaction with the target, whereas the neutron does not. The three-body kinematics is depicted in Fig. 11.2. Assuming the short-range interaction between neutron and target to be negligible, the asymptotic Schrödinger equation is written as

$$\left[K_R + V_{\text{pt}} \left(\vec{R} - \frac{1}{2}\vec{r} \right) + H_{\text{np}}(r) \right] \Psi^+(\vec{R}, \vec{r}\,) = E\Psi^+(\vec{R}, \vec{r}\,),$$

where K_R is the kinetic energy of the relative motion. Now, since the distance r between proton and neutron, which describes the deuteron internal motion, is a slow variable, whereas the distance R between the deuteron center of mass and the target is a fast variable, the deuteron wave function can be frozen in the ground state, and the last equation can be simplified in the form

$$[K_R + V_{\text{pt}}(\vec{R} - \vec{r}\,)]\Phi^+(\vec{R}) = (E - E_{\text{d}})\Phi^+(\vec{R}),$$

where E_{d} is the binding energy of the deuteron and Φ^+ is its wave function in the relative motion. The strong simplification means that a difficult three-body problem

---

[1] In the β-decay a neutron decays according to the reaction $n \rightarrow p + e^- + \bar{\nu}_e$, where $\bar{\nu}_e$ is the electronic anti-neutrino.

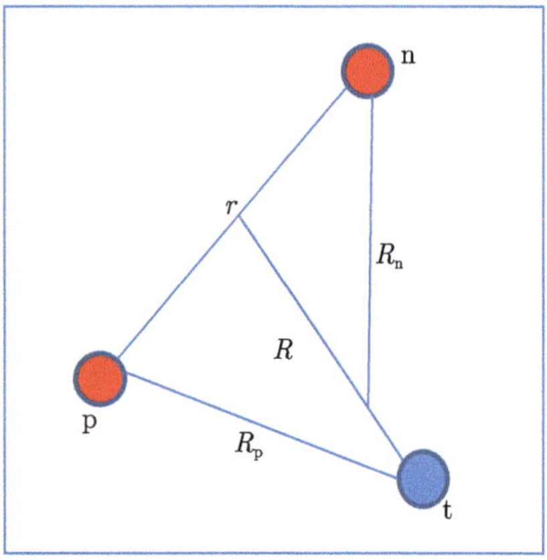

Fig. 11.2 Deuteron Coulomb breakup (d, pn): the proton (p) and the neutron (n) are depicted with red spots, the target (t) with a blue spot

is reduced to a two-body problem, with the variable r playing the role of an external parameter. This procedure has been extended to the Coulomb breakup of *halo* nuclei[2].

Last but not least, the process with slow and fast variables we want to address is the propagation of sound waves in a diluted Fermi gas that will be treated in Chap. 18. In diluted ^{3}He, two kinds of sounds have been observed. The propagation of acoustic waves is well known to occur in hydrodynamic regime, where the oscillation period T is much larger than the inter-atomic collision time τ. The condition $T \gg \tau$ ensures the propagation of the ordinary sound does not affect the local equilibrium of the medium (adiabatic process). At very low temperature, the so-called zero sound has been observed in collisionless regime, where $T < \tau$. The propagation of this kind of sound is due to the quantum behaviour of ^{3}He at a very low temperature. Here, the Pauli principle, blocking most atoms in frozen states, hinders the inter-particle collisions and thus no local equilibrium is possible.

11.5 Complement to Sect. 11.2: Interaction of Charged Particles with the Electromagnetic Field

In this section, the time-dependent perturbation due to the interaction between the atomic electrons and the electromagnetic field will be inferred. We start with the Lagrangian

$$\mathcal{L} = \mathcal{L}_0 - \frac{e}{c} A^i \frac{\mathrm{d}x_i}{\mathrm{d}t} = \frac{1}{2} m v^2 - e\phi + \vec{A} \cdot \vec{v},$$

[2] Halo nuclei are neutron-rich isotopes with one or two weakly bound neutrons, which are delocalized from the core.

where $\mathcal{L}_0$ is the Lagrangian of a free particle, ϕ is the scalar potential and $\vec{A}$ is the vector potential. In the non-relativistic limit, the first term can be replaced by the kinetic energy $\frac{1}{2}mv^2$. Moving to the Hamiltonian formalism through the equation

$$\vec{p} = \frac{\partial \mathcal{L}}{\partial \vec{v}} = \frac{e}{c}\vec{A} + m\vec{v},$$

we replace velocity with momentum to get the Hamiltonian

$$H = \vec{v} \cdot \frac{\partial \mathcal{L}}{\partial \vec{v}} - \mathcal{L} \tag{11.7}$$

and then

$$H = e\phi + \frac{1}{2m}\left(\vec{p} - \frac{e}{c}\vec{A}\right)^2,$$

Due to the gauge freedom, we can apply a suitable gauge transformation such that the scalar potential $\phi = 0$, and the vector potential fulfills the condition $\vec{\nabla} \cdot \vec{A} = 0$ (Lorentz condition). Afterwards, since the e.m. field strength is weak ($A^2 \ll A$), we can neglect A^2 in the preceding equation. Since the observables $\vec{\mathbf{p}}$ and $\vec{\mathbf{A}}(\vec{r}\,)$ do not commute in quantum mechanics, the Hamiltonian is written as

$$\mathbf{H} = \frac{1}{2m}\left(\vec{\mathbf{p}} - \frac{e}{c}\vec{\mathbf{A}}\right)^2 \approx \frac{\mathbf{p}^2}{2m} - \frac{e}{2mc}(\vec{\mathbf{p}} \cdot \vec{\mathbf{A}} + \vec{\mathbf{A}} \cdot \vec{\mathbf{p}}),$$

but, imposing the Lorentz condition, it follows:

$$\begin{aligned}
\langle \chi | \vec{\mathbf{p}} \cdot \vec{\mathbf{A}} | \psi \rangle &= \int \mathrm{d}\vec{r}\, \chi^*(\vec{r}\,) \frac{\hbar}{\mathrm{i}} \vec{\nabla} \cdot \vec{A} \psi(\vec{r}\,) \\
&= \int \mathrm{d}\vec{r}\, \chi^*(\vec{r}\,) \left(\frac{\hbar}{\mathrm{i}} \vec{\nabla} \cdot \vec{A}\right) \psi(\vec{r}\,) + \int \mathrm{d}\vec{r}\, \chi^*(\vec{r}\,) \vec{A} \cdot \left(\frac{\hbar}{\mathrm{i}} \vec{\nabla} \psi(\vec{r}\,)\right) \\
&= \langle \chi | \vec{\mathbf{A}} \cdot \vec{\mathbf{p}} | \psi \rangle,
\end{aligned}$$

for two arbitrary states of the system. The first two terms inside the gradient vanish. As a result, $\vec{\mathbf{p}}$ and $\vec{\mathbf{A}}$ in fact commute and the Hamiltonian takes the final form

$$\mathbf{H} = \frac{\mathbf{p}^2}{2m} - \frac{e}{mc}\vec{\mathbf{A}} \cdot \vec{\mathbf{p}} \tag{11.8}$$

In the case of plane monochromatic waves, the vector potential is

$$\vec{\mathbf{A}} = \mathcal{R}\vec{\mathbf{A}}_0 \mathrm{e}^{\mathrm{i}(\vec{k} \cdot \vec{r} - \omega t)}.$$

11.6 Notes and References

MES For a derivation of this time-energy uncertainty relation see also A. Messiah, *Quantum Mechanics*, North-Holland, 1966, Chap. XVII.8.

11.7 Problems

- Problem 1. An infinite potential well of width $2L$ is divided into two equal parts, left and right, by a separating wall. On the left, a particle is in the ground state. At the time t, the wall is abruptly removed to allow the particle to occupy the full potential well space. What is the probability for the particle to be in the new ground state?
- Problem 2. Consider the β-decay reaction $^3\mathrm{H}_1 \rightarrow {}^3\mathrm{He}_2 + e^- + \bar{\nu}_e$. Show that the condition for applying the sudden approximation (see Sect. 11.4) is satisfied, considering that the kinetic energy of the emitted electron e^- is about 6 keV.
- Problem 3. One of the two electrons of He atom in the ground state is emitted after a sudden collision. Calculate for the second electron its probability of being in the He^+ ion ground state.
- Problem 4. A particle is in the ground state of a one-dimensional infinite potential well of width L. During a time interval T a time-dependent perturbation $V(t) = c\mathbf{p}\cos(\omega t)$ acts on the particle (c and ω are constants). Calculate the transition probability to the first excited state at the first perturbative order.
- Problem 5. An electron is in the ground state of a one-dimensional infinite potential well of width L. During a time interval T a constant electrical field E acts on the electron. Calculate the transition probability to the first, second, and third excited states at the first order of the perturbation expansion.
- Problem 6. A particle of mass m in a one-dimensional harmonic oscillator potential, is perturbed by a constant weak interaction during a time T. For which excited states does the transition probability, calculated at the first perturbation order, not vanish? For which value of T is this probability the biggest? (Hint: use the creation and annihilation operators).

Chapter 12
Two-Body Problem: Bound States

The simplest quantum systems consist of two interacting particles: the hydrogen atom, made up of a proton and an electron, interacting via the electrostatic force, on the atomic scale; and the deuteron nucleus, made up of a proton and a neutron, interacting via the nuclear force, on the nuclear scale. The interactions responsible for the stability of these two systems have similarities and differences to highlight for a better understanding of their properties. In the non-relativistic limit, the two interactions are instantaneous, time-independent and invariant under space translations and rotations, hence $V(\vec{r}_1, t_1, \vec{r}_2, t_2) = V(|\vec{r}_1 - \vec{r}_2|)$, the positions of the two particles being $\vec{r}_1$ and $\vec{r}_2$. Since the two interactions are attractive, bound states are expected both for the hydrogen atom and for the deuteron nucleus. In fact, both of them exist in nature. However, the size and energy scales of the two systems are quite different: the hydrogen is 10^5 times greater than the deuteron and its binding energy is 10^{-6} times smaller. These significant differences must be traced back to the different nature of the two forces: the former is weak and long range and the latter is strong and short range. Although these properties have the same significance in both classical and quantum mechanics, their implications are quite different. Just to mention one, while the binding energy is negative in both cases, the classical values are continuous, whereas the quantum values are quantized. The same holds true for the angular momentum.

12.1 Central Potential

Consider the Hamiltonian of two particles interacting by means of the potential $V(\vec{\mathbf{r}}_1 - \vec{\mathbf{r}}_2)$

$$\mathbf{H} = \frac{\mathbf{p}_1^2}{2m_1} + \frac{\mathbf{p}_2^2}{2m_2} + V(\vec{\mathbf{r}}_1 - \vec{\mathbf{r}}_2),$$

Supplementary Information The online version contains supplementary material available at https://doi.org/10.1007/978-981-97-1670-8_12.

U. Lombardo et al., *Quantum Mechanics*, UNITEXT for Physics,
https://doi.org/10.1007/978-981-97-1670-8_12

where m_1 and m_2 are the masses of the two particles. Introducing the center-of-mass and relative motion coordinates

$$\vec{\mathbf{R}} = \frac{m_1\vec{\mathbf{r}}_1 + m_2\vec{\mathbf{r}}_2}{m_1 + m_2}, \qquad \vec{\mathbf{r}} = \vec{\mathbf{r}}_1 - \vec{\mathbf{r}}_2,$$

respectively, the Hamiltonian takes the form

$$\mathbf{H} = \frac{\mathbf{P}^2}{2M} + \frac{\mathbf{p}^2}{2m} + V(\vec{\mathbf{r}}), \tag{12.1}$$

where $M = m_1 + m_2$ and $m = (m_1 m_2)/M$ are total and reduced mass Eq. (7.12), respectively, and

$$\vec{\mathbf{P}} = \vec{\mathbf{p}}_1 + \vec{\mathbf{p}}_2, \qquad \vec{\mathbf{p}} = \frac{m_2\vec{\mathbf{p}}_1 - m_1\vec{\mathbf{p}}_2}{M}$$

are the total and relative momenta, respectively. From the new Hamiltonian, Eq. (12.1), we see that center-of-mass motion and relative motion are independent of each other so that the total wave function can be factorized as

$$\Psi(\vec{r}_1, \vec{r}_2) = \Phi(\vec{R}) \cdot \psi(\vec{r}\,) = (2\pi\hbar)^{-\frac{3}{2}} e^{i\vec{P}\cdot\vec{R}/\hbar}\psi(\vec{r}\,).$$

We see that the CM motion decouples from the relative motion, and its wave function $\Phi(\vec{R})$ is trivially that of a free particle. As a result, we are left with the study of the equation of the relative motion

$$\left(\frac{p^2}{2m} + V(\vec{\mathbf{r}})\right)\psi(\vec{r}\,) = E\psi(\vec{r}\,).$$

We assume now that the potential is spherically symmetric, i.e. depending only on the distance $r = \vec{r}_1 - \vec{r}_2$. In this case, the square modulus of orbital angular momentum $\mathbf{l^2}$ and its z-projection $\mathbf{l}_z$ commute with the Hamiltonian, so that they have a complete set of simultaneous eigenfunctions $\psi_{nlm}(\vec{r})$, where n, l, m are the quantum numbers identifying the eigenvalues of $\mathbf{H}$, $\mathbf{l^2}$, and $\mathbf{l}_z$, respectively. As in classical mechanics, the angular motion decouples from the radial motion owing to the isotropy of the interaction. Splitting the kinetic energy into the radial and centrifugal kinetic components, the Hamiltonian is divided into a radial term, including the radial kinetic and potential energy, and an angular term, including only the centrifugal term. As a result, the wave function of the relative motion factorizes as $\psi_{nlm}(\vec{r}\,) = \psi_{nl}(r) \cdot Y_{lm}(\Omega)$, where the spherical harmonics $Y_{lm}(\Omega)$ are eigenstates of $\mathbf{l^2}$ and $\mathbf{l}_z$[1]. Hence, after removing the spherical harmonics, the radial equation becomes

---

[1] In the case of non-spherical potential, the relative wave function will be a superposition of spherical harmonics.

$$\left(\frac{p_r^2}{2m} + \frac{\hbar^2 l(l+1)}{2mr^2} + V(r)\right)\psi_{nl}(r) = E_n\,\psi_{nl}(r), \tag{12.2}$$

where $\hbar^2 l(l+1)/2mr^2$ is the centrifugal potential. The expression of the square radial momentum is derived from the transformation of Cartesian coordinates into polar coordinates (see Problem 7). This is found to be

$$p_r^2 = -\hbar^2 \frac{1}{r}\frac{\mathrm{d}^2}{\mathrm{d}r^2} r.$$

The radial equation depends parametrically on the quantum number l, but it does not depend on the quantum number m with the consequence that each l state is $2l+1$ times degenerate.

The Hamiltonian of a free particle ($V(r) = 0$) falls into the class of rotationally invariant Hamiltonians. The Schrödinger equation of free particles has been studied in Sect. 4.4, using Cartesian coordinates, as its Hamiltonian is also translationally invariant. However, the polar coordinates are a viable alternative, whenever the partial wave analysis is requested, e.g. in scattering processes. The angular part of the solutions is given by spherical harmonics $Y_{lm}(\theta, \phi)$ and the radial part by Bessel's functions $j_l(kr)$, that is

$$\psi_{nlm}(r) \sim j_l(kr) Y_{lm}(\Omega), \tag{12.3}$$

where $k = \sqrt{2mE/\hbar^2}$ is the wave number and E the energy. Details on the Bessel functions are given in Appendix B. Simultaneous eigenstates of momentum and angular momentum cannot exist, as the corresponding operators do not commute, but we can express the plane-wave eigenfunctions as superposition of the eigenstates $Y_{lm}(\theta, \phi)$, because these form a complete set of orthogonal states. As shown in Appendix B, the result is

$$\mathrm{e}^{\mathrm{i}\vec{k}\cdot\vec{r}} = \sum_{l=0}^{\infty} \mathrm{i}^l (2l+1) j_l(kr) P_l(\cos\theta), \tag{12.4}$$

where $P_l(\cos\theta)$ are the Legendre polynomials (see Appendix B).

We proceed now to solve the radial equation Eq. (12.2) for bound states. We must impose suitable boundary conditions at $r = 0$ and $r = \infty$ to select only physically acceptable solutions. Consider first the behaviour at $r = 0$: near the origin the centrifugal barrier increases as $1/r^2$, whereas the potentials we will study increase less than $1/r^2$ or do not increase at all. Therefore, the radial equation, Eq. (12.2) can be approximated as it follows:

$$\left[-\frac{\hbar^2}{2m}\frac{1}{r}\frac{\mathrm{d}^2}{\mathrm{d}r^2} r + \frac{\hbar^2 l(l+1)}{2mr^2}\right]\psi_l(r \sim 0) \sim 0. \tag{12.5}$$

This equation has solutions of the form $r\psi(r \sim 0) = r^{l+1}$ at $r \sim 0$. Consider then a distance $r \sim \infty$: the centrifugal term can be neglected and the radial equation can be approximated

$$-\frac{\hbar^2}{2m}\frac{1}{r}\frac{\mathrm{d}^2}{\mathrm{d}r^2}r\psi_l(r \sim \infty) = (E - V(r \sim \infty))\psi_l(r \sim \infty). \tag{12.6}$$

The solution of this equation depends on the asymptotic behaviour of the potential. Since we want to deal with bound states, $E - V(r \sim \infty)$ must be negative; otherwise, the wave function would oscillate indefinitely at $r \sim \infty$. Most interactions vanish asymptotically, leading to bound states with negative energy spectrum and exponentially decreasing wave function. The spherical harmonic oscillator potential, which asymptotically increases as r^2, deserves a separate discussion (see Sect. 14.4).

12.2 Hydrogen Atom

We first solve the Schrödinger equation for the hydrogen atom, made up of a proton and an electron with equal charge e. The two particles interact via the Coulomb potential $V(r) = -e^2/r$, r being their distance. The reduced mass of the two particles is almost equal to the proton mass, the electron mass being three orders of magnitude smaller. With the Coulomb potential, the radial equation is written as

$$\left(\frac{p_r^2}{2m} + \frac{\hbar^2 l(l+1)}{2mr^2} - \frac{e^2}{r}\right)\psi_{nl}(r) = E_n\,\psi_{nl}(r), \tag{12.7}$$

The solution of the radial equation is featured by the completion between the repulsive centrifugal and the attractive Coulomb potentials, as shown in Fig. 12.1.

The boundary conditions at $r = 0$ and $r = \infty$, discussed in the last section, will select the physically acceptable solutions. At $r \sim 0$, the solutions must have the form

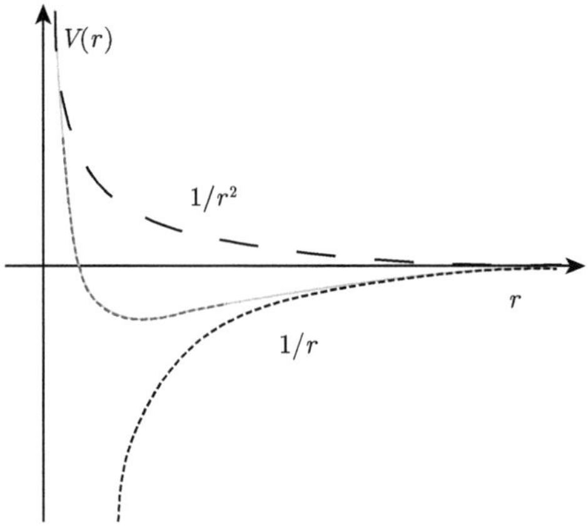

Fig. 12.1 Coulomb (short dashes), centrifugal (long dashes), and, in between, total potential

$r\psi(r \sim 0) = r^{l+1}$. At asymptotic distance $r \sim \infty$, both Coulomb and centrifugal fields can be neglected, and thus the radial equation can be approximated as follows:

$$\left(-\frac{\hbar^2}{2m}\frac{1}{r}\frac{\mathrm{d}^2}{\mathrm{d}r^2}r - E\right)\psi_l(r \sim \infty) = 0,$$

so that the solutions physically acceptable must have the form $r\psi_l(r \sim \infty) \sim \mathrm{e}^{-kr}$, where $k = \sqrt{-2mE/\hbar^2}$ $(E < 0)$. Therefore, the full solution of the radial equation Eq. (12.7) must have the form

$$r\psi_l(r) = r^{l+1}\mathrm{e}^{-kr} f_l(r), \tag{12.8}$$

with $f_l(r)$ a regular function in the range between $r = 0$ and $r \to \infty$. After the expression of $r\psi_l(r)$ from Eq. (12.8) has been plugged into the radial equation Eq. (12.7), the latter becomes a Laplace equation for $f_l(r)$. Introducing the dimensionless variable $x = kr$, we can rewrite the radial equation in the form

$$\left[x\frac{\mathrm{d}^2}{\mathrm{d}x^2} + 2(l+1-x)\frac{\mathrm{d}}{\mathrm{d}x} - 2\left(l+1-\frac{1}{ka}\right)\right] f_l(x) = 0, \tag{12.9}$$

where $a = \dfrac{\hbar^2}{me^2}$. By developing $f_l(x)$ in Taylor series around $r = 0$, the differential equation Eq. (12.9) transforms into an infinite set of coupled algebraic equations for the coefficients c_i $(c_0 = 1)$

$$(2l+2)c_1 = 2\left(l+1-\frac{1}{ka}\right), \tag{12.10}$$

$$2(2l+3)c_2 = 2\left(l+2-\frac{1}{ka}\right)c_1, \tag{12.11}$$

$$\ldots\ldots.$$

The solution is known as a confluent hypergeometric series. It is found that it diverges for $r \to \infty$ more rapidly than e^{-kr} converges. Hence, this solution cannot describe bound states. But for special values of the energy the series becomes a polynomial, which is damped by the exponential for $r \to \infty$. These kinds of solutions can describe bound states and the corresponding energy values range into a quantized spectrum, rather than continuous, as expected from the experimental atomic spectroscopy. Let us see, in detail, how the quantized energies arise from Eqs.(12.10) and (12.11). Choosing the energy E such that $ka = 1/(l+1)$, Eq. (12.10) gives $c_1 = 0$, so that Eq. (12.11) gives $c_2 = 0$ and the coefficients of the subsequent equations vanish as well. For this first value of energy E, the corresponding polynomial is simply $f_l = c_0 = 1$ and the full radial eigenfunction is

$$\psi_l(r) = r^l\mathrm{e}^{-kr}.$$

With the choice of the parameter E such as to fulfill the condition $ka = \dfrac{1}{l+2}$ it turns out that $c_1 \neq 0$, but c_2 as well as all coefficients of the successive equations are identically equal to zero. For this second value of E, the full radial eigenfunction is

$$\psi_l(r) = r^l e^{-kr}(1 + 2kr\, c_1).$$

Iterating this procedure, we get all bound-state eigenvalues and eigenfunctions with the condition $k_{n'}a = \dfrac{1}{l+1+n'}$, where $n' = 0, 1, 2, 3, \cdots$ As a result, the energy spectrum is

$$E_n = \frac{\hbar^2 k_n^2}{2m} = \frac{\hbar^2}{2m} \frac{1}{a^2(l+1+n')^2} \qquad (n' = 0, 1, 2, \cdots.),$$

where $n = l + 1 + n'$ is the principal quantum number. Each energy level n is degenerate, because the wave function $\psi_{nlm}(\vec{r}\,) = \psi_{nl}(r)Y_{lm}(\Omega)$ depends on three quantum numbers. For each value of n, $0 \leq l \leq n-1$ and $-l \leq m \leq l$, the degeneracy of the E_n energy level is

$$\mathcal{N} = \sum_{l=0}^{n-1}(2l+1) = n^2.$$

In conclusion, the eigenstates ψ_{nlm} of the hydrogen Hamiltonian are simultaneous eigenstates of $\mathbf{H}, \mathbf{l}^2$ and $\mathbf{l}_z$ labeled by the quantum numbers n, l and m; the energy eigenvalues E_n are degenerate with degeneracy equal to n^2. The quantum numbers of the ground state are $n = 1, l = 0, m = 0$ with energy $E_1 = \hbar^2/2ma^2$, where $a \approx 0.53$ Å is the Bohr radius, and $E_1 \approx -13.6$ eV. The whole energy spectrum is plotted in Fig. 12.2: by increasing n the levels rapidly approach $E = 0$, which is the border between bound states and scattering states. The value $E = 13.6$ eV is the ionization energy, that is the energy necessary to excite the electron from the ground state to an unbound state.

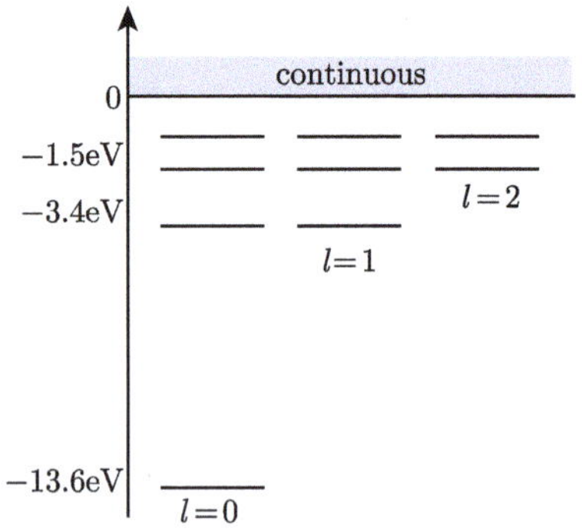

Fig. 12.2 Energy spectrum of the hydrogen atom

12.3 Isospin

The nucleus consists of two types of nucleons, namely protons and neutrons, the neutrons being uncharged. The nuclear interaction, plotted in Fig. 12.4, is responsible for the interaction between the two nucleons. It is a short-range force with an attractive component, but a quite repulsive component (hard core) at a very short distance. There is experimental evidence (see Table 13.1) that the nuclear force is independent of charge, which means that it does not distinguish neutrons from protons. Therefore, they have to be considered as two different states of the same particle, which defines a two-dimensional space, named isospin space. Hence, any single-nucleon state will be written $|\vec{r}, \vec{s}, \vec{t} >$ with $\vec{t}$ the isospin variable. The same as the spin case, the symmetry transformations in the isospin space belong to the class of SU(2) transformations.

The isospin variable, as generator of the SU(2) symmetry transformations, satisfies the same algebra as angular momentum.

Denoting the isospin vector as $\vec{\mathbf{t}}$, its magnitude will be $t = \frac{1}{2}$ and the eigenvalue of its projection $\mathbf{t}_z$ onto the z-axis of isospin space will be $t_z = +\frac{1}{2}$ for protons and $t_z = -\frac{1}{2}$ for neutrons. The corresponding states in the isospin two-dimensional space are

$$n = \begin{pmatrix} 0 \\ 1 \end{pmatrix}, \quad p = \begin{pmatrix} 1 \\ 0 \end{pmatrix},$$

and the components of the vector variable $\vec{\mathbf{t}}$, represented in terms of Pauli matrices (see Sect. 6.4), are

$$\tau_x = \frac{1}{2}\begin{pmatrix} 0 & 1 \\ 1 & 0 \end{pmatrix}, \quad \tau_y = \frac{1}{2}\begin{pmatrix} 0 & -\mathrm{i} \\ \mathrm{i} & 0 \end{pmatrix}, \quad \tau_z = \frac{1}{2}\begin{pmatrix} 1 & 0 \\ 0 & -1 \end{pmatrix}.$$

The Pauli matrices and the identity matrix form a complete basis in the 2×2 matrix space. Let us consider two nucleons with isospin $\vec{\mathbf{t}}_1$ and $\vec{\mathbf{t}}_2$, respectively, interacting via the nuclear force, assumed to be invariant in space, spin and isospin space. Since the total isospin $\vec{\mathbf{T}} = \vec{\mathbf{t}}_1 + \vec{\mathbf{t}}_2$ is a good quantum number along with the total spin S and total angular momentum L, because of the charge independence of the nuclear force, it is more significant to feature the two-nucleon state in the coupled representation. Using the Clebsch–Gordan coefficients, the coupled isospin two-nucleon state is

$$|\psi_{TT_z}\rangle = \sum_{TT_z} C\left(\frac{1}{2}, t_{z1}, \frac{1}{2}, t_{z2} | T T_z\right) \left|\frac{1}{2} t_{z1}, \frac{1}{2} t_{z2}\right\rangle,$$

where $T = 0, 1$. The same coupling procedure is applied to angular momentum and spin. Since nucleons are identical fermions, the two-nucleon state must be antisymmetrized, exchanging particle 1 and particle 2. This exchange involves a factor

$(-1)^L$,$(-1)^S$ and $(-1)^T$, for angular momentum[2], spin and isospin, respectively, owing to the exchange property of the Clebsch–Gordan coefficients. In conclusion, the anti-symmetrized two-nucleon state is written as

$$\Psi_{LST} = \Psi_{LST}(12) - \Psi_{LST}(21) = \Psi_{LST}(12)[1 - (-1)^{L+S+T}].$$

so that the full anti-symmetrization requires the sum $L + S + T$ to be an odd number, this rule being known as *generalized Pauli principle*.

For a system of A nucleons, the isospin operators $\mathbf{T}$ and $\mathbf{T}_z$ are defined as

$$\mathbf{T} = \sum_{i=1}^{A} \mathbf{t}_i, \qquad \mathbf{T}_z = \sum_{i=1}^{A} t_z^{(i)}.$$

Hence, the eigenvalue equation of the operator $\mathbf{T}_z$

$$\begin{aligned} \mathbf{T}_z|1, 2, \cdots, A\rangle &= \left(-\frac{1}{2} - \cdots - \frac{1}{2} + \frac{1}{2} + \cdots + \frac{1}{2}\right)|1, 2, \cdots, A\rangle \\ &= \frac{(N - Z)}{2}|1, 2, \cdots, A\rangle, \end{aligned} \tag{12.12}$$

where Z is the proton number and N the neutron number. As a result, for a system of A nucleons with $N \neq Z$ the eigenvalue of the $\mathbf{T}_z$ operator is proportional to the neutron excess. If $\mathbf{H}$ is the Hamiltonian of the system of A nucleons, the commutation relation

$$[\mathbf{H}, \vec{\mathbf{T}}] = 0$$

states the rotational invariance in the isospin space, which is a consequence of the charge independence of nuclear forces. Moreover, the rotational invariance with respect to the quantization axis

$$[\mathbf{H}, \mathbf{T}_z] = 0$$

is related to the charge conservation of the system. Indeed, since $A = N + Z$, we have $Z = \frac{A}{2} - T_z$. The $\mathbf{T}^2$ eigenvalue is a good total isospin quantum number: the nuclear states, which are associated with the same isospin eigenvalue, have the same structure, and, as a result, the same energy. The further inclusion of the Coulomb interaction changes the energy values of such a class of states, and we refer to these states as isospin multiplets. As an example, we consider the three isobars (nuclei with the same mass number) ^{14}C$(Z = 6, N = 8)$, ^{14}N$(Z = 7, N = 7)$, ^{14}O$(Z = 8, N = 6)$, which form a $T = 1$ isospin multiplet. In accordance with the properties of rotation operators, there are $2T + 1 = 3$ eigenvalues of the $\mathbf{T}_z$ operator which can be calculated using Eq. (12.12)

---

[2] The particle exchange in the relative motion amounts to the axis reflection, hence the parity $(-)^L$ changes sign (see Sect. 6.5).

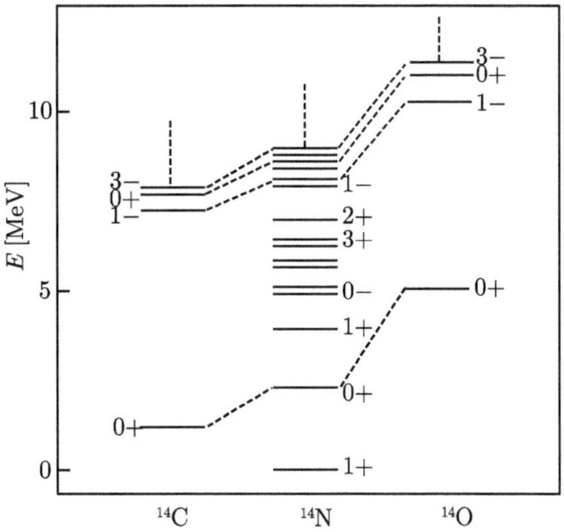

Fig. 12.3 Energy levels of the $A = 14$ isobars

$$T_z(C) = 1, \qquad T_z(N) = 0, \qquad T_z(O) = -1.$$

For $T = 1$, the three isobars form an isospin triplet: they have the same level structures. For $T = 0$, we have an isospin singlet related to the ground state of the ^{14}N. This means that the isospin of a nuclear ground state is the minimum T value related to the possible minimum T_z value:

$$T_{\text{groundstate}} = |T_z|$$

The energy differences among the analogous states of the three nuclei, which are displayed in Fig. 12.3, are due to the different Coulomb energies of the three systems.

12.4 Ground State of the Deuteron

The deuteron, made up of one proton and one neutron, is the only two-nucleon system that has a bound state. The neutron–neutron and proton–proton couples are not found to have any bound states. The minimum of the nuclear interaction, in the range of $r \sim 1$ fm (Fermi), as shown in Fig. 12.4, suggests the existence of bound states. The low-energy neutron–proton scattering data (see Chap. 13) tell us that the ground state is isotropic in the relative motion ($L = 0$), which has even parity ($P = (-1)^L = 1$) and is a triplet-spin state ($S = J = 1$)[3]. Imposing these experimental constraints, we

[3] We neglect the spin dependence of the nuclear interaction, except for the fact of privileging the triplet state over the singlet state.

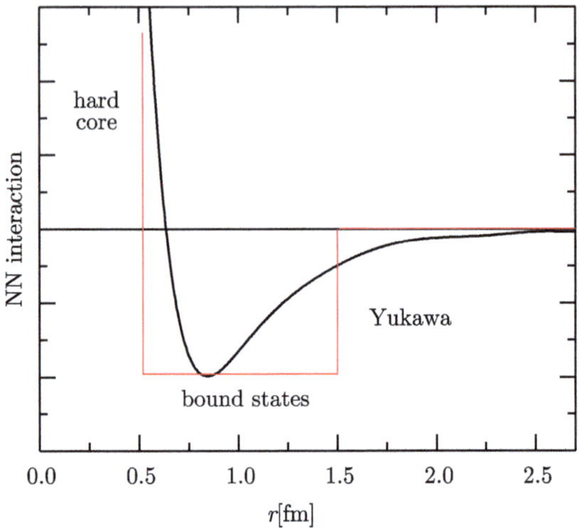

Fig. 12.4 Nucleon–nucleon potential

solve the deuteron Schrödinger equation with, the nuclear potential approximated by a square well (red curve in Fig. 12.4)

$$V(r) = \begin{cases} \infty & r < r_0 \\ V_0 & r_0 \le r \le r_0 + d \\ 0 & r > r_0 + d \end{cases} ,$$

where $V_0 = -35$ MeV and $d = 2.1$ fm are consistent with the experimental binding energy $E_B \approx -2.2$ MeV. Due to the short-range nature of the nuclear force ($r_0 \approx 1$ fm), the square well turns out not to be a bad approximation. Another approximation is to assume the solution to be the eigenstate with $L = 0$. It is in fact possible to show that, in the present model, bound states with $L > 0$ are not permitted because, as the deuteron is a weakly bound nucleus, a non-vanishing repulsive centrifugal potential would exceed the attractive nuclear potential (see Problem 1). Therefore, we neglect the centrifugal potential and hence, inside the well, the radial equation, Eq. (12.2), simplifies as follows:

$$\left(\frac{p_r^2}{2m} + V_0\right)\psi_0(r) = E\psi_0(r), \tag{12.13}$$

where the reduced mass is $m_p/2$, as the neutron mass m_n is approximately equal to the proton one m_p. After the integration, the solution $r\psi(r) = c\sin(kr)$ with $k = \sqrt{2m(E - V_0)/\hbar^2}$ fulfills the boundary condition $[r\psi(r)]_{r=0} = 0$ because for $r = 0$ it results $V(r = 0) = \infty$. Outside the well, for $r > d$, the radial equation is

$$\frac{p_r^2}{2m}\psi_0(r) = E\psi_0(r),$$

where $\kappa = \sqrt{-2mE/\hbar^2}$. The solution $r\psi(r) = const \cdot e^{-\kappa r}$ ensures the boundary condition $[r\psi(r)]_{r=\infty} = 0$. The continuity condition at $r = d$ requires that the logarithmic derivatives of the left and right solutions must be identical, i.e.

$$\frac{k\cos(kd)}{\sin(kd)} = -\kappa. \tag{12.14}$$

Equation (12.14) can only be solved numerically. The result of the binding energy is $E = -2.24$ MeV, which agrees well with the experimental value[4]. Despite the pretty good prediction of the binding energy, the wave function of the ground state cannot be an S-wave($L = 0$), because it entails the quadrupole moment Q to be equal to zero, at variance with the non-vanishing experimental value. The agreement can be obtained only by assuming that the ground state is a mixture of the S-wave with higher partial waves. The only possible state, which conserves the even parity, is the D-wave ($L = 2$). The coupling between different channels requires the addition of a non-central component to the nuclear potential.

The only possible force, which conserves the total spin, total angular momentum and parity is the tensor force

$$\mathbf{V}_{\text{tensor}} \equiv V_T(r)\,\mathbf{T}_{\text{np}} = V_T(r)\,3(\hat{r}\cdot\vec{s}_{\text{n}})(\hat{r}\cdot\vec{s}_{\text{p}}) - \vec{s}_{\text{n}}\cdot\vec{s}_{\text{p}},$$

where $\hat{r} = \vec{r}/r$. To get a physical insight into this force, one can resort to the picture of the nucleon's spin. From this point of view, neutron and proton look like two interacting electric current loops. A classical derivation of the tensor force is given in the Complement to this section. By adding the tensor force to the central force, the partial-wave expansion of the deuteron Schrödinger equation, Eq. (12.13), goes over into the coupled equations for the two components, ($S-$wave) and ($D-$wave), of the ground-state wave function, respectively

$$\left[\frac{p_r^2}{2m} + \frac{\hbar^2 L(L+1)}{2mr^2}\right]\psi_L(r) + \sum_{LL'}\int d^3r' V_{LL'}(r,r')\psi_{L'}(r') = E\psi_L(r). \tag{12.15}$$

This equation must be solved numerically, when employing a realistic potential[5], the Paris potential [PAR] as an example. Figure 12.5 displays the two components of the deuteron wave function, obtained with this potential. It is worth noting that the D component is much smaller than the S component, as expected from the small magnitude of the experimental quadrupole moment. Apart from the deuteron case, the role of the tensor force has been consolidated in the further developments of nuclear physics.

4 It is worth pointing out that the binding energy of the ^{4}He nucleus is -28.3 MeV.

5 With the name *realistic potential* one usually means a nuclear potential fitting the nucleon–nucleon scattering data and the deuteron binding energy.

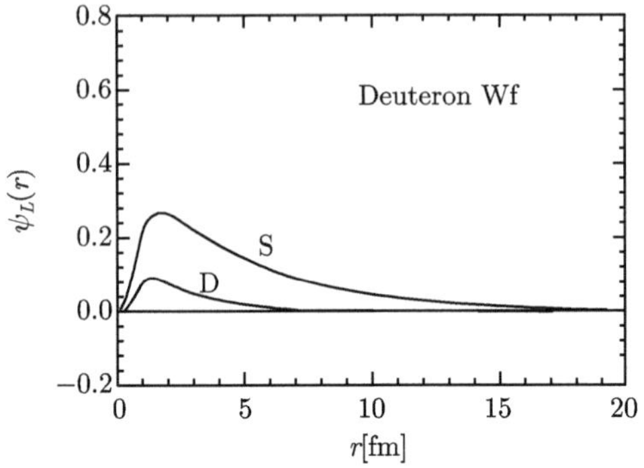

Fig. 12.5 S and D components of the deuteron wave function $\psi_L(r)$ using the Paris potential (arbitrary units)

12.5 Complement to Sect. 12.4: Tensor Interaction

According to the quark model of elementary particle physics, the quark configuration of the proton is (uud) and that of the neutron is (udd), where the quark *up* (u) has a charge equal to $\frac{2}{3}\,e$ and the quark *down* (d) a charge of $-\frac{1}{3}\,e$. Hence, the proton has a total charge equal to e and the neutron has a total charge equal to zero. Both of them have magnetic properties due to the currents generated by the charged quarks. Experimental evidence of these magnetic properties is the spin. The spin $\vec{s}_\tau$ of the nucleon τ (= n, p) classically looks like the angular momentum of an electric current loop, which creates a magnetic field

$$\vec{B}_\tau(\vec{r}\,) = \frac{\mu_0}{4\pi\, r^3}[3\hat{r}(\vec{\mu}_\tau\cdot\hat{r}) \;-\; \vec{\mu}_\tau], \tag{12.16}$$

where $\vec{\mu}_\tau \sim \vec{s}_\tau$ is the magnetic moment of the nucleon τ and $\hat{r} \equiv \vec{r}/r$. In turn, the electric loop, embedded inside the magnetic field $\vec{B}_\tau(\vec{r})$, acquires a potential energy

$$V_{\tau'}(\vec{r}\,) = \vec{\mu}_{\tau'}\cdot\vec{B}_\tau(\vec{r}\,), \tag{12.17}$$

where $\vec{\mu}_{\tau'} \sim \vec{s}_{\tau'}$ is the magnetic moment of the nucleon τ'. Therefore, the proton–neutron interaction due to their magnetic fields, equivalent to the loop–loop interaction, is

$$V(\vec{r}\,) = V_{\rm p}(\vec{r}\,)\cdot\vec{B}_{\rm n}(\vec{r}\,) + V_{\rm n}(\vec{r}\,)\cdot\vec{B}_{\rm p}(\vec{r}\,).$$

Using Eqs.(12.16) and (12.17) and replacing the proton and neutron magnetic moments with their respective spins, we obtain

$$V_{\rm tensor} \equiv V_T(r)S_{\rm np} = V_T(r)[3(\hat{r}\cdot\vec{s}_{\rm n})(\hat{r}\cdot\vec{s}_{\rm p}) - \vec{s}_{\rm p}\cdot\vec{s}_{\rm n}].$$

All constants are incorporated into $V_T(r)$, including those connecting the magnetic moments to the spin. In quantum mechanics, $V_{\rm tensor}$ is an operator, which does not commute with the total orbital angular momentum $\mathbf{L}^2$ so that the mixing of different L-values is supposed to occur in the wave functions of non-spherical systems. On the contrary, it commutes with the total spin $\mathbf{S}^2$ and hence it does not mix spin-singlet and spin-triplet states. It also conserves the parity.

12.6 Notes and References

PAR M. Lacombe et al., Physical Review C **81** (1981) 2405 [Paris, 1981].
DAV A. S. Davydov, *Quantum Mechanics*, Pergamon Press (1965), pg. 28.
YAM Y. Yamaguchi, Physical Review **95** (1954) 1628.
MES A. Messiah, *Quantum Mechanics* Vol.1 & 2, North-Holland Publishing Company Amsterdam (1961), pg. 564.
KAM A. Kamal, *Nuclear Physics*, Springer (2014), Chap. 5.

12.7 Problems

- Problem 1. Consider a hydrogen-like ion with Z protons and 1 electron. Calculate the ground-state energy correction due to the finite-size effect of the nucleus.
- Problem 2. Given two particles of equal mass m, which interact through the non-local potential [YAM]

$$V(r, r') = -\lambda v(r) v(r'),$$

where $\lambda > 0$ is the strength and $v(r) = \frac{e^{-kr}}{r}$, $(k > 0)$, is the form factor. Calculate the ground-state energy. [Hint: project the Schr ø“dinger equation onto momentum space].
- Problem 3. Calculate the ground-state wave function ψ_0 of a deuteron subject to the central square well potential of Fig. 14.4. What about excited states?
- Problem 4. Prove that the tensor force is zero for a two-body system with total spin $S = 0$.
- Problem 5. Determine the kinetic energy of the radial motion (see Eq. (12.7)). (Hint: Express the operator ∇^2 in polar coordinate. Take out the radial part, neglecting the terms with the angular derivatives.)

Chapter 13
Two-Body Problem: Scattering States

The experimental setup of atomic or nuclear scattering is an appropriate realization of the ideal setup depicted in Fig. 13.1. The beam is prepared by a particle accelerator in a state Ψ^{in}. Then it is sent to a target, where the beam particles interact with the target particles. The outgoing particles reach a detecting array, where each detector D_k is devised to detect particles in a definite state Ψ_k^{out}. The experimental data are the set of counts at each detector, giving the particle's probability of being in Ψ_k^{out}, expressed in terms of cross section.

In the theoretical approach, the key quantity is the transition operator, known as T-matrix, describing the interaction mechanism inside the target. It is obtained from the solution of the Schrödinger equation with positive energy eigenvalues. We refer to them as the continuum spectrum of eigenstates. At variance with the bound states with negative energy, the continuum states describe particles with finite probability of being found far from the interaction region. In the classical scattering theory, particles move along trajectories of infinite length. Each trajectory is determined by the impact parameter along with momentum direction. But, in quantum mechanics, the impact parameter cannot be assigned with a definite value in collisions, even if the range of values can be retrieved through the partial-wave analysis. This analysis is the main tool for obtaining information on the interaction from the experimental differential cross sections of nucleon–nucleon collisions.

Supplementary Information The online version contains supplementary material available at https://doi.org/10.1007/978-981-97-1670-8_13.

U. Lombardo et al., *Quantum Mechanics*, UNITEXT for Physics,
https://doi.org/10.1007/978-981-97-1670-8_13

13.1 Lippmann–Schwinger Equation

The same as we learned in the case of two-body bound states, the elastic collision between two particles can be reduced to the scattering of a particle from a central potential, with mass equal to the reduced mass. The process can be described either following the time evolution of the collision or as a stationary state with suitable boundary conditions. Normally, the second choice is adopted. Thus, the time-independent Schrödinger equation must be solved with a plane wave describing particles entering the interaction region plus those leaving without interacting, and a spherical wave describing the particles scattered in all directions. In the frame of the relative motion, the Schrödinger equations for the particles with energy E inside and outside the interaction area, see Fig. 13.1, are

$$\left(E - \frac{\mathbf{p}^2}{2m}\right)|\psi\rangle = V(\mathbf{r})|\psi\rangle, \tag{13.1}$$

$$\left(E - \frac{\mathbf{p}^2}{2m}\right)|\vec{p}\,\rangle = 0, \tag{13.2}$$

respectively. Subtracting Eq. (13.2) from Eq. (13.1), we obtain

$$\left(E - \frac{\mathbf{p}^2}{2m}\right)(|\psi\rangle - |\vec{p}\,\rangle) = V(\mathbf{r})|\psi\rangle. \tag{13.3}$$

Equation (13.3) can be recast in the following form:

$$|\psi_p\rangle = |\vec{p}\,\rangle + \frac{1}{E^{\pm} - \dfrac{\mathbf{p}^2}{2m}} V(\mathbf{r})|\psi_p\rangle,$$

where an imaginary infinitesimal term $\pm \mathrm{i}\epsilon$ has been added to the energy E, i.e. $E^{\pm} = E \pm \mathrm{i}\epsilon$, to avoid the singularity $E = \dfrac{p^2}{2m}$. This quantity has to be set equal to zero at the end of calculations. The double sign corresponds to choosing the physical boundary condition of outgoing (plus sign) or incoming (minus sign) spherical waves.

After projecting the Schrödinger equation into the coordinate space

$$\langle \vec{r}\,|\psi_p\rangle = \langle \vec{r}\,|\vec{p}\,\rangle + \int \mathrm{d}^3\vec{r}\,' \left\langle \vec{r} \left| \frac{1}{E^{\pm} - \dfrac{\mathbf{p}^2}{2m}} \right| \vec{r}\,' \right\rangle \langle \vec{r}\,' |\mathbf{V}|\psi_p\rangle, \tag{13.4}$$

we are left with an integral equation, known as the Lippmann–Schwinger equation.

13.2 Asymptotic Form of the Continuum States

The experimental information on the scattering mechanism is carried by the particles leaving the interaction area and reaching the detectors placed at a considerable distance. As a consequence, the asymptotic solution of Eq. (13.4), i.e. far from the interaction center, is of primary importance. Firstly, we calculate the kernel $G_0(\vec{r},\vec{r}\,')$, known as free-particle Green's function, which is defined as

$$G_0(\vec{r}\,,\vec{r}\,') = \left\langle \vec{r} \left| \frac{1}{E^{\pm} - \frac{\mathbf{p}^2}{2m}} \right| \vec{r}\,' \right\rangle = \left(\frac{1}{2\pi\hbar}\right)^3 \int \mathrm{d}^3\vec{p}\,' \frac{\mathrm{e}^{\mathrm{i}\vec{p}\,'\cdot(\vec{r}-\vec{r}\,')/\hbar}}{E^{\pm} - \dfrac{p'^2}{2m}}.$$

After integration in momentum space (see Appendix D.3), we obtain

$$G_0(\vec{r},\vec{r}\,') = -\frac{m}{2\pi\hbar^2}\frac{\mathrm{e}^{\pm\mathrm{i}p_r|\vec{r}-\vec{r}\,'|/\hbar}}{|\vec{r}-\vec{r}\,'|},$$

$\vec{p}_r$ being a vector with the modulus of $\vec{p}$, but the direction of $\vec{r}$. In order to calculate the asymptotic limit, i.e. $r \to \infty$, we use the approximations

$$\lim_{r\to\infty} \frac{1}{|\vec{r}-\vec{r}\,'|} \approx \frac{1}{r},$$

$$\lim_{r\to\infty} |\vec{r}-\vec{r}\,'| \approx r\left(1 - \frac{\vec{r}\cdot\vec{r}\,'}{r^2}\right).$$

So doing, Green's function takes the asymptotic form

$$G_0(\vec{r},\vec{r}\,') \approx -\frac{m}{2\pi\hbar^2}\frac{\mathrm{e}^{\pm\mathrm{i}p_r r/\hbar}}{r}\mathrm{e}^{\mp\mathrm{i}\vec{p}_r\cdot\vec{r}\,'/\hbar},$$

then the asymptotic solution of the Lippmann–Schwinger equation takes the form

$$\psi_p(\vec{r}\,) \approx \mathrm{e}^{\mathrm{i}\frac{\vec{p}\cdot\vec{r}}{\hbar}} + \int \mathrm{d}^3\vec{r}\,' G_0(\vec{r},\vec{r}\,')V(\vec{r}\,')\psi_p(\vec{r}\,') \tag{13.5}$$

$$= \mathrm{e}^{\frac{\mathrm{i}\vec{p}\cdot\vec{r}}{\hbar}} + f_{\vec{p}\vec{p}\,'}\frac{\mathrm{e}^{\pm\frac{\mathrm{i}pr}{\hbar}}}{r}, \tag{13.6}$$

where we have introduced the quantity

$$f_{\vec{p}\vec{p}\,'} = -\frac{m}{2\pi\hbar^2}\int \mathrm{d}^3\vec{r}\,'\mathrm{e}^{-\frac{\mathrm{i}\vec{p}\,'\cdot\vec{r}\,'}{\hbar}} V(\vec{r}\,')\psi_p(\vec{r}\,'), \tag{13.7}$$

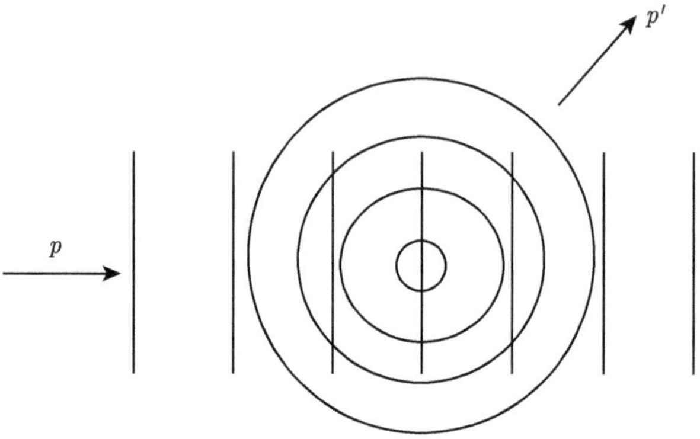

Fig. 13.1 Asymptotic form of the wave function in a scattering process

with $\vec{p}' = \vec{p}_r$. This quantity, known as *scattering amplitude*, plays a crucial role for the elastic collisions in quantum mechanics, similar to what the deflection function does in classical mechanics.

The physical interpretation of Eq. (13.6) is provided by Fig. 13.1. The incoming plane waves describe the incident particles with momentum $\vec{p}$. The outgoing plane waves correspond to those particles, which do not interact with the scattering center and proceed in the forward direction. The wave function of both plane waves is the first term on the right-hand side of Eq. (13.6). The outgoing spherical waves correspond to those particles that have interacted with the potential. Their wave function is the second term on the right-hand side of Eq. (13.6) with the plus sign on the exponential.[1] The particles leaving the interaction region are emitted in every direction with different probabilities given by the square modulus of $f_{pp}{}'$, as we will see in the next section.

The momentum representation is useful for several purposes. It can be obtained by introducing the Fourier transforms of the scattering wave function of both wave function and potential into the Lippmann–Schwinger equation. By doing so, it takes the form

$$\psi_p(\vec{k}) = \delta(\vec{p} - \vec{k}) + \frac{1}{E - k^2/2m + \mathrm{i}\epsilon} \int \mathrm{d}^3 q V(\vec{q}\,)\psi_p(\vec{k} - \vec{q}\,).$$

Combining this equation with Eq. (13.7), we obtain a new version of the Lippmann–Schwinger equation

$$\mathbf{T} = \mathbf{V} + \mathbf{V}\mathbf{G}_0\mathbf{T},$$

in terms of the so-called T-matrix defined as

[1] The minus sign corresponds to incoming spherical waves that will not be considered for the present physical situation.

$$\langle \vec{p}\,'|\mathbf{T}|\vec{p}\,\rangle = -\frac{2\pi\hbar^2}{m} f_{\vec{p}\,\vec{p}\,'}. \tag{13.8}$$

13.3 Solving the Lippmann–Schwinger Equation

In general, the Lippmann–Schwinger equation cannot be solved exactly, even if one adopts simple potentials, so that we must resort to approximate solutions. However, an exact solution would allow us to study some general properties of the scattering amplitude. The Yamaguchi potential [YAM] gives us a chance of reaching this goal. This is a rank-one separable potential of the form

$$V(\vec{r}, \vec{r}\,') = -\lambda\, u^*(\vec{r}\,)u(\vec{r}\,'), \tag{13.9}$$

where λ is the strength of the interaction and the functions $u(r)$ are the form factors.[2] With this potential the Lippmann–Schwinger equation for the T-matrix, Eq. (13.8), becomes

$$\langle \vec{k}\,'|\mathbf{T}(E)|\vec{k}\rangle = -\lambda u^*(\vec{k}\,')u(\vec{k}) - \lambda u^*(\vec{k}\,') \int \mathrm{d}\vec{q}\; u(\vec{q})\, G_0(q, E)\langle \vec{q}\,|\mathbf{T}(E)|\vec{k}\rangle,$$
$$G_0(q, E) = \frac{2m}{2mE - q^2 + \mathrm{i}\epsilon},$$

choosing the option of outgoing spherical wave. After solving this equation algebraically, the resulting T-matrix is found to have the same analytical expression as the potential, Eq. (13.9), with λ replaced by $\Lambda(E)$

$$\langle \vec{k}\,'|\mathbf{T}(E)|\vec{k}\rangle = -\Lambda(E)u^*(\vec{k}\,')u(\vec{k}),$$
$$\Lambda(E) = \frac{\lambda}{1 + \lambda \int \mathrm{d}\vec{q}\; G_0(q, E)u(\vec{q}\,)^2}.$$

The structure of this equation exhibits the following general features:

(1) If the strength λ is small ($\lambda^2 \ll \lambda$) then $\Lambda \sim \lambda$;
(2) If the strength is not small, as is the case of the nuclear force, Λ is sizably different from λ and the truncation of the λ expansion to some order might no longer be a good approximation;
(3) The two opposite ranges of the incident energy are worth of being examined. Firstly, at very high energy the first-order expansion in the strength λ might be a reasonable approximation; this is the Born approximation that will be discussed in the next section. Conversely, at very low energy, the cross section could show

[2] This potential belongs to the class of the non-local potentials, such as the tensor force introduced to explain the non-vanishing quadrupole moment of the deuteron (see Sect. 12.3).

a resonant structure, if the system of two interacting particles has a bound state with energy E_B close to zero but slightly negative,—as is the case of the deuteron.

The same potential can be used to calculate the bound states ($E < 0$). Considering for simplicity the form factors to be only dependent on the $\vec{r}$ modulus, then the solution will be restricted to $l = 0$ s-wave only. The Schrödinger equation, projected onto the momentum space, can be algebraically solved the same way as the Lippmann–Schwinger equation. It is easily found that the binding energy $E_B < 0$ is the solution of the secular equation

$$1 - \lambda \int \mathrm{d}\vec{q}\, G_0(q, -E_B) u(\vec{q}\,)^2 = 0.$$

13.4 Elastic Scattering Cross Section

Collision experiments concern a beam of particles with the same kinetic energy which is sent onto a target at rest. These experiments aim to measure a quantity called cross section, which is defined as

$$\sigma_{\mathrm{exp}}(\Omega)\mathrm{d}\Omega = \frac{\text{number of particles scattered into solid angle d}\Omega\text{ unit time}}{\text{number of particles incident on unit surface/unit time}}$$

From the experimental cross section, one can extract information on the structure of matter and on the interaction mechanisms. From a theoretical point of view, we define the cross section as the ratio of the outgoing flow in the $\vec{p}\,'$ direction per unit solid angle to the incident flow per unit surface in the $\vec{p}$ direction

$$\sigma(\Omega) = \frac{\dfrac{\mathrm{d}\Phi_{\mathrm{out}}}{\mathrm{d}\Omega}}{\dfrac{\mathrm{d}\Phi_{\mathrm{in}}}{\mathrm{d}S}} = \frac{\dfrac{\mathrm{d}\Phi_{\mathrm{out}}}{\mathrm{d}S}}{\dfrac{\mathrm{d}\Phi_{\mathrm{in}}}{\mathrm{d}S}} r^2,$$

where $\mathrm{d}S = r^2 \mathrm{d}\Omega$ is the surface element within the solid angle $\mathrm{d}\Omega$. The incoming flow in the $\hat{p}$ direction is equal to the current flowing in the $\hat{p}$ direction itself

$$\frac{\mathrm{d}\Phi_{\mathrm{in}}}{\mathrm{d}S} = \hat{p} \cdot \vec{J}_{\mathrm{in}} = \hat{p} \cdot \left[\frac{\hbar}{2im} (\psi^* \nabla \psi - \psi \nabla \psi^*) \right] = \frac{p}{m} = v. \tag{13.10}$$

Equation (13.10) includes only the incident plane wave term of the wave function, that is $\psi \sim \mathrm{e}^{\mathrm{i}\frac{\vec{p}\cdot\vec{r}}{\hbar}}$. The outgoing current in the $\vec{p}\,'$ direction is

$$\frac{d\Phi_{out}}{dS} = \hat{p}\,' \cdot \vec{J}_{out} = \hat{p}\,' \cdot \left[\frac{\hbar}{2im}(\psi^* \nabla \psi - \psi \nabla \psi^*)\right]$$
$$= \frac{\hbar}{2im}\left(\psi^* \frac{\partial}{\partial r}\psi - \psi \frac{\partial}{\partial r}\psi^*\right) = \frac{v}{r^2}|f_{\vec{p}\vec{p}\,'}|^2. \tag{13.11}$$

In this case, we have only included the outgoing spherical wave term of the wave function, that is $\psi \sim f_{\vec{p}\vec{p}\,'}\frac{e^{\frac{ipr}{\hbar}}}{r}$. For the calculation of Eq. (13.11), we have used the property $\hat{p}\,' \cdot \nabla = \frac{\partial}{\partial r}$ and neglected the terms of the order $\frac{1}{r^4}$. The ratio between the two currents is

$$\sigma(\Omega) = |f_{\vec{p}\,\vec{p}\,'}|^2. \tag{13.12}$$

The experimental cross section is then connected to the theoretical scattering amplitude. Such a result highlights the importance of studying the asymptotic behaviour of the wave function. The unit of measurement for the cross section is

$$1\,\text{barn} = 10^{-28}\text{m}^2.$$

13.4.1 Born Approximation for the Elastic Scattering Cross Section

The scattering amplitude can be written as a power expansion of the **V** potential by repeatedly applying the Lippmann–Schwinger equation Eq. (13.5) upon the wave function that appears on the right-hand side of Eq. (13.7)

$$\begin{aligned} f_{\vec{p}\vec{p}\,'} = & -\frac{m}{2\pi\hbar^2}\int d^3\vec{r}\,' e^{-i\frac{\vec{p}\,'\cdot\vec{r}\,'}{\hbar}} V(\vec{r}\,') \\ & \cdot \left[e^{i\frac{\vec{p}\cdot\vec{r}\,'}{\hbar}} + \int d^3\vec{r}\,'' G_0(\vec{r}\,', \vec{r}\,'') V(\vec{r}\,'') \psi_p(\vec{r}\,'')\right] \\ = & -\frac{m}{2\pi\hbar^2}\left[\int d^3\vec{r}\,' e^{-i\frac{\vec{p}\,'\cdot\vec{r}\,'}{\hbar}} V(\vec{r}\,') e^{i\frac{\vec{p}\cdot\vec{r}\,'}{\hbar}}\right. \\ & \left. + \int d^3\vec{r}\,' \int d^3\vec{r}\,'' e^{-i\frac{\vec{p}\,'\cdot\vec{r}\,'}{\hbar}} V(\vec{r}\,') G_0(\vec{r}\,', \vec{r}\,'') V(\vec{r}\,'') e^{i\frac{\vec{p}\cdot\vec{r}\,''}{\hbar}} + \cdots\right]. \end{aligned}$$

For a weak enough interaction, we expect the higher-than-first-order terms to be negligible. The first-order truncation of the expansion is called the Born approximation: the exact wave function of Eq. (13.7) is approximated to its asymptotic form, that is the plane wave. The Born approximation for the scattering amplitude (13.7) is then

$$f_{\vec{p},\vec{p}\,'} \approx -\frac{m}{2\pi\hbar^2}\int d\vec{r}\,' e^{-i\frac{(\vec{p}\,'-\vec{p})\cdot\vec{r}\,'}{\hbar}} V(\vec{r}\,') = -\frac{m}{2\pi\hbar^2}\tilde{V}(\vec{p}\,' - \vec{p}).$$

The function $\tilde{V}(\vec{q}\,)$ represents the Fourier transform of the potential with respect to the momentum transfer $\vec{p}\,' - \vec{p}$. It follows that the cross section within the Born approximation is

$$\sigma(\Omega) = \frac{m^2}{4\pi^2\hbar^4}|\tilde{V}(\vec{p}\,' - \vec{p}\,)|^2. \tag{13.13}$$

The weak potential strength does not guarantee the validity of the Born approximation. The requirement of a second condition is discussed in Sect. 13.3 for the special case of a solvable model. In general, the condition that the scattering wave function is close to the free wave function must be met.

13.4.2 Nuclear and Coulomb Potential

According to quantum electrodynamics, the interaction between charged particles is caused by the exchange of massless particles, i.e. the Einstein photons. In analogy to this theory, Yukawa suggested that the nuclear interaction is due to the exchange of particles with mass $M_\pi = 140$ MeV, known as pions [YUK]. The Yukawa potential

$$V(r) = V_0\frac{\mathrm{e}^{-\mu r}}{r} \tag{13.14}$$

is the static limit of the pion-exchange model. The parameter $\mu = M_\pi c/\hbar \sim 0.7\,\mathrm{fm}^{-1}$ determines the damping rate, in other words, the range of interaction, 1.4×10^{-15} m. The nuclear cross section in Born approximation, Eq. (13.13), is obtained from the Fourier transform of Eq. (13.14). It results to be

$$\tilde{V}(\vec{p}\,' - \vec{p}) = \frac{4\pi V_0}{(\vec{p}\,' - \vec{p}\,)^2/\hbar^2 + \mu^2} = \frac{4\pi V_0}{\left[2(p/\hbar)\sin\left(\frac{\theta}{2}\right)\right]^2 + \mu^2}.$$

At the limit for $\mu \to 0$, the Yukawa potential tends to $\frac{V_0}{r}$, it follows that the Yukawa potential coincides formally with the Coulomb potential as a limiting case, except for the strength. Such a feature is consistent with the picture that the Coulomb interaction is mediated by the exchange of massless particles, and hence it has an infinite range. Therefore, taking the limit $\mu = 0$, the Coulomb differential cross section in the Born approximation turns out to be

$$\sigma_R(\theta) = \left(\frac{V_0}{2E}\right)^2 \frac{1}{4\sin^4\left(\frac{\theta}{2}\right)}, \tag{13.15}$$

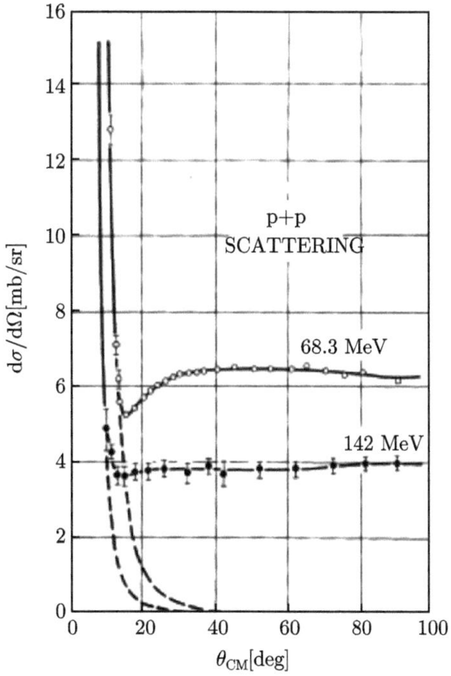

Fig. 13.2 Differential cross sections for protons with kinetic energies in the laboratory frame of 68 and 142 MeV scattered by protons. The dashed curves refer to the Rutherford cross section. The bars regard the experimental uncertainties. [Data from J. M.Blair et al., Phys. Rev. 74 (1948) 553]

where $E = \frac{p^2}{2m}$. This formula is known as the Rutherford cross section [RUT]. We note that σ_R has the same form as the classical form, but this is a purely random coincidence. At variance with the nuclear interaction, the Born approximation works well with the Coulomb scattering, because the strength of Coulomb force, known as fine structure constant, is $\alpha = \frac{e^2}{4\pi\hbar c} \approx \frac{1}{137}$, much smaller than 1.

The Born approximation for the nuclear as well as the Coulomb interaction can be probed with the experimental proton–proton scattering at high energy, whose differential cross section is plotted in Fig. 13.2. In a wide angular range, the latter turns out to be rather flat, driven by the attractive nuclear force, whereas a strong enhancement occurs at small angles due to the rapid increase of the Coulomb repulsion. A minimum marks the competition between the two forces. Furthermore, the expected energy reduction of the cross section is confirmed by two curves at quite different incident energies plotted in the figure.

13.4.3 Electron Scattering and Nuclear Density

An interesting application of the Coulomb scattering within Born approximation is the scattering of electrons from nuclei. In this experiment, the electrons probe the proton distribution inside the nucleus, and hence the nuclear charge density.

In electron scattering, the potential of the electron–nucleus system is the sum of the Coulomb interaction of each proton of the nucleus with the incident electron, i.e.

$$V(r) = -\sum_{k=1}^{Z} \frac{e^2}{|\vec{r} - \vec{r}_k|} = \int \mathrm{d}^3\vec{r}\,'\rho(\vec{r}\,')\frac{Ze^2}{|\vec{r} - \vec{r}\,'|}. \tag{13.16}$$

In Born approximation the scattering amplitude can be evaluated by means of the convolution theorem, that is, the Fourier transform of the potential (13.16) is equal to the product of the Fourier transforms of the Coulomb potential and charge density (see Appendix D)

$$\begin{aligned} f_{\vec{p},\vec{p}'} &\approx -\frac{m}{2\pi\hbar^2}\int \mathrm{d}^3\vec{r}\mathrm{e}^{-\mathrm{i}\frac{\vec{q}\cdot\vec{r}}{\hbar}}\int \mathrm{d}^3\vec{r}\,'\rho(\vec{r}\,')\frac{Ze^2}{|\vec{r} - \vec{r}\,'|} \\ &= f_0(\theta)\int \mathrm{d}^3\vec{r}\rho(\vec{r})\mathrm{e}^{-\mathrm{i}\frac{\vec{q}\cdot\vec{r}}{\hbar}} \equiv f_0(\theta)F(q), \end{aligned}$$

where $f_0(\theta)$ is the Coulomb scattering amplitude for point-like particles

$$f_0(\theta) = -\frac{2m}{\hbar^2}\frac{Ze^2}{q^2},$$

and $q = |\vec{p} - \vec{p}\,'| = 2p\sin(\theta/2)$ is the momentum transfer. The resulting cross section, including the form factor is

$$\sigma(\theta) = |f_0(\theta)|^2|F(q)|^2 = \sigma_R(\theta)|F(q)|^2,$$

The Fourier transform of the form factor, i.e. $\tilde{F}(r)$, is the charge density $\rho(r)$ of the nucleus.

Figure 13.3 shows some electron-nucleus experimental cross sections together with the theoretical fit. The form factor can be extracted from these distributions. Afterwards, the Fourier transform of the form factor provides the behaviour of the nuclear charge distributions as a function of the distance from the center of the nucleus. Looking at Fig. 13.4, we observe the gradual quenching to zero of the charge distribution which indicates the existence of the nuclear surface. We also observe that for the heavier systems, from ^{40}Ca to ^{209}Bi, the charge distribution is almost constant. This behaviour is an indication of the saturation of the nuclear force.

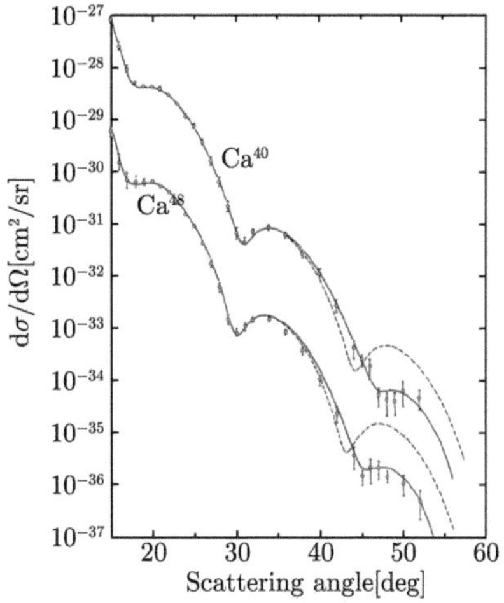

Fig. 13.3 Differential cross sections of electrons scattered by ^{16}O and ^{12}C nuclei. The full and open symbols refer to experimental data, and the lines refer to theoretical calculations. Courtesy of American Physical Soc

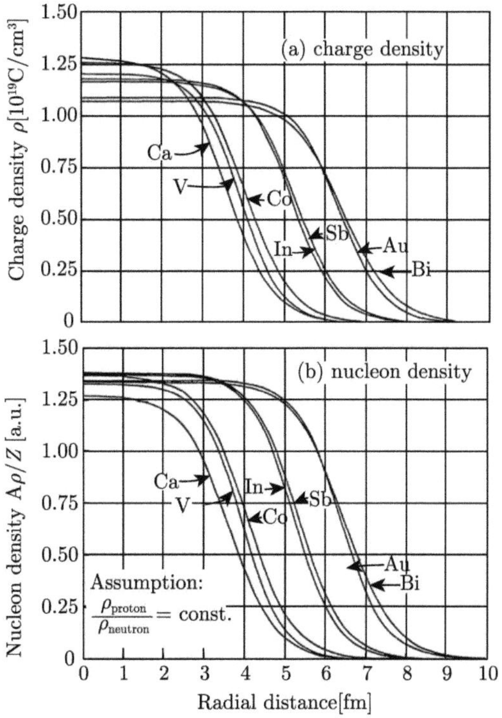

Fig. 13.4 Experimental charge density distributions as a function of the distance from the center r(fm)

13.5 Partial-Wave Analysis

In the classical scattering theory, the cross section depends on the impact parameter (or equivalently on the angular momentum), in addition to momentum. In quantum mechanics, these two variables are incompatible, and hence all possible angular momenta contribute, each one with its own weight. The partial-wave analysis gives information on the angular momentum dependence of the differential cross section and, occasionally, on the relation with the classical impact parameter.

In order to derive the partial-wave expansion of the asymptotic wave function, we consider the asymptotic plane-wave and scattering amplitude expressed as a series of Legendre polynomials (see Appendix C)

$$\mathrm{e}^{\mathrm{i}\vec{k}\cdot\vec{r}} \sim \sum_{l=0}^{\infty} \mathrm{i}^l(2l+1)\frac{\mathrm{e}^{\mathrm{i}\left(kr-l\frac{\pi}{2}\right)}-\mathrm{e}^{-\mathrm{i}\left(kr-l\frac{\pi}{2}\right)}}{2\mathrm{i}kr}P_l(\cos\theta), \tag{13.17}$$

$$f(\theta)=\sum_{l=0}^{\infty}(2l+1)f_l P_l(\cos\theta), \tag{13.18}$$

where the polar axis is chosen in the direction of the incident wave so that $\hat{k}\cdot\hat{r}=\cos\theta$. Thus, the wave function $\psi_E(r,\theta,\phi)$ will be a function only of r and θ. After introducing these two expressions into Eq. (13.6), we get

$$\psi_{\vec{k},\vec{k}'}(r,\theta) \sim \sum_{l}(2l+1)\left[\mathrm{i}^l\frac{\mathrm{e}^{\mathrm{i}\left(kr-l\frac{\pi}{2}\right)}-\mathrm{e}^{-\mathrm{i}\left(kr-l\frac{\pi}{2}\right)}}{2\mathrm{i}kr}+f_l\frac{\mathrm{e}^{\mathrm{i}kr}}{r}\right]P_l(\cos\theta). \tag{13.19}$$

Separating the incoming from the outgoing spherical waves, we get

$$\sum_{l}(2l+1)\left[-\frac{1}{2\mathrm{i}k}\frac{\mathrm{e}^{-\mathrm{i}(kr-l\pi)}}{r}+\left(\frac{1}{2\mathrm{i}k}+f_l\right)\frac{\mathrm{e}^{\mathrm{i}kr}}{r}\right]P_l(\cos\theta), \tag{13.20}$$

since $i^l=e^{il\pi/2}$. Applying the continuity equation on any spherical surface centered around the scattering center, the total flow outgoing from that surface must be zero, which means that the two coefficients in front of the incoming and outgoing partial waves, respectively, must be equal, i.e. $|S_l|=|1+2\mathrm{i}kf_l|=1$. In other words, the interaction can only induce a phase-shift between the incoming and outgoing waves in elastic scattering [Sak]. Introducing the *scattering phase-shifts* δ_l with the definition $S_l(k)=\mathrm{e}^{2\mathrm{i}\delta_l(k)}$, the partial-wave expansions of the asymptotic wave function and scattering amplitude take the final form

$$\psi_{\vec{k},\vec{k}'} \sim \frac{1}{2\mathrm{i}k}\sum_{l=0}^{\infty}(2l+1)\left[\mathrm{e}^{2\mathrm{i}\delta_l}\frac{\mathrm{e}^{\mathrm{i}kr}}{r} - \frac{\mathrm{e}^{-\mathrm{i}(kr-l\pi)}}{r}\right]$$
$$= \sum_{l=0}^{\infty}\mathrm{i}^l(2l+1)\mathrm{e}^{\mathrm{i}\delta_l}\sin\left(kr - l\frac{\pi}{2} + \delta_l\right)P_l(\cos\theta), \quad (13.21)$$
$$f(\theta) = \sum_{l=0}^{\infty}(2l+1)\frac{(\mathrm{e}^{2\mathrm{i}\delta_l}-1)}{2\mathrm{i}k}P_l(\cos\theta)$$
$$= \frac{1}{k}\sum_{l=0}^{\infty}(2l+1)\mathrm{e}^{\mathrm{i}\delta_l}\sin\delta_l P_l(\cos\theta).$$

The differential cross section is then

$$\sigma(\theta) = |f(\theta)|^2$$
$$= \frac{1}{k^2}\sum_{l,l'}(2l+1)(2l'+1)\mathrm{e}^{\mathrm{i}(\delta_l-\delta_{l'})}\sin\delta_l\sin\delta_{l'}P_l(\cos\theta)P_{l'}(\cos\theta),$$

and the total cross section is

$$\sigma_{\mathrm{tot}} = \int_{\Omega}\sigma(\theta)\mathrm{d}\Omega = 2\pi\int_{-1}^{1}|f(\theta)|^2\mathrm{d}(\cos\theta) = \frac{4\pi}{k^2}\sum_l(2l+1)\sin^2\delta_l = \sum_l\sigma_l. \quad (13.22)$$

For a potential with finite range R: $\sigma_{\mathrm{tot}} = \sum_l^{l_{\max}}\sigma_l$, where $l_{\max} = kR$. The total cross section can be calculated from the "forward" scattering amplitude

$$f(\theta=0) = \frac{1}{k}\sum_l(2l+1)\mathrm{e}^{\mathrm{i}\delta_l}\sin\delta_l P_l(1).$$

The imaginary part of $f(\theta=0)$ is connected to the total cross section (see Eq. (13.22))

$$\sigma_{\mathrm{tot}} = \frac{4\pi}{k}\mathrm{Im}\,f(\theta=0).$$

Such a result is known as the optical theorem.

The partial-wave analysis allows us to compare the theoretical and experimental cross sections through the phase-shifts extracted from the measurements of differential cross sections.

13.6 Low-Energy Scattering and Bound States

We discuss first the case of short-range interaction by addressing the low energy neutron–proton scattering. Let us consider in the laboratory frame a beam of neutrons with kinetic energy $E_{\text{lab}} = \dfrac{p^2}{2m_{\text{n}}}$ impinging upon a proton-rich target. Given b as the neutron–proton impact parameter, the classical value of the neutron angular momentum is $L = pb$. From quantum mechanics, we know that the $\mathbf{L}^2$ eigenvalues are $L^2 = l(l+1)\hbar^2$; therefore, from the relation $pb = \sqrt{l(l+1)}\hbar$, we get an estimation of the impact parameter

$$b \sim \frac{\sqrt{l(l+1)}}{k},$$

where $k = \dfrac{p}{\hbar}$ is the wave number. It follows that only those neutrons with $b \leqslant r_{\text{np}}$ will be scattered, r_{np} being the range of the neutron–proton interaction. In accordance with this semi-classical reasoning, the only partial waves involved in the scattering process are those for which $l(l+1) < k^2r^2$, with $r \leqslant r_{\text{np}}$. For incident energies less than or equal to 10 MeV only $l = 0$, partial wave is involved, and hence we can restrict the resolution of the Schrödinger equation to the s-wave channel. In the center-of-mass reference frame, the Schrödinger equation with the neutron–proton potential $V_{\text{np}}(r)$ in the relative motion is written as

$$\frac{\mathrm{d}^2u_0}{\mathrm{d}r^2} + [k^2 - V_{\text{np}}(r)]u_0 = 0, \tag{13.23}$$

where $k = \sqrt{\dfrac{\mu E}{\hbar^2}}$ is the wave number, $\mu = \dfrac{m_{\text{n}}m_{\text{p}}}{m_{\text{n}} + m_{\text{p}}} \sim \dfrac{m_{\text{n}}}{2}$ is the reduced mass; and E the energy. Notice that the centrifugal potential vanishes in s-wave ($l = 0$). For $r > r_{\text{np}}$, the interaction potential is equal to zero and Eq. (13.23) turns into

$$\frac{\mathrm{d}^2u_0'}{\mathrm{d}r^2} + k^2u_0' = 0,$$

which is fulfilled by the wave function (see also Eq. (13.21))

$$u_0' = \frac{A_0}{k}\sin(kr + \delta_0), \tag{13.24}$$

In the extreme limit of zero energy, the wave function must remain finite beyond the interaction range and the phase-shift $\delta_0 \sim k + O(k^2)$. Then, taking the limit for $k \to 0$, we get

$$\lim_{k\to 0} u_0' = \lim_{k\to 0}\left(r\cos\delta_0\frac{\sin kr}{kr} + \frac{\cos kr}{k}\sin\delta_0\right) \sim (r-a), \tag{13.25}$$

where $\cos\delta_0 \sim 1$ and a is the finite limit

$$a = -\frac{\sin\delta_0}{k}, \tag{13.26}$$

which is known as *scattering length*. Moreover, the slope of the wave function is related to the wave function itself by

$$\lim_{k\to 0}\frac{du_0'/dr}{u_0'} \sim \frac{1}{r-a}. \tag{13.27}$$

As we will see later, it mainly characterizes the elastic scattering at low energy and provides significant information on the nature of interaction. First of all, in the low energy limit Eq. (13.22) implies that

$$\begin{aligned}\sigma_{k\to 0} &= \frac{4\pi}{k^2}\sin^2\delta_0 = 4\pi\frac{a^2}{1+k^2a^2},\\ \sigma_0 &= \sigma_{k=0} = 4\pi a^2.\end{aligned} \tag{13.28}$$

It should be noted that the cross section of classical scattering from the hard sphere with radius a is πa^2 (see Problem 1). The enhancement of a factor 4 over the classical value in Eq. (13.28) must be attributed to the quantum diffractive effect, that cannot occur with classical trajectories. Thus, the measurements of the scattering cross sections provide the values of the phase-shift and of the scattering length.

In addition to the scattering length, the low-energy scattering gives information on the interaction range. To prove that, we consider the elastic scattering from a square potential with radius $r = r_{\rm np}$, described by Eq. (13.23). The logarithm derivative of its solution, Eq. (13.24) evaluated at $r = r_{\rm np}$ gives us

$$\left[\frac{1}{u_0'(r)}\frac{du_0'}{dr}\right]_{r=r_{\rm np}} = k\cot(kr_{\rm np}+\delta_0). \tag{13.29}$$

The right-hand side of Eq. (13.29) can be expressed as a power expansion in k^2

$$k\cot(kr_{\rm np}+\delta_0) = \alpha + \beta k^2 + \gamma k^4 + \cdots .$$

For low-energy scattering, i.e. neutrons with kinetic energies up to 10 MeV, we can neglect the terms of order higher than k^2. From the definition of the scattering length, Eq. (13.26), we write the relation between phase-shift and scattering length.

$$k\cot(kr_{\rm np}+\delta_0) = -\frac{1}{a} + \frac{r_{\rm np}}{2}k^2. \tag{13.30}$$

Accordingly, we modify the s-wave scattering cross section (13.28) as

$$\sigma_0 = \frac{4\pi}{k^2}\frac{1}{1+\cot^2\delta_0} = \frac{4\pi}{k^2} + \left(\frac{1}{a} - \frac{r_{np}}{2}k^2\right)^2.$$

The real value of the nuclear potential range is not well defined. Nevertheless, the procedure just described is still valid with the parameter r_{np}, known as *effective range*, to be determined by a fit with the experimental cross section [PRE].

The knowledge of a and r_{np} is not enough to obtain information on the shape of the neutron–proton potential. Indeed, the constants of a potential function of the form $V_{np}(r) = V_0 f(r)$ can be adjusted to obtain the experimental values of a and r_{np} no matter what the shape of the $f(r)$ function is. As a result, low-energy scattering studies do not provide information on the shape of the neutron–proton potential. We might also say that the low-energy properties of the neutron–proton system are independent of the potential shape. The connection of scattering length and effective range with the arising of a bound state comes about as follows. The asymptotic form of the wave function for a bound state is $u \sim e^{-\gamma r}$, which goes to zero for $r \to \infty$. The related energy eigenvalue, the binding energy, has to be negative

$$E_B = \frac{\hbar^2 k^2}{2\mu} = -\frac{\hbar^2\gamma^2}{2\mu},$$

where $k = i\gamma$. In such a case the relation Eq. (13.30) between phase-shift and scattering length takes the form

$$k\cot\delta_0 = -\gamma,$$

while the power expansion (13.30) corresponds to

$$\gamma = \frac{1}{a} + \frac{r_0}{2}\gamma^2.$$

The positive value of a is consistent with the positive value of γ, for the bound state to be an exponentially decreasing wave function. The binding energy is then obtained once the scattering length and the effective range are determined. The scattering length a cannot give any information on the shape of the interaction; however, its sign tells us whether the interaction is repulsive or attractive. In the absence of interaction, the wave function is a plane wave and a is zero. Turning the interaction on, a takes a positive value for repulsive potentials and negative values for attractive potentials. This is illustrated in Fig. 13.5, where the asymptotic s-wave function u_0' from a square potential with radius R is drawn in the $k \to 0$ limit. According to Eq. (13.25), the scattering length a is the intercept of the wave function with the r axis. In the case of repulsive potential (case 1), $0 < a < R$ and the slope of u_0' is large (see Eq. (13.27)), which makes the probability that the particle will be found inside the potential unlikely. In the case of attractive potential (case 2), a takes a negative value, and the slope of u_0' is smaller, so that the probability

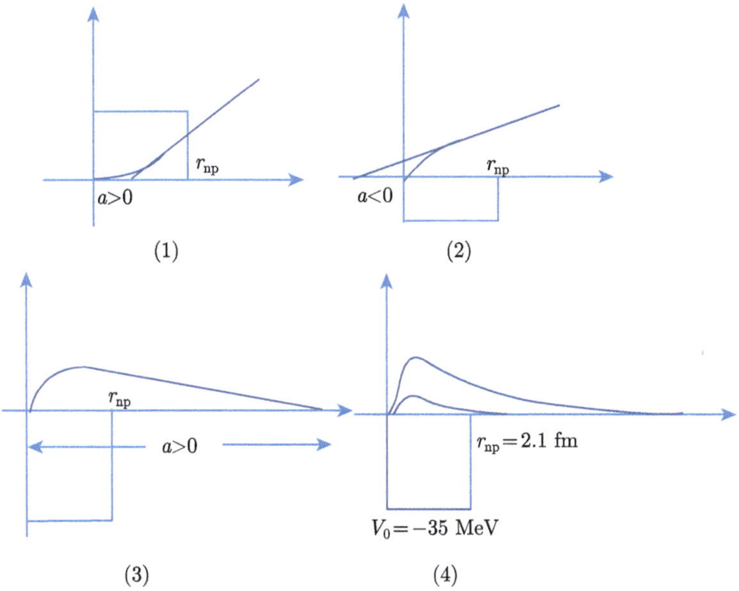

Fig. 13.5 $u_0'(r)$ as a function of r for a potential barrier (1), a square well (2) and a deeper square well (3), deuteron ground state (4). The intercept a represents the corresponding scattering length for the three cases

that the particle will be found inside the potential is greater. Enhancing the strength of the attractive potential, the scattering length goes over again into positive and significantly large magnitude with negative slope of u_0'. This behaviour marks the transition of the wave function from the oscillatory shape of the scattering wave function to the exponentially decreasing shape of a bound state (case 4).

Let us consider now the low energy elastic scattering with a long-range interaction, addressing the Coulomb collision between protons as an example. In this case, the Schrödinger equation can be solved without any approximation, even though the procedure is rather cumbersome. The exact solution is however preferable because, in the physically significant cases, the system is simultaneously submitted to a short-range potential in addition to the Coulomb potential, and it is convenient to adopt the approximations required by the short-range potential on top of exact Coulomb distortions.

The Coulomb potential introduces a distortion on the asymptotic wave function (13.6). According to Eq. (13.32), the new asymptotic form is

$$\psi_{\text{Coul}} \underset{r\to\infty}{\sim} \mathrm{e}^{\mathrm{i}[kz+\eta\ln k(r-z)]} + f(\theta)\frac{\mathrm{e}^{\mathrm{i}(kr-\eta\ln 2kr)}}{r},$$

where $\eta = \dfrac{\hbar^2 k}{2m_p c^2}$; m_p is the proton mass, and the scattering amplitude is

$$f(\theta) = -\frac{1}{2ik}\sum_{l=0}^{\infty}(2l+1)[e^{2i(\delta_l+\sigma_l)} - 1]P_l(\cos\theta),$$

where σ_l is the phase-shift due to the Coulomb potential, and δ_l is the phase-shift due to the nuclear potential. The scattering amplitude $f_C(\theta)$ for pure Coulomb interaction is obtained from the previous equations for $\delta_l = 0$. Since the two protons are identical particles, the Coulomb scattering cross section takes the form (see Sect. 5.4)

$$\sigma_C(\theta) = |f_C(\theta) - f_C(\pi - \theta)|^2$$
$$= \left(\frac{e^2}{4E}\right)^2 \left[\frac{1}{\sin^4\dfrac{\theta}{2}} + \frac{1}{\cos^4\dfrac{\theta}{2}} - \frac{\cos\left(\eta \ln\tan\dfrac{\theta}{2}\right)}{\sin^2\dfrac{\theta}{2}\cos^2\dfrac{\theta}{2}}\right], \tag{13.31}$$

where $E = \dfrac{\hbar^2 k^2}{2m}$. This equation is known as the Mott cross section. We observe that, while the first two terms represent the Rutherford scattering cross section evaluated for θ and for $\pi - \theta$, the last term is an interference term (for comparison see Sect. 5.4) giving rise to oscillations around the non-symmetrized scattering curve, as shown in Fig. 5.4 for two identical atoms. As an example, $\eta = 0.05$ for proton–proton collisions at energy of 10 MeV in the laboratory reference frame.

The behaviour of the experimental angular distribution is reproduced by the theoretical calculations based on formula (13.31) with respect to the calculations based on the Rutherford formula for the Coulomb scattering cross section.

Due to the infinite range of the Coulomb potential, no semiclassical consideration can impose a cutoff to the number of partial waves and hence the partial-wave expansion itself is not useful. The relation between bound and unbound states enables us to determine the asymptotic form of the scattering wave function from the same procedure as for the scattering from short-range potentials. In the s-wave case, the asymptotic wave function is $u_0 \sim kre^{-kr}$ with $k = \sqrt{2\mu|E|/\hbar^2}$, the eigenvalue E being negative. Moving from negative to positive energy, we must replace k with ik obtaining the asymptotic scattering wave function

$$u'_0 \sim ikre^{-ikr} = i(kr)^2\frac{e^{-ikr}}{kr} = \frac{i}{k}\frac{e^{-ikr+2\ln(kr)}}{r}, \tag{13.32}$$

that is a distorted wave function at variance with the spherical shape of that shown in Fig. 13.1.

13.7 Nuclear Interaction from Nucleon–Nucleon Scattering

Assuming that isospin is a good quantum number, the generalized Pauli principle dictates that the scattering state of two nucleons, interacting via the nuclear force, must fulfill the rule $L + S + T =$ odd, L being the orbital angular momentum of the relative motion, S and T the total spin and isospin, respectively. In the case of low-energy scattering, the $L = 0$ component is the dominant wave so that $S + T$ must be odd and hence either ($S = 0,\ T = 1$) or ($S = 1,\ T = 0$). The spin-singlet and isospin-triplet state describes neutron–neutron and neutron–proton states, which are symmetric in the isospin channel, i.e. $\psi_\mathrm{n}(1)\psi_\mathrm{p}(2) + \psi_\mathrm{n}(2)\psi_\mathrm{p}(1)$. The spin triplet and isospin singlet state describes neutron–proton states, which are anti-symmetric in the isospin channel, i.e. $\psi_\mathrm{n}(1)\psi_\mathrm{p}(2) - \psi_\mathrm{n}(2)\psi_\mathrm{p}(1)$. Concerning the study of the charge independence of the nuclear force, the experimental investigation has been mostly concentrated on the low-energy scattering between nucleons so as to only take into account the s-wave ($l = 0$) contribution to the cross section.

Let us firstly consider the neutron–proton scattering with a neutron beam at low energy (the limit $E \to 0$ for thermal neutrons). If the neutron beam is not spin-polarized, the total cross section is given by the sum of the spin-singlet and spin-triplet squared scattering lengths weighted on their respective spin multiplicities, according to Eq. (13.28), i.e.

$$\sigma_0 \sim (a_s^2 + 3a_t^2). \tag{13.33}$$

In the case of scattering of thermal neutrons ($E \sim 10^{-2}$ eV) on a hydrogen target, there is interference between the two components so that the total cross section is given by the squared sum of the spin-singlet and spin-triplet scattering lengths weighted on their respective spin multiplicities, according to the uncertainty principle, i.e.

$$a_H = \left(\frac{a_s}{2} + \frac{3}{2}a_t\right), \tag{13.34}$$

$$\sigma_0 \sim |a_s^2 + 3a_t^2|^2.$$

From the experimental determination of the two total cross sections,[3] the singlet and triplet scattering lengths are determined. The singlet one, which is of interest to us, is shown in Table 13.1.

The negative a_s sign tells us that neutron–proton spin singlet states cannot be bound states. As a result, the neutron–proton bound state, the deuteron, could be found only in the triplet spin state. The measurements of the total angular momentum, $J = 1$, confirm this assumption. The effective range of the neutron–proton interaction in the spin-triplet state $r_{0t} = 1.749 \pm 0.008$ fm is obtained from Eq. (13.6). In order to calculate the effective range for the spin singlet state, we have to express the total cross section Eq. (13.33) as a function of the scattering lengths a_s and a_t, as well as the effective ranges r_{0s} and r_{0t}

[3] Data taken from [PRE].

Table 13.1 Experimental scattering lengths and effective radii in nucleon–nucleon spin triplet state [NNN]

	Neutron–proton	Proton–proton	Neutron–neutron
a_s [fm]	−23.74	−17.30	−19.90
r_s [fm]	2.77	2.85	2.75

$$\sigma_{\rm tot} = \pi \left\{ \left[k^2 + \left(\frac{1}{a_s} - \frac{r_{0s}}{2} k^2 \right)^2 \right]^{-1} + 3 \left[k^2 + \left(\frac{1}{a_t} - \frac{r_{0t}}{2} k^2 \right)^2 \right]^{-1} \right\}.$$

In order to extract the parameters of the nuclear force from the experimental proton–proton scattering, we have to disentangle the effects of the Coulomb interaction. As shown in Sect. 13.6, the Coulomb potential introduces a distortion on the asymptotic wave function, Eq. (13.6).

Once the nuclear potential is included, the total phase-shift is $\delta_l = \delta_l^{\rm C} + \delta_l^{\rm N}$, being $\delta_l^{\rm C}$ and $\delta_l^{\rm N}$ the Coulomb and nuclear phase-shifts, respectively, and the partial wave expansion of the full scattering amplitude is written as

$$f(\theta) = -\frac{1}{2ik} \sum_l (2l+1)[e^{i(\delta_l^{\rm C}+\delta_l^{\rm N})} - 1] P_l(\cos\theta) \equiv f_{\rm C}(\theta) + f_{\rm N}(\theta) + f_{\rm NC}(\theta), \tag{13.35}$$

with obvious notations. Notice the interference term $f_{\rm NC}$ between the Coulomb and the nuclear interaction. From Eq. (13.35), we can calculate the proton–proton scattering cross section. In the further investigation of low-energy scattering, the exact Coulomb contribution is taken into account, whereas the nuclear contribution is approximated to the low-energy value, $\sigma_{\rm N}^{\rm pp} \sim \dfrac{\sin^2 \delta_0^{\rm N}}{k^2}$. Figure 13.6 shows the low-energy proton–proton scattering cross section data. The main features are clearly seen: the highly repulsive Coulomb effect is prevalent at small angles, the flat nuclear effect is prevailing at large angles, and a minimum, due to the interplay between them, appears at medium angles.

In order to extract the scattering length one needs significantly low-energy scattering data, that are not available below the Coulomb barrier threshold. Hence, one should resort to a nuclear potential, fitting the data with this. The results with the Argonne V18 potential, with the Coulomb effect removed, are reported in Table 13.1 compared to the neutron–proton data [NN].

For proton–proton collisions at incident energies from 1 MeV to 10 MeV, the comparison between experimental and theoretical cross sections has given the following values for the scattering length and the effective range of proton–proton interactions

$$a_{\rm pp} = -7.821 \pm 0.004 {\rm fm}, \qquad r_{0\rm pp} = 2.79 \pm 0.02 {\rm fm}.$$

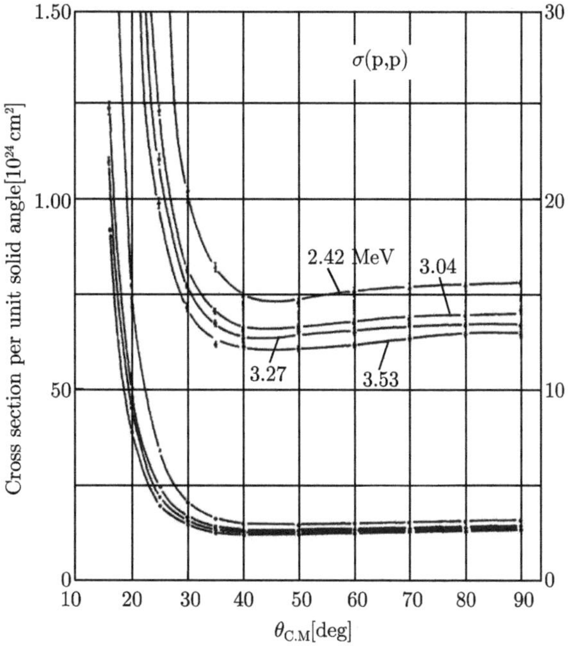

Fig. 13.6 Low-energy differential cross section for proton–proton scattering, [BLA] The y-axis scale for lower (higher) curves is on the right (left)

The quantities a_{pp} and r_{0pp} are connected to the phase-shift δ through an expansion analogous to Eq. (13.30) and depend, as we have seen for the neutron–proton interaction, on the nuclear potential. It has been possible to employ nuclear potentials to fit experimental data up to 350MeV and obtain

$$a_{pp} = -17.1 \pm 0.2\text{fm}, \qquad r_{0pp} = 2.81 \pm 0.02\text{fm}.$$

Lastly, we discuss how to determine the spin singlet parameters of the nuclear interaction from the low energy (s-wave) neutron–neutron scattering. So far, measurements from direct scattering are not available at very low energy, because neutrons, as neutral particles, are not charged particles. Indirect methods only exist, which involve studying the nuclear reactions with three particles in the exit channel, two of them being neutrons. A good candidate for this study is the reaction

$$\pi^- + d \rightarrow \gamma + 2n,$$

where π^- is a negative charged pion and γ is a gamma photon. In this reaction, the two neutrons in the exit channel do not experience nuclear interaction with any other particles, in particular with the γ photon.[4] In addition, the formation of a two-neutron bound state must be excluded, otherwise there would be a unique energy value of the emitted photon. In accordance with the energy and momentum conservation laws, the

[4] The effect of the neutron magnetic moment can be neglected.

γ-ray energy distribution would peak around 131 MeV, most of the rest-mass energy of the pion, $m_\pi c^2 = 139.6$ MeV, being used for the reaction. The real scenario of the final state corresponds to a two-neutron scattering state, and a continuous gamma ray energy distribution with a peak of around $E_\gamma = 131$ MeV. The kinematic condition, in which the relative momentum between the two neutrons is almost equal to zero, corresponds to such a peak. As a result, within this kinematic region, zero energy is almost reached for the neutron–neutron system. The relation between phase-shift and neutron–neutron scattering length is

$$\sin\delta_0 = \frac{\hbar k a_{\rm nn}}{\sqrt{1+\hbar^2 k^2 a_{\rm nn}^2}}.$$

The fits of the experimental data have given the value of the scattering length reported in Table 13.1. The simultaneous measurements of the scattering length and the effective range are made possible thanks to the comparison of the three-particle exit channels of the same reaction

$$^3{\rm He} +^3{\rm He} \to \alpha + 2{\rm p}, \qquad ^3{\rm He} +^3{\rm He} \to \alpha + {\rm n} + {\rm p}, \qquad ^3{\rm He} +^3{\rm He} \to \alpha + 2{\rm n}.$$

The known scattering lengths $a_{\rm np}$ and $a_{\rm pp}$ are used in turn to fit the parameters of a theoretical model. Afterwards, the same model is used to study the $\alpha + 2n$ exit channel with the final result for the effective range, reported in Table 13.1. In conclusion, the effective radius values obtained in the neutron–proton, proton-proton, and neutron–neutron scattering confirm the presumed charge independence of the nuclear force, but significant disagreement is found for the scattering length. A further investigation is directed towards sub-nucleonic degrees of freedom.

13.8 Notes and References

YAM Y.Yamaguchi, Physical Review **95** (1954) 1628.

YUK I. Yukawa, Proc. Phys. Math Soc.(Japan) **17** (1935) 48.

WEI S. Weinberg, Physica **A96** (1979) 327.

RUT The large angle scattering of α particles from gold target suggested to Rutherford (1911) the interpretation of a collision between pointlike particles, according to the Eq. (13.15). Hence, the protons of the gold atom had to be concentrated in a nucleus much smaller than the atomic scale of 10^{-10} m.

SAK J.J. Sakurai, *Modern Quantum Mechanics*, Addison-Wesley Publishing Company (1994), Chap. 7.2.

PRE M.A. Preston and R.K. Bhaduri, *Structure of the Nucleus*, Westview Press (1975), Chap.1–3.

NNN For a review see: A. Gårderstig, Journal of Physics G **36** (2009) 053001.

BLA Data from J.M.Blair et al., Physical Review 74 (1948) 553.

13.9 Problems

- Problem 1. Calculate in Born approximation the nuclear form factor for electron scattering on a nucleus with constant charge density, i.e. $\rho(r) = \rho_0$ for $r \leqslant R$ and zero for $r > R$.
- Problem 2. Calculate, within the Born approximation, the differential cross section for elastic scattering of a particle with charge e from the Coulomb potential generated by two point-like particles with charge q and $-q$, which are located on x-axis in the positions $(a, 0, 0)$ and $(-a, 0, 0)$, respectively. Assume the distance of the charge e from the charged particles to be $r >> a$.
- Problem 3. Calculate the classical total cross section of a particle for elastic scattering from a hard sphere with radius R. (Hint: The particle elastically bounces off the sphere in the same way as in the reflection of a light ray.)
- Problem 4. Calculate the total cross section of a particle for elastic scattering from a hard sphere with radius R. Compare the result with the classical cross section $\sigma_{cl} = \pi r^2$.
- Problem 5. Study the s-wave phase-shifts for the scattering from a square potential of depth V_0 and radius R. Compare the two cases of attractive $V_0 < 0$ and repulsive $V_0 > 0$ potential.
- Problem 6. Consider the high-energy Coulomb break up of deuteron (see Fig.11.2) assuming the neutron–proton r to be a slow parameter and the proton–target distance R_{pt} a fast parameter, determine within the sudden approximation the asymptotic form of the Schrodinger equation. [Hint: assume the deuteron to be "frozen" in the ground state].

Chapter 14
Many-Body Systems

In the last chapter, we showed how the system of two interacting particles can be exactly reduced to a system of one particle in a potential. This is not possible in the case of many-body interacting systems. However, an approximation, in which the particles move in a common self-consistent potential, was devised. This is the Hartree–Fock approximation, extensively used in atomic and nuclear structures, as discussed in this chapter. An extension of the Hartree–Fock approximation, called Brueckner–Hartree–Fock approximation, was developed to remove the hard-core divergences appearing in the bare interaction. With many-nucleon systems, from the simplest nucleus ^{3}He to nuclear matter, the problem of introducing many-body forces arises, because neither the experimental binding energy of ^{3}He nor the empirical saturation energy of nuclear matter can be reproduced by calculations using only bare or effective two-body forces. Several calculations in the last few decades have shown that two-plus three-body forces can do this. The part missing from the Hartree–Fock or Brueckner–Hartree–Fock approximation, known as residual interaction, is responsible for the so-called screening phenomenon due to medium polarization. A typical example is the screening of the electric interaction in water: in the presence of a Coulomb field the water molecules H_2O polarize and, in turn, weaken the original field. The net result is a significant reduction of the dielectric constant in the Coulomb interaction law. Another important phenomenon appearing in many-body systems is the collective motion due to a departure of the system from stability. It has already been treated with the lattice oscillations in crystals (see Chap. 8), and it will also be with the giant resonances in nuclei (see Chap. 17).

The original version of the chapter has been revised: A correction to this chapter can be found at https://doi.org/10.1007/978-981-97-1670-8_19.

Supplementary Information The online version contains supplementary material available at https://doi.org/10.1007/978-981-97-1670-8_14.

U. Lombardo et al., *Quantum Mechanics*, UNITEXT for Physics,
https://doi.org/10.1007/978-981-97-1670-8_14

14.1 Systems of Identical Particles

Identical particles are particles with the same observable physical properties, such as mass, charge, spin, and half-life. Examples are composite systems like atoms or nuclei and particles like electrons, protons, or neutrons. In Sect. 5.4, we saw that the scattering of two identical particles can be interpreted only if we treat them as indistinguishable particles. In this chapter, we discuss systems of many identical particles in a general context. In classical mechanics, once a particle is labeled, one can track its path since its coordinates and momentum can be measured simultaneously with absolute precision at any instant, independently of experimental uncertainties. As a result, one can always distinguish one particle from another. On the contrary, in quantum mechanics, coordinates and momentum of the particles cannot be measured simultaneously, so that it becomes impossible to track the particle. Particles lose their individuality. Due to its very significant implications, this is one of the most fundamental principles of quantum mechanics: identical particles cannot be distinguished from one another. Indeed, this principle dictates the symmetrization of wave function of identical particles.

Let us consider a system of N identical particles. Its state is described by the wave function $\psi(q_1, q_2, \ldots, q_N)$, where q_i denotes the number of degrees of freedom of the ith particle, like space coordinates and spin coordinates. When we exchange any two particles, e.g. the ith particle and the jth particle, the wave function $\psi(q_1, \ldots, q_i, \ldots, q_j, \ldots, q_N)$ becomes $\psi(q_1, \ldots, q_j, \ldots, q_i, \ldots, q_N)$. Here we can define the permutation operator $\mathbf{P}_{ij}$

$$\mathbf{P}_{ij}\psi(q_1, \ldots, q_i, \ldots, q_j, \ldots, q_N) \equiv \psi(q_1, \ldots, q_j, \ldots, q_i, \ldots, q_N).$$

Since they describe the same state, $\mathbf{P}_{ij}\psi(q_1, \ldots, q_i, \ldots, q_j, \ldots, q_N)$ and $\psi(q_1, \ldots, q_i, \ldots, q_j, \ldots, q_N)$ must be equal within a phase factor λ

$$\mathbf{P}_{ij}\psi(q_1, \ldots, q_i, \ldots, q_j, \ldots, q_N) = \lambda\psi(q_1, \ldots, q_i, \ldots, q_j, \ldots, q_N).$$

By repeating the exchange, the wave function returns to its original state, so that

$$\begin{aligned}
\mathbf{P}_{ij}^2\psi(q_1, \ldots, q_i, \ldots, q_j, \ldots, q_N) &= \mathbf{P}_{ij}\lambda\psi(q_1, \ldots, q_i, \ldots, q_j, \ldots, q_N) \\
&= \lambda^2\psi(q_1, \ldots, q_i, \ldots, q_j, \ldots, q_N) \\
&= \psi(q_1, \ldots, q_i, \ldots, q_j, \ldots, q_N),
\end{aligned}$$

which leads to $\lambda^2 = 1$, $\lambda = \pm 1$. This means that, when two particles are exchanged, the wave function ψ either changes or does not change sign, i.e. it is antisymmetric or symmetric. Particles described as antisymmetric wave functions, called fermions, obey the Fermi–Dirac statistics, whereas particles described as symmetrical wave functions, called bosons, obey the Bose–Einstein statistics (see Chap. 16). Experiments show that the statistics obeyed by particles is uniquely related to the spin. Particles with half-integer spin ($s = \hbar/2, 3\hbar/2, 5\hbar/2, \ldots$) are fermions, e.g. electrons, protons, and neutrons, and those with integer spin ($s = 0, \hbar, 2\hbar, \ldots$) are bosons, e.g. photons and mesons.

Let us consider a system of two non-interacting identical fermions. Particle 1 occupies the state ϕ_{k_1} with energy ϵ_1, and particle 2 occupies the state ϕ_{k_2} with energy ϵ_2. Here ϕ_k are orthonormalized wave functions. Then $\phi_{k_1}(q_1)\phi_{k_2}(q_2)$, $\phi_{k_1}(q_2)\phi_{k_2}(q_1)$ or any linear superposition of them is an eigenfunction of the total Hamiltonian, with eigenvalue $\epsilon_1 + \epsilon_2$. Since the wave function of two fermions must be anti-symmetrized, we take the linear combination

$$\begin{aligned}\phi^A_{k_1k_2}(q_1, q_2) &= \frac{1}{\sqrt{2}}(1 - P_{12})\phi_{k_1}(q_1)\phi_{k_2}(q_2)\\ &= \frac{1}{\sqrt{2}}[\phi_{k_1}(q_1)\phi_{k_2}(q_2) - \phi_{k_1}(q_2)\phi_{k_2}(q_1)]\\ &= \frac{1}{\sqrt{2}}\begin{vmatrix}\phi_{k_1}(q_1) & \phi_{k_1}(q_2)\\ \phi_{k_2}(q_1) & \phi_{k_2}(q_2)\end{vmatrix}.\end{aligned}$$

This form is called a Slater determinant. When $k_1 = k_2$, the wave function $\phi^A_{k_1k_2}(q_1, q_2)$ vanishes, meaning that two identical fermions cannot occupy the same state (k is a complete set of quantum numbers defining the state). This is the statement of the Pauli principle. For bosons, the wave function has to be symmetrized, then

$$\phi^S_{k_1,k_2}(q_1, q_2) = \frac{1}{\sqrt{2}}[\phi_{k_1}(q_1)\phi_{k_2}(q_2) + \phi_{k_1}(q_2)\phi_{k_2}(q_1)].$$

The symmetrization can be generalized to systems of N non-interacting identical particles. In fermion systems, due to the Pauli principle, each state k_i, can be occupied by a single particle. Let $k_1, k_2, \ldots, k_N$ denote the occupied single-particle states defining a given many-body state of N particles. There are in total $N!$ possible permutations of the N particles among the N occupied states. The permutation achieved by the operator **P** is called even or odd according to whether the number of exchanges is an even or odd number. We introduce the parameter δ_P, where $\delta_P = 1$ if **P** is an even permutation, and $\delta_P = -1$ if **P** is an odd permutation. The normalized anti-symmetrized wave function is then

$$\begin{aligned}\phi^A_{k_1,k_2,\ldots,k_N}(q_1, q_2, \ldots, q_N) &= \frac{1}{\sqrt{N!}}\sum_P \delta_P P[\phi_{k_1}(q_1)\phi_{k_2}(q_2)\ldots\phi_{k_N}(q_N)]\\ &= \frac{1}{\sqrt{N!}}\begin{vmatrix}\phi_{k_1}(q_1) & \phi_{k_1}(q_2) & \cdots & \phi_{k_1}(q_N)\\ \phi_{k_2}(q_1) & \phi_{k_2}(q_2) & \cdots & \phi_{k_2}(q_N)\\ \vdots & \vdots & \cdots & \vdots\\ \phi_{k_N}(q_1) & \phi_{k_N}(q_2) & \cdots & \phi_{k_N}(q_N)\end{vmatrix}.\end{aligned}\tag{14.1}$$

The Pauli principle lies at not only the base of the electron distribution in atoms but also that of the behavior of Fermi gases (see Sect. 16.7) and other quantum systems.

In the case of boson systems, several particles can occupy the same state. Suppose there are N bosons and n_1 of them occupy the state k_1, n_2 occupy the state $k_2, \cdots,$ and n_M occupy the state k_M. There are totally $\frac{N!}{\prod_{i=1}^{N} n_i!}$ distinct permutations. Notice here we only permutate particles in states different from each other. The normalized symmetrized wave function is then

$$\phi^S_{n_1 n_2 \cdots n_M}(q_1, q_2, \ldots, q_N) = \sqrt{\frac{\prod\limits_{i=1}^{N} n_i!}{N!}} \sum_P P[\phi_{k_1}(q_1) \cdots \phi_{k_1}(q_{n_1})$$
$$\phi_{k_2}(q_{n_1+1}) \cdots \phi_{k_2}(q_{n_1+n_2}) \cdots$$
$$\phi_{k_M}(q_{N-n_M+1}) \cdots \phi_{k_M}(q_N)]. \quad (14.2)$$

At low temperature, most particles of a boson gas occupy the lowest energy single-particle state. This property is the origin of the superfluidity and superconductivity phenomena.

14.2 The Hartree–Fock Approximation

In Chap. 9, the variational principle is introduced to deal with the many-body system with interactions, which cannot be strictly solved. In principle, the variational principle is exact and equivalent to the Schrödinger equation. However, in practice, one needs to restrict the trial wave function in a subspace of the whole Hilbert space, which introduces the approximation. One way to achieve the variational principle is the Ritz variational method, which was introduced in Chap. 9, and the other commonly used way is the Hartree–Fock method.

14.2.1 The Hartree-Fock Approximation: Stationary Case

In the Hartree–Fock approximation, it is assumed that each particle moves independently in an average (or mean field) potential provided by other particles as well as the external field, if there is one. Under this idea, the trial wave function is the product of single-particle wave functions, i.e.:

$$\Psi(\mathbf{r}_1, \mathbf{r}_2, \ldots, \mathbf{r}_N) = \phi_{k_1}(\mathbf{r}_1)\phi_{k_2}(\mathbf{r}_2) \ldots \phi_{k_N}(\mathbf{r}_N) \quad (14.3)$$

for a system with N particles, where $\phi_k(\mathbf{r})$ is the single-particle wave function. However, in quantum mechanics, these particles are identical particles, so they have to obey their statistics, according to whether they are fermions or bosons. The above wave function has to be symmetrized for bosonic systems and anti-symmetrized for fermionic systems. Running over the space of these trial wave functions, we will find wave functions with the lowest expectation value of the Hamiltonian.

Considering a system of N fermions interacting with each other via a two-body interaction $V(\mathbf{r}, \mathbf{r}')$ each via an external field $v_{\text{ext}}(\mathbf{r})$, the Hamiltonian of the system is

$$\mathbf{H} = \sum_{i=1}^{N}\left[-\frac{\hbar^2}{2m}\nabla_i^2 + v_{\text{ext}}(\mathbf{r}_i)\right] + \frac{1}{2}\sum_{ij} V(\mathbf{r}_i, \mathbf{r}_j).$$

The trial wave function in the Hartree–Fock method is the product of independent particle wave functions, which is a generic Slater determinant for fermion systems, i.e.:

$$\Psi(1, 2, \ldots, N) = \frac{1}{\sqrt{N!}}\begin{vmatrix} \phi_1(\mathbf{r}_1, \sigma_1) & \phi_1(\mathbf{r}_2, \sigma_2) & \cdots & \phi_1(\mathbf{r}_N, \sigma_N) \\ \phi_2(\mathbf{r}_1, \sigma_1) & \phi_2(\mathbf{r}_2, \sigma_2) & \cdots & \phi_2(\mathbf{r}_N, \sigma_N) \\ \vdots & \vdots & \cdots & \vdots \\ \phi_N(\mathbf{r}_1, \sigma_1) & \phi_N(\mathbf{r}_2, \sigma_2) & \cdots & \phi_N(\mathbf{r}_N, \sigma_N) \end{vmatrix},$$

where ϕ_i is the orthonormal single-particle wave function, with position $\mathbf{r}$ and spin σ. The expectation value of the Hamiltonian on this wave function is

$$\begin{aligned}\overline{H} &= \langle\Psi|\mathbf{H}|\Psi\rangle \\ &= \sum_{i\sigma}\int d\mathbf{r}\phi_i^*(\mathbf{r}, \sigma)\left[-\frac{\hbar^2}{2m}\nabla^2 + v_{\text{ext}}(\mathbf{r})\right]\phi_i(\mathbf{r}, \sigma) \\ &+ \frac{1}{2}\sum_{ij,\sigma\sigma'}\int\int d\mathbf{r}d\mathbf{r}'\phi_i^*(\mathbf{r}, \sigma)\phi_j^*(\mathbf{r}', \sigma')V(\mathbf{r}, \mathbf{r}')\left[\phi_i(\mathbf{r}, \sigma)\phi_j(\mathbf{r}', \sigma')\right. \\ &\left.-\phi_j(\mathbf{r}, \sigma)\phi_i(\mathbf{r}', \sigma')\right].\end{aligned} \tag{14.4}$$

With the constraint

$$\langle\phi_i|\phi_i\rangle = 1, \qquad i = 1, 2, \ldots$$

we minimize the energy $\overline{H}$ with respect to arbitrary variations of ϕ_i and ϕ_i^*, as follows:

$$\delta\left(\overline{H} - \sum_i \epsilon_i(\phi_i, \phi_i)\right) = 0, \tag{14.5}$$

where ϵ_i are Lagrangian multipliers.

$$\begin{aligned}\delta\overline{H} &= \sum_{i\sigma}\int d\mathbf{r}\delta\phi_i^*(\mathbf{r}, \sigma)\left[-\frac{\hbar^2}{2m}\nabla^2 + v_{\text{ext}}(\mathbf{r})\right]\phi_i(\mathbf{r}, \sigma) \\ &+ \frac{1}{2}\sum_{ij,\sigma\sigma'}\int\int d\mathbf{r}d\mathbf{r}'\left[\delta\phi_i^*(\mathbf{r}, \sigma)\phi_j^*(\mathbf{r}', \sigma') + \phi_i^*(\mathbf{r}, \sigma)\delta\phi_j^*(\mathbf{r}', \sigma')\right] \\ &\cdot V(\mathbf{r}, \mathbf{r}')\left[\phi_i(\mathbf{r}, \sigma)\phi_j(\mathbf{r}', \sigma') - \phi_j(\mathbf{r}, \sigma)\phi_i(\mathbf{r}', \sigma')\right] \\ &+ \text{ complex conjugate terms}\end{aligned}$$

$$
\begin{aligned}
&= \sum_{i\sigma} \int \mathrm{d}\mathbf{r} \delta\phi_i^*(\mathbf{r},\sigma) \left[-\frac{\hbar^2}{2m} \nabla^2 + v_{\text{ext}}(\mathbf{r}) \right] \phi_i(\mathbf{r},\sigma) \\
&\quad + \sum_{ij,\sigma\sigma'} \int\int \mathrm{d}\mathbf{r}\mathrm{d}\mathbf{r}' \delta\phi_i^*(\mathbf{r},\sigma)\phi_j^*(\mathbf{r}',\sigma') \\
&\quad \cdot V(\mathbf{r},\mathbf{r}') \left[\phi_i(\mathbf{r},\sigma)\phi_j(\mathbf{r}',\sigma') - \phi_j(\mathbf{r},\sigma)\phi_i(\mathbf{r}',\sigma') \right] \\
&\quad + \text{ complex conjugate terms.}
\end{aligned} \tag{14.6}
$$

Plugging Eq. (14.6) into Eq. (14.5), we get

$$
\begin{aligned}
&\sum_{i\sigma} \int \mathrm{d}\mathbf{r} \delta\phi_i^*(\mathbf{r},\sigma) \left[-\frac{\hbar^2}{2m} \nabla^2 + v_{\text{ext}}(\mathbf{r}) \right] \phi_i(\mathbf{r},\sigma) \\
&\quad + \sum_{ij,\sigma\sigma'} \int\int \mathrm{d}\mathbf{r}\mathrm{d}\mathbf{r}' \delta\phi_i^*(\mathbf{r},\sigma)\phi_j^*(\mathbf{r}',\sigma') V(\mathbf{r},\mathbf{r}') \\
&\quad \cdot \left[\phi_i(\mathbf{r},\sigma)\phi_j(\mathbf{r}',\sigma') - \phi_j(\mathbf{r},\sigma)\phi_i(\mathbf{r}',\sigma') \right] - \sum_{i\sigma} \epsilon_i \int \mathrm{d}\mathbf{r} \delta\phi_i^*(\mathbf{r},\sigma)\phi_i(\mathbf{r},\sigma) \\
&\quad + \text{ complex conjugate terms} = 0.
\end{aligned} \tag{14.7}
$$

Since $\delta\phi_i$ and $\delta\phi_i^*$ are arbitrary, we have the following equations:

$$
\begin{aligned}
&\left[-\frac{\hbar^2}{2m} \nabla^2 + v_{\text{ext}}(\mathbf{r}) \right] \phi_i(\mathbf{r},\sigma) \\
&\quad + \sum_{j,\sigma'} \int \mathrm{d}\mathbf{r}' \phi_j^*(\mathbf{r}',\sigma') V(\mathbf{r},\mathbf{r}') \left[\phi_i(\mathbf{r},\sigma)\phi_j(\mathbf{r}',\sigma') - \phi_j(\mathbf{r},\sigma)\phi_i(\mathbf{r}',\sigma') \right] \\
&= \epsilon_i \phi_i(\mathbf{r},\sigma).
\end{aligned} \tag{14.8}
$$

These are the Hartree–Fock self-consistent mean-field equations. We notice that when $j = i$ the term in the summation vanishes. It is convenient to introduce the density matrix in the coordinate representation

$$
\rho(\mathbf{r},\sigma;\mathbf{r}',\sigma') = \sum_j \phi_j(\mathbf{r},\sigma)\phi_j^*(\mathbf{r}',\sigma').
$$

Its diagonal term

$$
\rho(\mathbf{r},\sigma) = \sum_j |\phi_j(\mathbf{r},\sigma)|^2
$$

is the density of particles with spin σ. The function

$$
\begin{aligned}
U(\mathbf{r}) &= \sum_{j,\sigma'} \int \mathrm{d}\mathbf{r}' |\phi_j(\mathbf{r}',\sigma')|^2 V(\mathbf{r},\mathbf{r}') \\
&= \sum_{\sigma'} \int \mathrm{d}\mathbf{r}' \rho(\mathbf{r}',\sigma') V(\mathbf{r},\mathbf{r}'),
\end{aligned}
$$

defines the Hartree potential, which is local, and the function

$$U_{\rm exc}(\mathbf{r},\mathbf{r}') = \sum_j \phi_j^*(\mathbf{r}',\sigma')\phi_j(\mathbf{r},\sigma)V(\mathbf{r},\mathbf{r}')$$
$$= \rho(\mathbf{r},\sigma;\mathbf{r}',\sigma')V(\mathbf{r},\mathbf{r}'),$$

defines the exchange potential, which is non-local. With these definitions, Eq. (14.8) is written as

$$\left[-\frac{\hbar^2}{2m}\nabla^2 + v_{\rm ext}(\mathbf{r})\right]\phi_i(\mathbf{r},\sigma) + U(\mathbf{r})\phi_i(\mathbf{r},\sigma)$$
$$-\sum_{\sigma'}\int d\mathbf{r}' U_{\rm exc}(\mathbf{r},\mathbf{r}')\phi_i(\mathbf{r}',\sigma') = \epsilon_i\phi_i(\mathbf{r},\sigma). \tag{14.9}$$

This is a set of integro-differential equations, whose solutions are the single-particle energies ϵ_i and the corresponding single-particle wave functions ϕ_i of the N particles moving in the common Hartree–Fock potential. This set of equations has to be solved by iteration, since the Hartree–Fock potential is a self-consistent field, i.e. depending on the solution itself, consisting of the set of N wave functions ϕ_i. Starting from N known trial wave functions $\phi_i^{(0)}$, we calculate the zero-order potentials $U^{(0)}$ and $U_{\rm exc}^{(0)}$. Thus, we can solve the Hartree–Fock equations Eq. (14.9) getting a first-order set of $\phi_i^{(1)}$. We then repeat this procedure many times until the convergence is achieved, namely $\phi_i^{(n-1)} = \phi_i^{(n)}$. The Slater determinant obtained with the single-particle wave functions of the last iteration is the Hartree–Fock ground state. Its energy, derived from Eqs. (14.4) and (14.8), is

$$E = \langle\Psi_{\rm HF}|\mathbf{H}|\Psi_{\rm HF}\rangle$$
$$= \sum_i^N \epsilon_i - \frac{1}{2}\sum_{ij,\sigma\sigma'}\int\int d\mathbf{r}d\mathbf{r}'\phi_i^*(\mathbf{r},\sigma)\phi_j^*(\mathbf{r}',\sigma')V(\mathbf{r},\mathbf{r}')$$
$$\cdot\left[\phi_i(\mathbf{r},\sigma)\phi_j(\mathbf{r}',\sigma') - \phi_j(\mathbf{r},\sigma)\phi_i(\mathbf{r}',\sigma')\right]$$
$$= \sum_i^N \epsilon_i - \frac{1}{2}\sum_{ij=1}^N(\langle ij|V|ij\rangle - \langle ij|V|ji\rangle).$$

We notice that the ground-state energy is not equal to the sum of all occupied single-particle energies, because the interaction energy between particles i and j is counted twice, when we sum the single-particle energies.

If we do not consider the exchange contribution of the potential matrix elements, namely we use only the trial wavefunction Eq. (14.3), we get the Hartree equation

$$\left[-\frac{\hbar^2}{2m}\nabla^2 + v_{\rm ext}(\mathbf{r}) + U(\mathbf{r})\right]\phi_i(\mathbf{r},\sigma) = \epsilon_i\phi_i(\mathbf{r},\sigma).$$

In this equation, only the direct local potential $U(\mathbf{r})$ appears in the self-consistent Hartree–Fock equation. $U(\mathbf{r})$ is called Hartree potential. The non-local exchange potential $U_{\text{exc}}(\mathbf{r})$ increases the complexity of solving the Hartree–Fock equations. However, there is a cancellation between $U(\mathbf{r})$ and $U_{\text{exc}}(\mathbf{r})$ when $j = i$, which means that the particle described by the wavefunction ϕ_i feels the potential generated by the other $N - 1$ particles, without including itself. On the contrary, in the Hartree approximation, this unphysical interaction of a particle with itself is present.

The second quantization (see Sect. 14.5) furnishes a more appropriate formalism to the theory of many-body systems. It already gives a simpler derivation of the HF approximation than the formalism of Slater determinants. Starting with an arbitrary complete and orthonormal set of single-particle states ϕ_i of energy ε_i, we introduce the creation operator $\mathbf{c}_i^\dagger$ and annihilation operator $\mathbf{c}_i$ of a particle in the state ϕ_i. In this formalism an arbitrary state of N independent fermions, as an alternative to the Slater determinant, is written

$$|\Phi\rangle = \prod_{i=1,N} c_i^\dagger |0\rangle,$$

where $|0\rangle$ is the vacuum state. To derive the HF equations, we consider a trial wave function $|\text{HF}\rangle$ of the form Eq. (14.2.1), where the N particles occupy the lowest energy levels. In the selected basis the single-particle density matrix is diagonal[1], i.e.

$$\rho_k = \begin{cases} 1 & \text{for occupied states}(k \leqslant N) \\ 0 & \text{for empty states}(k > N) \end{cases}$$

Hereafter in this section, we use the indices i,j for occupied states, m,n for empty states and k,l for generic states. Therefore, the one-body density is a projection operator, $\rho^2 = \rho$, with eigenvalue 1 for occupied states and 0 for unoccupied states. In the second quantization, according to Eq. (14.28) and (14.29) the antisymmetrized Hamiltonian of two-body interaction, in any arbitrary basis, is written as

$$H = \sum_{l_1 l_2} t_{l_1 l_2} c_{l_1}^\dagger c_{l_2} + \frac{1}{4} \sum_{l_1 l_2 l_3 l_4} \bar{v}_{l_1 l_2 l_3 l_4} c_{l_1}^\dagger c_{l_2}^\dagger c_{l_4} c_{l_3},$$

where $\bar{v}_{l_1 l_2 l_3 l_4} = v_{l_1 l_2 l_3 l_4} - v_{l_1 l_2 l_4 l_3}$. Applying the algebra of the fermion creation and annihilation operators (see Sect. 14.5), the energy functional $E(\rho)$ is written as

$$E = \langle\Phi|\mathbf{H}|\Phi\rangle = \sum_{l_1 l_2} t_{l_1 l_2} \rho_{l_2 l_1} + \frac{1}{2} \sum_{l_1 l_2 l_3 l_4} \rho_{l_3 l_1} \bar{v}_{l_1 l_2 l_3 l_4} \rho_{l_4 l_2}.$$

In the HF basis, it becomes

$$E = \langle\text{HF}|\mathbf{H}|\text{HF}\rangle = \sum_{i=1}^{N} t_{ii} + \frac{1}{2} \sum_{i,j=1}^{N} \bar{v}_{ijij}.$$

[1] Otherwise, it can be diagonalized by a suitable unitary transformation leaving the density invariant

In order to determine the HF equations, we have to minimize the energy by the variational procedure on the one body density ρ, with $\rho^2 = \rho$, corresponding to the subspace of Slater determinants. Hence

$$\delta E = E[\rho + \delta\rho] - E[\rho] = \sum_{kk'} h_{kk'} \delta\rho_{kk'} = 0,$$

where

$$h_{kk'} \equiv \frac{\partial E[\rho]}{\partial \rho_{k'k}} = t_{kk'} + \sum_{i=1}^{N} \bar{v}_{kl'k'l} \rho_{ll'}. \tag{14.10}$$

The variation of the constraint $\rho^2 = \rho$ with respect to density gives

$$\rho\delta\rho + \delta\rho\rho = \delta\rho,$$

from which the following conditions arise

$$\rho\delta\rho\rho = \sigma\delta\rho\sigma = 0,$$

where $\sigma = 1 - \rho$. This means that the density variation between particle states $\delta\rho_{pp'} = 0$ as well as between hole states $\delta\rho_{hh'} = 0$, so that particle-hole excitations, i.e. $\delta\rho_{ph}$ and $\delta\rho_{hp}$ only are allowed. It stems that the variation principle takes the simplified form Eq. (14.2.1),

$$\delta E = E[\rho + \delta\rho] - E[\rho] = \sum_{hp} \left(h_{hp} \delta\rho_{hp} + h_{ph} \delta\rho_{ph} \right) = 0,$$

so that

$$h_{hp} = h_{ph} = 0$$

$\delta\rho_{hp}$ and $\delta\rho_{ph}$ being non vanishing arbitrary variations. In the basis, where the density matrix is diagonal, the previous equation can be globally written

$$[h, \rho] = 0 \tag{14.11}$$

Since the submatrices h_{pp} and h_{hh} can be diagonalized simultaneously with ρ, it is not a lose of generality to require the latter to be diagonal, that is, $(h_{pp})_{mn} = \epsilon_m \delta_{mn}$ and $(h_{hh})_{ij} = \epsilon_i \delta_{ij}$. We end up with

$$h_{kk'} = t_{kk'} + \sum_{l=1,N} \bar{v}_{klk'l} = \epsilon_k \delta_{kk'},$$

and the ground state will be the state with the lowest single-particle energies occupied by the N particles. These are the HF equations in momentum representation, in keeping with the equivalent Eqs. (14.9) in coordinate representation.

14.2.2 The Hartree-Fock Approximation: Time-Dependent HF Equations

Let us consider a system of N interacting fermions in a non-stationary state $|\Psi(t)\rangle$ at the instant t and calculate the time evolution of the one-body density

$$\begin{aligned} i\hbar\partial_t\rho_{ki}(t) &= \langle\Psi(t)|c_i^\dagger c_k|i\hbar\partial_t\Psi(t)\rangle + \langle i\hbar\partial_t\Psi(t)|c_l^\dagger c_k|\Psi(t)\rangle \\ &= \langle\Psi(t)|c_l^\dagger c_k\mathbf{H}|\Psi(t)\rangle - \langle\Psi(t)|\mathbf{H}c_l^\dagger c_k|\Psi(t)\rangle \\ &= \langle\Psi(t)|[c_l^\dagger c_k, \mathbf{H}]|\Psi(t)\rangle, \end{aligned}$$

where $\mathbf{H} = \mathbf{T} + \mathbf{V}$ is the Hamiltonian of the system. Using the anticommutation rules between creation and annihilation operators, we obtain the equation

$$i\hbar\partial_t\rho_{kl}(t) = \sum_l^{\prime}[t_{kj}\rho_{l'l}(t)_{-}\rho_{kl'}(t)t_{l'l}] + \frac{1}{2}\sum_{jlm}[\bar{v}_{kl'lm}\rho_{lmil'}(t) - \bar{v}_{lmij}\rho_{kjlm}(t)].$$

This is the time-dependent equation of motion for one body density. The time-dependent two-body density matrix ρ_{lmij} measures the correlations between two particles generated by the two-body interaction. As matter of fact, this function contains two parts: the first one describing the two particles as if they were independent of each other, the second describing the influence of one particle on another, due to the interaction. In the independent particle model, there are no correlations, hence the two-body density matrix only keeps the first part, that is the anti-symmetrized product of two one-body densities

$$\begin{aligned} \rho_{ijkl}(t) &= \langle\Psi(t)|c_i^\dagger c_j^\dagger c_k c_l|\Psi(t)\rangle \\ &= \langle\Psi(t)|c_i^\dagger c_l|\Psi(t)\rangle\langle\Psi(t)|c_j^\dagger c_k|\Psi(t)\rangle - \langle\Psi(t)c_i^\dagger c_k|\Psi(t)\rangle\langle\Psi(t)|c_j^\dagger c_l|\Psi(t)\rangle \end{aligned}$$

Plugging this expression into the equation of motion, we obtain

$$\begin{aligned} i\hbar\partial_t\rho_{ki}(t) &= \sum{}_j[t_{kl}\rho_{li}(t) - \rho_{kl}(t)t_{li}] \\ &\quad + \sum{}_{jlm}[\bar{v}_{kjlm}\rho_{li}(t)\rho_{mj}(t) - \bar{v}_{lmkj}\rho_{kl(t)}\rho_{jm}(t)] \\ &= [t + U_{HF}(\rho), \rho]_{\mathrm{ki}} \end{aligned} \tag{14.12}$$

This is the time dependent Hartree-Fock (TDHF) equation in a generic single particle basis, In this form, i.e. expressed in terms of the one body density matrix, it will be extensively used in next chapters.

14.3 Atomic Structure

Atoms are complex systems with a nucleus of Z protons surrounded by an electron cloud made of Z electrons. Each electron interacts with the attractive Coulomb field of the nucleus and with the repulsive Coulomb field of the other electrons of the cloud. Thus, the Hamiltonian is

$$\mathbf{H} = \sum_{i=1}^{Z} \left[-\frac{\hbar^2}{2m} \nabla_i^2 - \frac{Ze^2}{r_i} \right] + \frac{1}{2} \sum_{i,j=1}^{Z} \frac{e^2}{|\mathbf{r}_i - \mathbf{r}_j|},$$

In the non-relativistic approximation, the electron cloud of atoms is described by the Schrödinger equation. In general, this equation can be solved only with numerical methods, except for the hydrogen H and the hydrogen-like ions He^+, Li^{++}, $\cdots$, which have only one electron. In these cases, the Schrödinger equation can be exactly solved, as shown in Sect. 12.2. Here we only recall the energy spectrum (see Eq. (12.12))

$$E_n = -\frac{mZ^2e^4}{2\hbar^2} \frac{1}{n^2},$$

where $m = m_e M/(m_e + M)$, M being the nuclear mass and m_e the electron mass, respectively. Each energy level E_n is degenerate with degeneracy $f_n = 2\sum_{l=0}^{n-1}(2l + 1) = 2n^2$, where the factor of two comes from the electron spin.

In the Hartree–Fock approximation, each electron is subjected to the self-consistent potential created by the cloud, in addition to the nuclear potential. The Hartree–Fock equations are

$$\left[-\frac{\hbar^2}{2m_e} \nabla^2 - \frac{Ze^2}{r_i} \right] \phi_i(\mathbf{r}, \sigma) + U(\mathbf{r}) \phi_i(\mathbf{r}, \sigma) - \sum_{\sigma'} \int d\mathbf{r}' U_{exc}(\mathbf{r}, \mathbf{r}') \phi_i(\mathbf{r}', \sigma')$$
$$= \epsilon_i \phi_i(\mathbf{r}, \sigma),$$

with

$$U(\mathbf{r}) = \sum_{\sigma'} \int d\mathbf{r}' \rho(\mathbf{r}', \sigma') \frac{e^2}{|\mathbf{r} - \mathbf{r}'|},$$
$$U_{\text{exc}}(\mathbf{r}, \mathbf{r}') = \rho(\mathbf{r}, \sigma; \mathbf{r}', \sigma') \frac{e^2}{|\mathbf{r} - \mathbf{r}'|}.$$

By self-consistently solving these equations, a set of electron single-particle (s.p.) states with energy ϵ_i and wave function ϕ_i is obtained. Analogously to the hydrogen atom, each state is characterized by the principal quantum number n and orbital quantum number l ($n = l + 1, l + 2, \ldots$) because the total Coulomb potential is still central. At variance with the hydrogen atom, where the energy level n is degenerate, in complex atoms, this degeneracy is broken. This is because the effective potential acting on the electron occupying the level n is no longer just the Coulomb potential

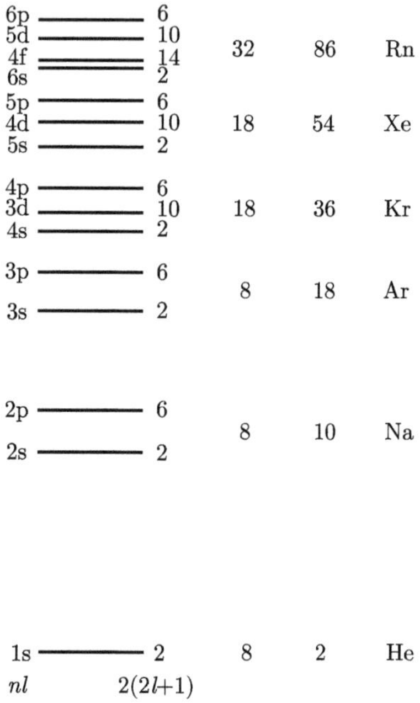

Fig. 14.1 The shell structure for electrons in an atom

of the nucleus only but also the potential created by the surrounding electrons. This latter affects in different ways orbitals with different angular momenta l: the electron in an orbital with larger l is further from the nucleus, and hence the screening effect of electrons in inner shells is larger. As a result, the energy level for larger l is higher in comparison to smaller l. The single-particle levels of electrons with the screening effect on Coulomb nuclear potential are plotted in Fig. 14.1. The orbital angular momentum quantum number l is denoted with letters

$$l = 0, 1, 2, 3, 4, 5, 6, 7, 8, 9, 10, \ldots$$
$$\text{s, p, d, f, g, h, i, k, l, m, n}, \ldots$$

For given n and l, electrons still have associated projections of orbital angular momentum and spin on z-axis, which are denoted as m and σ. For a given l, there can be $2l + 1$ choices of m, i.e. $m = -l, -l + 1, \ldots, l$, and two choices of σ, i.e. $\sigma = \pm 1/2$. There are thus $2(2l + 1)$ different states for a given n and l. According to the Pauli principle, each state can only be occupied by one fermion, so that the maximum electron occupancy of each orbital is $2(2l + 1)$. The interplay between energy quantization and Pauli principle can explain the existence of the so-called magic numbers. Indeed, passing from one atom with Z electrons to the next, the added electron acquires an energy slightly higher than the previous one, as long as both are in the same shell. But, when the shell is full, the next electron passes to the upper shell, in which the energy is much higher. Table 14.1 lists the electron shell structure and magic numbers in the periodic table. In terms of the binding energy, passing from one atom,

Table 14.1 The electron shell structure and magic numbers in the periodic table

Shell number	s.p. states	Accommodated electrons	Accumulated electrons	"Magic number"
1	1s	2	2	2
2	2s, 2p	8	10	10
3	3s, 3p	8	18	18
4	4s, 4p, 3d	18	36	36
5	5s, 5p, 4d	18	54	54
6	6s, 6p, 5d, 4f	32	86	86
7	7s, 7p, 6d, 5f	32	118	–

corresponding to shell closure (e.g. ^{2}He) to the next one, (^{3}Li), the binding energy of the added electron turns out to be much smaller than that of the previous one. $Z = 2$ is one (also the lowest) magic number. Atoms with a number of electrons equal to a magic number plus 1 are called hydrogenoids, e.g. Li, Na, K, Rb, Cs, Fr. The extra electron above the fully filled electron shells is called the valence electron. The chemical properties as well as the optical spectrum of the atom are determined by this electron. Due to the screening effect of the inner-shell electrons, this electron feels an effective charge of the nucleus Z^*. For electron state with small l, $Z^* > 1$; conversely, for orbitals with large l, $Z^* \sim 1$, in which case, the energies are very close to those of the hydrogen atom. In general, the energy levels of hydrogenoids are similar to those of the hydrogen atom with a corrective term Δ_l, which is called Rydberg's correction.

$$E_{nl} = -\frac{m_e e^4}{2\hbar^2}\frac{1}{(n-\Delta_l)^2}.$$

Δ_l is a positive number. When l is large, Δ_l is small, while when l is small, Δ_l is relatively large.

To get a complete description of an atom, one needs to know the states of all electrons as well as the total L, S, J coupled by all electrons. For the total L, S, J, one uses $^{2S+1}L_J$ to label the state of the atom, while, for the state of a single electron, one uses nl to label it. Here the capital letters are used to denote $(L)l$. For example, 1s 2p $^3P_0$ denotes a state of the helium atom in which $L = 1$, $S = 1$, and $J = 0$, and two electrons are in 1s and 2p states. If m electrons are in states with the same l and n, nl^m can be used, e.g. the symbol 1p^2 denotes the 1p state with two electrons. The distribution of the electrons on states with n and l is called the electron configuration. For a given electron configuration, there are different possible atom states. If electrons are in different n and l states, the possible values of L and S are determined by the rule of angular momentum coupling. For example, with the configuration of 1p and 2p, $L = 0, 1, 2$ and $S = 0, 1$, so that $^{1,3}S$, $^{1,3}P$, $^{1,3}D$ are possible atomic states. However, for electrons with the same n and l, there are fewer possibilities due to the Pauli principle. For example, for two electrons staying on 1p state, there are six possible states for each electron, denoted by m and σ, i.e. (1, 1/2), (0, 1/2),

$(-1, 1/2)$, $(1, -1/2)$, $(0, -1/2)$, and $(-1, -1/2)$. These two electrons can occupy two of the above six possible states, so there are 15 combinations. The possible $M_L = \sum m$ and $M_S = \sum \sigma$. (M_L, M_S) can be

$$(2,0)\ (1,0)\ (0,0)\ (-1,0)\ (-2,0)$$
$$(1,1)\ (0,1)\ (-1,1)\ (1,0)\ (0,0)\ (-1,0)\ (1,-1)\ (0,-1)\ (-1,-1)$$
$$(0,0)$$

The first line forms ^1D state; the second line forms ^3P state, and the last line forms ^1S state.

Before closing this section, we must discuss the interpretation of the level splitting, known as fine structure, observed, e.g. in the spectra of hydrogen-like atoms. This gives us the opportunity of introducing the spin–orbit interaction (see Chap. 15) due to the orbital motion of the nucleus in the rest frame of the electron. This is a little effect in atoms,[2] but a great effect in nuclei. Spin–orbit interaction breaks down the rotational invariance in coordinate space and spin space, separately. It is however recovered in a more general space including space and spin space coordinates together. As a result, the total angular momentum is conserved and l_z is to be replaced by j_z as a good quantum number. The electron states will be labeled by the quantum numbers l^2, s^2, j^2, j_z. The spin–orbit interaction of the hydrogen-like atom is

$$H_{so} = \frac{Ze^2}{2m^2c^2r^3}\vec{\mathbf{s}}\cdot\vec{\mathbf{l}} = \frac{Ze^2}{2m^2c^2r^3}\frac{1}{2}[\mathbf{j}^2 - \mathbf{l}^2 - \mathbf{s}^2]$$

according to Eq. (15.11).

To calculate the spin–orbit correction we resort to the first order of the perturbation theory, in degenerate case. However, since the unperturbed states $\psi_{nlsjj_z}(\vec{r})$ $(j_z = l \pm 1/2)$ are also eigenstates of the spin–orbit operator, the spin–orbit energy splitting is the difference between the two states that turns out to be (see Problem 3)

$$\begin{aligned}
&\int \mathrm{d}^3\vec{r}\ [\langle\psi_{nlsj,l+1/2}(\vec{r}\,)|\mathbf{H}_{\mathbf{so}}|\psi_{nlsj,l+1/2}(\vec{r}\,)\,\rangle \\
&- \langle\psi_{nlsj,l-1/2}(\vec{r}\,)\,|\mathbf{H}_{\mathbf{so}}|\psi_{nlsj,l-1/2}(\vec{r}\,)\,\rangle] \\
&= \frac{Ze^2\hbar^2}{4m^2c^2}(2l+1)\left\langle\frac{1}{r^3}\right\rangle_{nl} .
\end{aligned}$$

In Fig. 14.2, we report the calculation of the spin–orbit splitting in 2p orbital of the Na alkali atom.

---

[2] There is an even smaller splitting due to the interaction between the electron and the nuclear spin, called hyperfine-structure (see Problem 3).

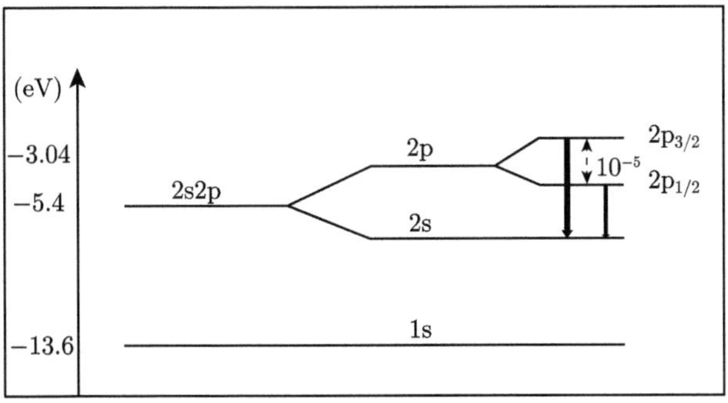

Fig. 14.2 Spin–orbit effect in Na atom

14.4 Nuclear Structure

The atomic nucleus is a quantum many-body system made up of A nucleons, of which Z are protons and N are neutrons. As discussed in Chap. 12, the nucleons are fermions and their free-particle states are characterized by momentum, spin, and isospin. The nuclear interaction is responsible for the stability of nuclei, against the Coulomb repulsion between protons.

There are two alternative phenomenological models to describe the nuclear structure. On one side, the nuclear spectroscopy inspired the *Nuclear Shell Model*, which strongly resembles the atomic shell model. According to this model, the nucleons occupy the single-particle states $\psi_{l,s,\tau}(r)$ determined by a common mean field. On the other side, scattering experiments on the nuclear density and the short-range nature of the nuclear force inspired the *Liquid Drop Model*. According to this model, the nucleus is a quantum liquid drop with volume, surface, and density. Simultaneously, quantum–mechanical microscopic models were developed with a main scope of reconciling the two phenomenological models.

14.4.1 The Nuclear Shell Model

The idea of the nuclear shell model has been borrowed from atomic physics [SMT]. Indeed, some experimental evidence shows that nuclei exhibit close analogies with phenomena occurring in atoms. In particular, we focus on the enhancement of the nucleon separation energies related to the so-called magic numbers of neutrons and protons: $N, Z = 2, 8, 20, 28, 50, 82, 126$. The nucleon separation energy is the amount of energy required to remove the least bound proton or neutron from a nucleus. For most nuclei, the separation energies are $\approx$8 MeV. However, strong enhancements can be observed for the aforementioned magic (N or Z magic) nuclei and also doubly magic (N and Z both magic) nuclei. Such phenomenon is analogous to the enhancement of the electron separation energies in noble gases. This analogy leads us to consider a single-nucleon subjected to the nuclear mean-field potential exactly as an electron subjected to the atomic Coulomb potential. Going further, we have to introduce an ad hoc nuclear potential. A reasonable approximation for the shape of the nuclear potential is the three-dimensional isotropic harmonic oscillator model (the infinite square well has also been employed in shell model calculations). Applying the empirical law $R = 1.15\,A^{\frac{1}{3}}$ fm for the nuclear radius (see Sect. 3.2), the potential is written as

$$V(r) = \begin{cases} -V_0\left[1 - \left(\frac{r}{R}\right)^2\right] & r \leqslant R \\ 0 & r > R. \end{cases}$$

The vanishing of the potential beyond the surface is not a crude approximation, because it can be proved that the bound states are almost insensitive to the potential tail. The angular frequency is $\omega = \sqrt{\dfrac{2V_0}{mR^2}}$ with $V_0 \approx 50$ MeV and the nucleon mass $m \approx 938$ MeV.

Neglecting for a while the spin of nucleons, the energy spectrum of the three-dimensional isotropic harmonic potential, Eq. (8.10), easily extends to the nucleus rescaling values by V_0 with the result

$$E_N = \left(N + \frac{3}{2}\right)\hbar\omega + V_0. \tag{14.13}$$

The degree of degeneracy for a fixed number of quanta N is

$$G(N) = \frac{1}{2}(N+2)(N+1),$$

where $N = 2(n-1) + l$ with $n = 1, 2, \ldots$ and $l = 0, 1, 2, \ldots$. We call *shells* the E_n energy levels. The structure of the nucleus is obtained by filling the shells with $G(N)$ neutrons and $G(N)$ protons starting from the lowest energy level in accordance with the Pauli principle, which dictates two particles to occupy the same state. Table 14.2 shows the distributions of the numbers of protons and neutrons along the oscillator shells. The last column is the total number of both neutrons and protons that have been gathered to fill the considered shell. We notice that the model can predict the first three magic numbers $N, Z = 2, 8, 20$ but not the higher ones $N, Z = 28, 50, 82, 126$.

So far, we have not considered the spin effects. In the atomic case the electron spin–orbit coupling generates, as explained in Sect. 14.3, the spin–orbit interaction, which is responsible for the fine structure of the electron energy spectrum. We exploit this concept by including an analogous nucleon spin–orbit interaction term in the shell model Hamiltonian, which has to be traced back to the nucleon magnetic moment (see Sect. 12.3). It can be written as

$$V_{\rm so}(r)\mathbf{L}\cdot\mathbf{S}.$$

The shape of the spin–orbit potential $V_{\rm so}(r)$ can be determined from the fit of experimental data in scattering experiments. Once the spin operator $\mathbf{s}$ is included, the

Table 14.2 Nucleon distribution along the harmonic oscillator shells. The last column refers to the sum of the numbers of nucleons counted up to that shell

N	$\frac{N}{\hbar\omega}$	l	Shell	Number of nucleons (neutrons or protons)	Total
0	$\frac{3}{2}$	0	1s	2	2
1	$\frac{5}{2}$	1	1p	6	8
2	$\frac{7}{2}$	0, 2	2s, 1d	12	20
3	$\frac{9}{2}$	1, 3	2p, 1f	20	40
4	$\frac{11}{2}$	0, 2, 4	3s, 2d, 1g	30	70
5	$\frac{13}{2}$	0, 2, 4, 6	3p, 2f, 1h	42	112
6	$\frac{15}{2}$	0	4s, 3d, 2g, 1i	56	168

energy eigenstates have to be modified from $|n, l, m\rangle$ to $|n, s, l, j, m\rangle$, with j representing the eigenvalue of the total angular momentum operator $\mathbf{j} = \mathbf{l} + \mathbf{s}$. The energy spectrum Eq. (14.13) has to be corrected for the expectation value of the spin–orbit operator $\mathbf{l} \cdot \mathbf{s}$, to simplify matters, we assume $V_{\text{so}}(r) = const$ and, since $\mathbf{l} \cdot \mathbf{s} = \frac{1}{2}(\mathbf{j}^2 - \mathbf{l}^2 - \mathbf{s}^2)$ we have

$$\begin{aligned} E_{jls} &= \langle n, s, l, j, m \left| V_{\text{so}}(r) \frac{\mathbf{l} \cdot \mathbf{s}}{2} \right| n, s, l, j, m\rangle \\ &= \frac{1}{2} V_{\text{so}}(r)[j(j+1) - l(l+1) - s(s+1)]\hbar^2 \\ &= \frac{1}{2} V_{\text{so}}(r) \left[\left(l \pm \frac{1}{2} \right) \left(l \pm \frac{1}{2} + 1 \right) - l(l+1) - \frac{3}{4} \right] \hbar^2. \end{aligned}$$

Nucleons have half-integer spin $s = \frac{1}{2}$, thus we have $s(s+1) = \frac{3}{4}$ with $2s + 1 = 2$ possible orientations along the quantization axis, and

$$E_{jls} = V_{\text{so}}(r)\hbar^2 \frac{l}{2} \quad \text{for} \quad j = l + \frac{1}{2}, \tag{14.14}$$

$$E_{jls} = -V_{\text{so}}(r)\hbar^2 \frac{(l+1)}{2} \quad \text{for} \quad j = l - \frac{1}{2}. \tag{14.15}$$

Hence, the spin–orbit interaction removes the degeneracy in the j quantum number. According to Eqs. (14.14) and (14.15), the energy difference between the j-degenerate levels is

$$\Delta E(l) = V_{so}(r)\hbar^2\frac{1}{2}[l - (-l - 1)] = V_{so}(r)\hbar^2\frac{1}{2}(2l + 1).$$

Assuming an attractive spin–orbit potential, i.e. $V_{so}(r) < 0$, we have the $l + \frac{1}{2}$ energy level to be lower than the corresponding $l - \frac{1}{2}$ level. The effect of the spin–orbit interaction on the level structure is shown in Fig. 14.3. The first column in brackets refers to the number of nucleons, either neutrons or protons, that occupy the level; the second column refers to the total number of nucleons counted up to that level. The numbers outside the brackets are the total numbers of nucleons for the shell closure. In particular reference to this column, we can highlight how the inclusion in the Hamiltonian of the spin–orbit interaction leads to a correct description of the magic numbers. Finally, we can observe that this level structure is not unique but it can change according to the shape of the spin–orbit potential.

14.4.2 *Liquid Drop Model and Nuclear Matter*

The liquid drop model (LDM) is the first model proposed for the nucleus structure. The underlying idea of considering the nucleus as a quantum liquid drop stems from the following facts: (1) the nuclear forces exhibit saturation properties, meaning that the nuclear binding energy per nucleon $BE(A, Z)/A$ is roughly constant ~ -8 [MeV/nucleon] as seen in Fig. 14.4, instead of scaling with A when a non-saturating force, like that of electrons in atoms, is operating; (2) the nucleus has a low compressibility and so a well-defined surface. However, it is a quantum fluid which obeys Fermi statistics, so the Pauli principle prevents the nucleons from getting too close to each other, and the collisions are rare events in contrast to classical fluids. This explains why the average distance between nucleons is ~2.4 fm instead of the expected distance ~0.7 fm corresponding to the minimum of the inter-particle force. This average distance determines the so-called *saturation density* $\rho_0 = 0.16$ fm^{-3}. This value, roughly constant in the interior of nuclei, is experimentally estimated by electron scattering on nuclei (see Fig. (13.4)). So, the nuclear radius R can be estimated at $\frac{4}{3}\pi R^3 = A\frac{4}{3}\pi r_0^3$, where $2r_0 \sim 2.4$ fm is the average distance between nucleons, which is 2.4 fm. It leads to $R = 1.2A^{1/3}$. Based on the idea of the liquid drop model, Bethe and Weizsäcker proposed the following semi-empirical mass formula

$$B(N, Z) = a_V A + a_S A^{2/3} + a_C\frac{Z^2}{A^{1/3}} + a_I\frac{(N - Z)^2}{A} - \delta(A)$$

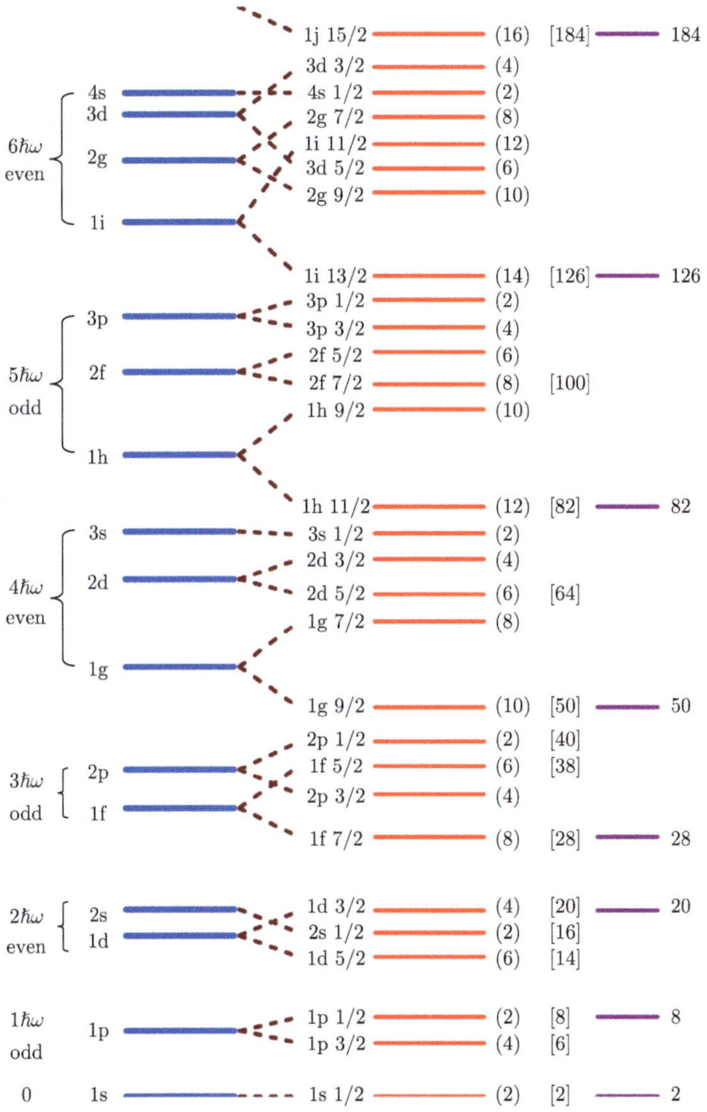

Fig. 14.3 Shell model level structure with the inclusion of the spin–orbit interaction. First column of numbers in brackets: number of nucleons, i.e. neutrons or protons, that fill the level. Second column of numbers in brackets: sum of the number of nucleons up to the considered level. Numbers outside the brackets: total numbers of nucleons at shell closure

to evaluate the binding energy in nuclei. Fitting the experimental nuclear binding energies, the coefficients take the values

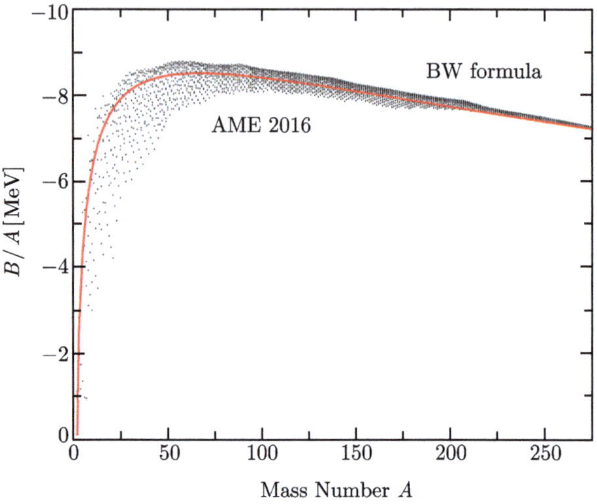

Fig. 14.4 The binding energies per nucleon from AME2016 atomic mass evaluation [WUA] and the binding energies per nucleon of stable nuclei from Bethe and Weizsäcker mass formula

$$a_V \sim -16\text{ MeV}, \quad a_S \sim 18.5\text{ MeV}, \quad a_C = 0.71\text{ MeV}, \quad a_I \sim 20 - 30\text{ MeV}, \tag{14.16}$$

and

$$\delta(A) = \begin{cases} 34A^{-3/4} & \text{even} - \text{even nuclei} \\ 0 & \text{even} - \text{odd nuclei} \\ -34A^{-3/4} & \text{odd} - \text{odd nuclei.} \end{cases}$$

The first term, called the volume term, assumes that all nucleons participate in the total energy with the same contribution on account of the saturation of the nuclear force. The second term, called the surface term, is a correction due to the surface nucleons, because these nucleons ($\sim A^{2/3}$) contribute with a smaller quantity owing to the lower density. The third term represents the Coulomb repulsion among protons. Assuming the charges are uniformly distributed over a sphere, the Coulomb energy is proportional to Z^2 and inversely proportional to the radius $R \propto A^{1/3}$. The fourth term is the symmetry energy due to the imbalance $\beta = \dfrac{N-Z}{A}$ between the number of neutrons and the number of protons. This term exists, despite the nuclear force being independent of charge. Indeed, if it were not for the repulsive Coulomb force between protons, the nuclear force would give the greatest stability to nuclei with $N = Z$ according to the Pauli principle. In fact, light nuclei tend to have the number of protons equal to the number of neutrons. But, in heavier and heavier nuclei, the Coulomb force increases and then a neutron excess for a nucleus with A nucleons is needed to guarantee the stability against the increasing Coulomb force. This neutron excess implies a cost that the binding energy must pay. Indeed, on the one hand, the excess neutrons go to occupy higher momentum levels than protons, due to the Pauli blocking, (kinetic energy effect), on the other hand, the ratio between pairs of unlike nucleons and pairs of like nucleons is reduced (potential effect)[BOU]. As a result, in nuclei with atomic numbers A and $N > Z$, the increase of kinetic energy is larger and that of nuclear potential energy is smaller than in nuclei with the same A, but $N = Z$. The combination of these two effects gives rise to the symmetry energy. The last term reflects the odd–even staggering effect on the nuclear binding energy due to the pairing correlations [PAI].

In Fig. 14.4, the experimental binding energies per nucleon from AME2016 atomic mass data [WAN] are shown, compared to the Bethe and Weizsäcker mass formula predictions. One can see this semi-empirical formula gives an overall good description of the experimental binding energies. Deviations are mostly due to shell effects.

A further development of the liquid drop model has led to the concept of *nuclear matter*, firstly introduced by Hans Bethe with the purpose of understanding the many-body structure of nuclei in terms of the underlying nuclear interaction [BET]. Later, it proved to be extremely useful for the study of neutron stars and heavy-ion collisions. Indeed, the inner core of neutron stars and the composite system formed during collisions between two heavy ions realize conditions closely similar to nuclear matter.

Roughly speaking, nuclear matter is an infinite homogenous drop of nucleons interacting via the nuclear force and no Coulomb force.[3] The main parameter characterizing the nuclear matter is the nucleon density defined as the ratio between the number of nucleons A and the volume V of the drop at the thermodynamic limit.

A phenomenological approach to nuclear matter stems from the extrapolation of the Bethe-Weizsäcker semi-empirical mass formula to any density ρ, other than the saturation density of finite nuclei. In the case of nuclear matter, the mass formula significantly simplifies owing to the absence of surface and Coulomb energy (the pairing contribution can be neglected in the present context). As a result, the binding energy per nucleon becomes

$$\epsilon(\rho, \beta) = a_V(\rho) + a_I(\rho)\beta^2, \tag{14.17}$$

where the N vs. Z imbalance is measured by the asymmetry parameter $\beta = (N - Z)/A = (\rho_n - \rho_p)/\rho$, ρ_n and ρ_p being the neutron and proton densities, respectively. The density dependence of the parameters in Eq. (14.17) has been emphasized, because the values, reported in Eq. (14.16), are those at the saturation density ρ_0. How to theoretically determine and how to experimentally constrain the values at $\rho \neq \rho_0$ is known as *the issue of the equation of state of nuclear matter.* Several microscopic calculations have been performed with various interactions [GIA]. A calculation with Skyrme forces is presented in the next section.

A common feature of these calculations is that the resulting equations of state have a van der Waals shape, which reflects the common property of the various interactions to be long-range attractive and short-range repulsive. A typical equation of state is shown in Fig. 14.5 for different asymmetries, from symmetric nuclear matter with $\beta = 0$ (symmetric nuclear matter) to asymmetric nuclear matter with $\beta = 1$ (pure

---

[3] The presence of the Coulomb force would give an energy divergence owing to its long-range nature.

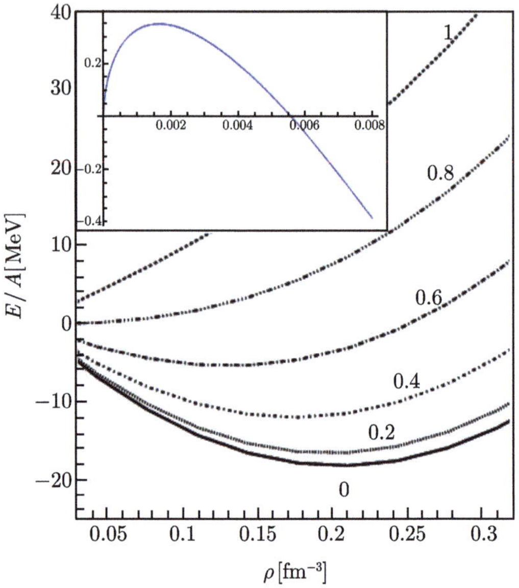

Fig. 14.5 The binding energies in nuclear matter at $T = 0$ for various asymmetries from symmetric nuclear matter to pure neutron matter

neutron matter). The energy minimum marks the saturation density ρ_0. The curvature of $E/A(\rho)$, evaluated at ρ_0, measures the compressibility of nuclear matter, defined as

$$K_\infty = 9\rho_0^2 \left[\frac{\partial^2 E/A(\rho)}{\partial \rho^2}\right]_{\rho=\rho_0} .$$

In the high-density range $\rho \gg \rho_0$, the interaction is more and more repulsive, because it corresponds to a smaller and smaller average distance between two nucleons. Conversely, in the low-density range $\rho \ll \rho_0$, the interaction is more and more attractive. In the domain of very low density, which is shown in the inset of Fig. 14.5, a kink appears featured by the kinetic energy, where the interaction is negligible because the two particles are a large distance apart. Increasing the isospin asymmetry, the system becomes less and less bound until the minimum disappears and it enters the region of full instability of very neutron-rich matter. The stability can only be restored by an attractive force of enough strength, as in the case of massive astrophysical objects such as neutron stars (see Sect. 16.8).

14.4.3 *Microscopic Approaches*

The Hartree–Fock method is the starting point for ab initio descriptions of the nuclear structure, i.e. a model in which the nucleon–nucleon interaction is obtained by fitting the experimental data from the in-vacuum scattering between nucleons. But in the nucleus, this interaction becomes very repulsive when the nucleons come close to each other, as shown in Fig. 12.4. As a consequence, the two-body matrix elements entering the Hartree–Fock potential become infinite at short distance. To extend the application of the Hartree–Fock method to the nucleus, two solutions have been proposed. One is the Brückner–Hartree–Fock method [BRU], and the other one is the Hartree–Fock method based on density functional theory [KOH,HAT]. In the Brückner–Hartree–Fock method, the bare nucleon–nucleon interaction is renormalized to include short-range correlations with the result of suppressing the hard-core repulsion. Extensive calculations, based on this model, have been performed to get the equation of state of nuclear matter, but its application to finite nuclei is rather cumbersome, making it difficult to accurately predict basic nuclear properties like mass and radii for finite nuclei.

The Hartree–Fock method based on density functional theory is not only a more phenomenological but also a simpler and more accurate method compared to the Brückner–Hartree–Fock theory. The density functional theory tells us that for any interacting many-body system, a universal energy density functional $E(\rho)$ exists, whose minimization gives the ground state energy, and the Kohn–Sham theorem tells us that one can always construct an auxiliary system of independent particles that has the same energy functional. Hence, having established that it is impossible to get the good independent particle picture of the nucleus from bare nucleon–nucleon interaction, one can always find a Hamiltonian with effective interactions that makes the single-particle picture to be a good approximation. This means that the Slater determinant serves as a good trial wave function and it makes the Hartree–Fock method applicable again. The most popular effective interaction is the Skyrme force [VAU]. Neglecting for simplicity the finite-range corrections, the corresponding Hamiltonian is

$$\mathbf{H} = -\sum_{i=1}^{A} \frac{\hbar^2}{2m}\nabla_i^2 + \frac{1}{2}\sum_{i,j=1}^{A} \Big\{ [t_0(1 + x_0\mathbf{P}_\sigma) + \frac{1}{6}t_3(1 + x_3\mathbf{P}_\sigma)\rho^\alpha(\vec{R}_{ij})]\delta(\vec{r}_{ij}) + \mathrm{i}W_0(\vec{\sigma}_i + \vec{\sigma}_j)\cdot[\vec{k}' \times \delta(\vec{r}_{ij})\vec{k}] \Big\}, \tag{14.18}$$

where $\vec{R}_{ij} = \dfrac{\vec{r}_i + \vec{r}_j}{2}$, $\vec{r}_{ij} = \vec{r}_i - \vec{r}_j$, $\vec{k} = \dfrac{1}{2\mathrm{i}}(\vec{\nabla} - \vec{\nabla}')$ and $\vec{k}'$ is the complex conjugate of $\vec{k}$ (acting on the left). The parameters t_0 and t_3 are the strengths of attractive and repulsive volume terms, respectively; the products t_0x_0 and t_3x_3 are the spin–spin term ($P_\sigma = \dfrac{1 + \vec{\sigma}_i \cdot \vec{\sigma}_j}{2}$ is the spin exchange operator) and the parameter W_0 is the strength of the surface spin–orbit term. The force parameters t_i, x_i for the central part, the power α of the density dependence, and the spin–orbit strength W_0, are free parameters which are adjusted according to some basic properties of finite nuclei and nuclear matter. Once these parameters are fixed, this effective Hamiltonian can be used within the Hartree–Fock approximation for calculations of nuclear properties. The Skyrme Hartree–Fock equations are

$$-\frac{\hbar^2}{2m_q}\nabla^2\phi_i(\vec{r})+\Big[U_q(\vec{r})-\mathrm{i}\vec{W}_q(\vec{r})\cdot(\vec{\nabla}\times\vec{\sigma})\Big]\phi_i(\vec{r})=\epsilon_i\phi_i(\vec{r}),$$

where the central potential U and the spin–orbital potential $\vec{W}$ are defined as

$$\begin{aligned}
U_q(\vec{r}) &= t_0\Big\{(1+\frac{1}{2}x_0)\rho-(x_0+\frac{1}{2})\rho_q\Big\}\\
&\quad+\frac{1}{12}t_3\Big\{\Big(1+\frac{1}{2}x_3\Big)(\alpha+2)\rho^{\alpha+1}\\
&\quad-\Big(x_3+\frac{1}{2}\Big)\big[\alpha(\rho_p^2+\rho_n^2)+2\rho\rho_q\big]\rho^{\alpha-1}\Big\}\\
&\quad-\frac{1}{2}W_0\vec{\nabla}(\vec{r})\cdot\Big(\vec{J}+\vec{J}_q\Big)+\left[\frac{e^2}{2}\int \mathrm{d}^3\vec{r}'\,\frac{\rho_q(\vec{r}')}{|\vec{r}-\vec{r}'|}-e^2\Big(\frac{3}{\pi}\Big)^{1/3}\rho_q^{1/3}\right]\delta_{qp},\\
\vec{W}_q(\vec{r}) &= \frac{1}{2}W_0\Big(\vec{\nabla}\rho+\vec{\nabla}\rho_q\Big).
\end{aligned}$$

Here, the density ρ_q and the spin current vector density $\vec{J}_q$ are defined as

$$\rho_q(\vec{r})=\sum_{i\sigma_z}|\phi_i(\vec{r},\sigma_z,q)|^2, \tag{14.19}$$

$$\vec{J}_q(\vec{r})=-\mathrm{i}\sum_{i\sigma_z'\sigma_z}\phi_i^*(\vec{r},\sigma_z',q)\big[\vec{\nabla}\phi_i(\vec{r},\sigma_z,q)\times\langle\sigma_z'|\vec{\sigma}|\sigma_z\rangle\big], \tag{14.20}$$

where q is the third component of isospin which distinguishes protons (p) and neutrons (n). The densities without the subscript q indicate the total densities as the sum of neutron and proton contributions. By solving the Skyrme Hartree–Fock equations iteratively, one obtains the single-particle energies ϵ_i and the single-particle wave functions ϕ_i. The Slater determinant of all occupied ϕ_i gives the nucleus wave function, with which one calculates all ground state properties of the nucleus. Besides the basic terms shown above, the complete Skyrme density functional also has momentum-dependent terms corresponding to the non-local part of the interaction. It has been shown that the Skyrme Hartree–Fock theory can successfully reproduce the binding energies and radii all over the periodic table. In Table 14.3, there are some examples. Table 14.3 lists the total binding energies and charge radii of ^{208}Pb, ^{40}Ca and ^{48}Ni for some typical Skyrme parameter sets SLy5, SkM*, and SAMi.

From the Skyrme density functional, one can also calculate the properties of nuclear matter. Nuclear matter being infinite and homogeneous, the single-particle wave functions are plane waves in Skyrme Hartree–Fock approximation. The expectation value of the Skyrme Hamiltonian Eq. (14.18) furnishes the energy of nuclear matter[4]

[4] The spin–orbit term is zero in nuclear matter (see Sect. 15.4).

Table 14.3 The total binding energies and charge radii of ^{208}Pb, ^{40}Ca, and ^{48}Ni by Skyrme parameter sets SLy5, SkM*, and SAMi

	^{208}Pb		^{40}Ca		^{48}Ni	
	E(MeV)	r_c(fm)	E(MeV)	r_c(fm)	E(MeV)	r_c(fm)
exp	−1636.43	5.501	−342.05	3.478	−348.72	–
SkM*	−1636.47	5.510	−341.12	3.518	−356.64	3.781
SLy5	−1636.08	5.508	−344.08	3.508	−352.72	3.791
SAMi	−1636.26	5.520	−347.03	3.483	−352.56	3.788

$$\begin{aligned}\frac{E}{A}(\rho_{\rm n},\rho_{\rm p}) &= \frac{3\hbar^2}{10m}\left(\frac{3\pi^2\rho}{2}\right)^{2/3}H_{5/3}+\frac{1}{8}t_0\rho\Big[2(x_0+2)-(2x_0+1)H_2\Big]\\ &\quad+\frac{1}{48}t_3\rho^{\alpha+1}\Big[2(x_3+2)-(2x_3+1)H_2\Big],\end{aligned} \tag{14.21}$$

where m is the nucleon mass, and $H_n = 2^{n-1}\left[\left(\frac{\rho_{\rm p}}{\rho}\right)^n+\left(\frac{\rho_{\rm n}}{\rho}\right)^n\right]$. With a simple fit of the basic properties of nuclear matter, one gets the following values of the Skyrme parameters (let us call them SK0):

$$t_0=-1867.0\,\text{MeV}\cdot\text{fm}^3,\ t_3=13523\,\text{MeV}\cdot\text{fm}^{3\alpha+3},\ W_0=105\,\text{MeV}\cdot\text{fm}^5,\ \alpha=\frac{1}{3}.$$

From the functional derivatives of energy density with respect to density, evaluated at ρ_0, one calculates the compression modulus K_∞, the symmetry energy a_I and its slope L

$$\begin{aligned}K_\infty &= \frac{3\hbar^2}{m}\left(\frac{3\pi^2\rho_0}{2}\right)^{2/3}H_{5/3}+\frac{9}{4}t_0\rho_0\Big[2(x_0+2)-(2x_0+1)H_2\Big]\\ &\quad+\frac{3}{16}t_3(\alpha+1)(\alpha+2)\rho_0^{\alpha+1}\Big[2(x_3+2)-(2x_3+1)H_2\Big],\\ a_I &= \frac{\hbar^2}{6m}\left(\frac{3\pi\rho_0}{2}\right)^{2/3}-\frac{1}{8}t_0(2x_0+1)\rho_0-\frac{1}{48}t_3(2x_3+1)\rho_0^{\alpha+1},\\ L &= \frac{\hbar^2}{3m}\left(\frac{3\pi\rho_0}{2}\right)^{2/3}-\frac{3}{8}t_0(2x_0+1)\rho_0-\frac{1}{16}t_3(2x_3+1)(\alpha+1)\rho_0^{\alpha+1},\end{aligned}$$

respectively. Table 14.4 lists the calculated basic properties of nuclear matter, including saturation density ρ_0, the binding energy per nucleon E/A, the compressibility K_∞, the symmetry energy a_I, and the slope of symmetry energy L, all calculated at the saturation density.

Table 14.4 The saturation density ρ_0 with the binding energy per nucleon E/A, the incompressibility K_∞, the symmetry energy a_I and the slope of a_I, calculated at ρ_0, using the Skyrme parameter set SK0

ρ_0 (fm^{-3})	E/A (MeV)	K_∞ (MeV)	a_I (MeV)	L (MeV)
0.157	−16.51	241.71	31.3	33.81

14.5 Complement to Sect. 14.2: Second Quantization

There is an alternative method of dealing with N identical particles: instead of defining the many-body state from the distribution of distinguishable particles in the various single-particle states and then symmetrizing the wave function as dictated by the non-distinguishability principle, we can define the state assigning the number of particles n_k (occupation number) to each single-particle state, so refraining from distinguishing the particles using labels. In this may, we are introducing the occupation numbers representation. The occupation numbers for each single-particle state will be any number from zero to N for bosons and a number either zero or one for fermions, the latter due to the Pauli principle. Of course, $\sum n_k = N$, where the sum runs over all single-particle states. In this representation, the state for a system is written as

$$|n_1 n_2 \ldots n_k \ldots\rangle,$$

which corresponds to the fermion state of Eq. (14.1) with $n_k = 0, 1$ and to the boson state of Eq. (14.2) with $n_k = 0, 1, 2, \ldots, N$, but the new forms are much simpler.

In Chap. 8, these states were built introducing the creation and annihilation operators a and $a^\dagger$ as excitation quanta of the harmonic oscillator. In that case, an arbitrary number of quanta could occupy the vacuum state of the system and satisfy the commutation rules. The excitation quanta behave like bosons, hence we can extend these operators to states of any identical bosons as follows:

$$|n_1 n_2 \cdots n_k \cdots\rangle = \frac{1}{\sqrt{\prod_i n_i!}} (a_1^\dagger)^{n_1} (a_2^\dagger)^{n_2} \cdots (a_k^\dagger)^{n_k} \cdots |0\rangle,$$

where $n_i = 0, 1, 2, \ldots, N$ and $\sum n_k = N$. Since the state must be symmetric under exchange of any two particles, the following commutation rules are fulfilled:

$$[a_i, a_k] = [a_i^\dagger, a_k^\dagger] = 0, \tag{14.22}$$

$$[a_i, a_k^\dagger] = \delta_{ik}, \tag{14.23}$$

By applying such rules, it is easy to show that

$$a_\alpha^\dagger |n_1 n_2 \ldots n_\alpha \ldots\rangle = \sqrt{n_\alpha + 1}\, |n_1 n_2 \ldots (n_\alpha + 1) \ldots\rangle, \tag{14.24}$$

$$a_\alpha|n_1n_2\ldots n_\alpha\ldots\rangle = \sqrt{n_\alpha}|n_1n_2\ldots(n_\alpha-1)\ldots\rangle, \tag{14.25}$$

$$\langle n_1n_2\ldots n_\alpha\ldots|a_\alpha = \sqrt{n_\alpha+1}\langle n_1n_2\ldots(n_\alpha+1)\ldots|, \tag{14.26}$$

$$\langle n_1n_2\ldots n_\alpha\ldots|a_\alpha^\dagger = \sqrt{n_\alpha}\langle n_1n_2\ldots(n_\alpha-1)\ldots|.$$

The state of N identical fermions is written in the same form

$$|n_1n_2\ldots n_k\ldots,\ldots\rangle = \frac{1}{\sqrt{N!}}(a_1^\dagger)^{n_1}(a_2^\dagger)^{n_2}\ldots(a_k^\dagger)^{n_k}\ldots|0\rangle, \tag{14.27}$$

but now $n_i = 0, 1$ for the Pauli principle and $\sum n_k = N$. Since the state must be antisymmetric under exchange of any two particles, the following anti-commutation rules are fulfilled:

$$\{a_i^\dagger, a_k^\dagger\} = \{a_i, a_k\} = 0,$$
$$\{a_i, a_k^\dagger\} = \delta_{ik},$$

where $\{a, b\} \equiv ab + ba$. The operators can also be expressed by the creation and annihilation operators for both fermions and bosons. For the one-body operator $\mathbf{F} = \sum_{a=1}^N \mathbf{f}(a)$, representing the sum of N single-particle operator $\mathbf{f}(a)$, e.g. the total momentum, angular momentum, particle number of the system or kinetic energy, it is

$$\mathbf{F} = \sum_{\alpha\beta} f_{\alpha\beta}a_\alpha^\dagger a_\beta, \tag{14.28}$$

where $f_{\alpha\beta} = \langle\alpha|\mathbf{f}|\beta\rangle = (\phi_\alpha, \mathbf{f}\phi_\beta)$, and for two-body operator $\mathbf{G} = \sum_{a<b}^N \mathbf{g}(a, b)$, e.g. the two-body interaction between particles, it is

$$\mathbf{G} = \frac{1}{2}\sum_{\alpha'\beta'\alpha\beta} g_{\alpha'\beta',\alpha\beta}a_{\alpha'}^\dagger a_{\beta'}^\dagger a_\beta a_\alpha, \tag{14.29}$$

where

$$g_{\alpha'\beta',\alpha\beta} = \langle\alpha'\beta'|\mathbf{g}|\alpha\beta\rangle = (\phi_{\alpha'}(q_1)\phi_{\beta'}(q_2), \mathbf{g}(1,2)\phi_\alpha(q_1)\phi_\beta(q_2)).$$

A typical two-body operator is the interaction between two particles.

14.6 Complement to Sect. 14.4: Isotropic Three-Dimensional Harmonic Oscillator

The isotropic three-dimensional oscillator potential is a central potential, so that the Hamiltonian Eq. (8.8) commutes with the angular momentum operators $\mathbf{L}^2$ and $\mathbf{L}_z$

$$[\mathbf{H}, \mathbf{L}^2] = [\mathbf{H}, \mathbf{L}_z] = 0.$$

As an alternative to Sect. 8.3, we search for the simultaneous eigenstates of $\mathbf{H}$, $\mathbf{L}^2$, and $\mathbf{L}_z$. We refer to them as $|n, l, m\rangle$, with $\left(n + \frac{3}{2}\right)\hbar\omega$, $l(l+1)\hbar^2$ and $m\hbar$ eigenvalues of $\mathbf{H}$, $\mathbf{L}^2$ and $\mathbf{L}_z$, respectively. Following the procedure developed in Sect. 12.1 to treat central potentials, we consider the Schrödinger equation in spherical coordinates

$$\left[-\frac{\hbar^2}{2m}\frac{1}{r}\frac{\partial^2}{\partial r^2}r + \frac{l(l+1)\hbar^2}{2mr^2} + \frac{1}{2}m\omega^2 r^2\right]\Psi_{nlm}(r, \theta, \varphi) = E_{nl}\Psi_{nlm}(r, \theta, \varphi) \tag{14.30}$$

with the wave functions of the form

$$\Psi_{nlm}(r, \theta, \varphi) = R_{nl}(r)Y_{lm}(\theta, \varphi), \tag{14.31}$$

where the angular components Y_{lm} are the eigenfunctions of $\mathbf{L}^2$ and $\mathbf{L}_z$, represented by the spherical harmonics (see Appendix B.2). The physical constraints at $r \sim 0$ and $r \sim \infty$ must be imposed to the radial wave function R_{nl}. According to the analysis of Sect. 12.1, at $r \sim 0$ $rR(r \sim 0) \sim r^{l+1}$. At $r \sim \infty$, R_{nl} must be a solution of the asymptotic radial equation

$$\left[-\frac{\hbar^2}{2m}\frac{1}{r}\frac{\partial^2}{\partial r^2}r + \frac{1}{2}m\omega^2 r^2\right]\Psi_{nlm}(r, \theta, \varphi) = E_{nl}\Psi_{nlm}(r, \theta, \varphi), \tag{14.32}$$

where the centrifugal term is asymptotically negligible. This equation can be easily integrated, giving $R(r \sim \infty) \sim \mathrm{e}^{-m\omega r^2/2\hbar}$. As a result, the full radial wave function is written as

$$r\,R_{nl} \sim r^{l+1}\mathrm{e}^{-\frac{m\omega r^2}{2\hbar}} f_{nl}(r), \tag{14.33}$$

where f_{nl} must be a regular function in the interval between $r = 0$ and $r = \infty$. Plugging this expression into the Schrödinger equation (14.31), we obtain the equation for f_{nl}

$$r^2\frac{\mathrm{d}^2 f_{nl}}{\mathrm{d}r^2} + 2r\left[(l+1) - \frac{m\omega}{\hbar}r^2\right]\frac{\mathrm{d}f_{nl}}{\mathrm{d}r} - \frac{m\omega}{\hbar}(2l+3)f_{nl} = -\frac{2mE_{nl}}{\hbar^2}r^2 f_{nl}. \tag{14.34}$$

Expanding the function f_{nl} in a series of r powers, $f_{nl} = \sum_{k=0}^{\infty} c_k r^k$, the differential equation transforms into an infinite system of algebraic equations for the coefficients c_k of the expansion. It is easy to show that these coefficients are related by the following recurrence relation:

$$(k+2)(k+2l+3)c_{k+2} = \left[(2k+2l+3)\frac{m\omega}{\hbar} - \frac{2mE_{nl}}{\hbar^2}\right]c_k. \tag{14.35}$$

From the constraint at $r = 0$ we have $c_0 \neq 0$ and $c_1 = 0$. The series is explicitly written as

$$2(2l+3)c_2 = \left[(2l+3)\frac{m\omega}{\hbar} - \frac{2mE_{nl}}{\hbar^2}\right]c_0, \tag{14.36}$$

$$3(2l+4)c_3 = \left[(2l+5)\frac{m\omega}{\hbar} - \frac{2mE_{nl}}{\hbar^2}\right]c_1, \tag{14.37}$$

$$4(2l+5)c_4 = \left[(2l+7)\frac{m\omega}{\hbar} - \frac{2mE_{nl}}{\hbar^2}\right]c_2, \tag{14.38}$$

$$4(2l+5)c_5 = \left[(2l+9)\frac{m\omega}{\hbar} - \frac{2mE_{nl}}{\hbar^2}\right]c_3. \tag{14.39}$$

$$\cdots\cdots.$$

We see that all odd-power coefficients vanish because $c_1 = 0$. Afterwards, with a procedure similar to that for the hydrogen bound states, we show that the energy quantization stems from imposing the convergence of the eigenfunctions at $r \to \infty$. Indeed, if

$$(2l+3)\frac{m\omega}{\hbar} = \frac{2mE_{nl}}{\hbar^2}, \tag{14.40}$$

then $c_2 = c_4 = \cdots = 0$ and $f_{nl} = c_0$; if

$$(2l+7)\frac{m\omega}{\hbar} = \frac{2mE_{nl}}{\hbar^2}, \tag{14.41}$$

then $c_4 = c_6 = \cdots = 0$ and $f_{nl} = c_0 + c_2 r^2$. Iterating this procedure, we can conclude that, for quantized values of the energy, the function f_{nl} becomes a polynomial, and hence the radial wave function decreases exponentially at $r \to \infty$ according to Eq. (14.33), as a bound state should do. Fixing l, the energy spectrum is

$$E_n = \hbar\omega\left(k + l + \frac{3}{2}\right) = \hbar\omega\left(n + \frac{3}{2}\right), \tag{14.42}$$

where $n = k + l$ ($n = 0, 1, 2, \ldots$) and even integer $k \geqslant 0$ ($k = 0, 2, 4, \ldots$). Therefore, the spectrum is degenerate with $(n+1)(n+2)/2$ angular momentum eigenvalues belonging to the same E_n, and $(2l+1)$ z-projection angular momentum eigenvalues belonging to the same l.

Figure 14.6 displays the first five energy levels of the isotropic three-dimensional harmonic oscillator, the letters: s, p, d, f, and g refer to the values of the angular momentum quantum number l.

Comparing with the hydrogen atom spectrum of Fig. 12.2, we note that the structure of the two spectra is the same, as a consequence of the common central symmetry of the two potentials. However, the energy levels are equally spaced in the harmonic oscillator spectrum, reminiscent of the one-dimensional case.

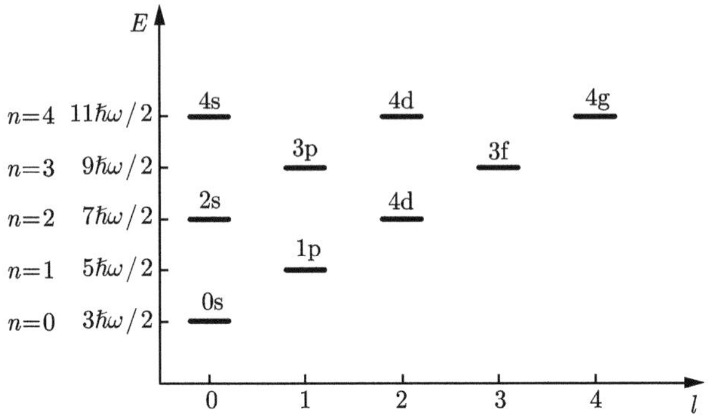

Fig. 14.6 Energy levels of the three-dimensional isotropic harmonic oscillator

14.7 Notes and References

SMT Goeppert Mayer, M.: Phys. Rev. **75**, 1969 (1949); **78**, 16 (1950).

SMT Axel, O., Jensen, J.H.D., Suess, H.E.: Phys. Rev. **55**, 1766 (1949); Z. Phyzik **128**, 295 (1950).

BET Bethe, H.: Theory of nuclear matter. Ann. Rev. Nuclear Sci. **21**, 93–244 (1971).

BOU Due to the spin dependence of the nuclear force (see Chaps. 12.3 and 13.12) the $T = 0$ ($V_{\rm np}$) component of the nuclear force is more attractive than the $T = 1$ ($V_{\rm nn} = V_{\rm pp} = V_{\rm np}$) component. Thus, a decrease in neutron–proton ratio reduces the nuclear binding energy of a nucleus with atomic number A.

WAN Wang, M., Audi, G., Kondev, F.G., Huang, W.J., Naimi, S., Xu, X.: Chin. Phys. C **41**, 030003 (2017).

GIA A review of applications in nuclear physics and astrophysics is reported in Giuliani, G., Zheng, H., Bonasera, A.: Progress Part. Nucl. Phys. **76**, 116 (2014).

BRU In the Brueckner theory the nuclear-matter energy, originally expressed in terms of the *in vacuum* interaction, is recast in terms of an effective interaction, the so-called G-matrix, which incorporates short-range correlations such as to prevent two nucleons from being a very short distance apart, as a consequence of the Pauli principle. A review of this theory is in: Brückner, K.A.: Theory of the nuclear structure. In: De Witt, C. (Ed.), The Many Body Problem, p. 47. Wiley (1959).

KOH Kohn, W., Sham, L.J.: Phys. Rev. **140**, A1133 (1965).

HAT Liang, H., Niu, Y.F., Hatsuda, T.: Phys. Lett. B **779**, 436 (2018).

VAU Vautherin, D., Brink, D.M.: Hartree–Fock calculations with Skyrme's interaction. I. Spherical nuclei. Phys. Rev. C **5**, 626 (1972).

14.8 Problems

- Problem 1. Calculate the exchange electron–electron interaction in the ^{4}He atom.
- Problem 2. Calculate the angular part of the matrix elements of the spin–orbit interaction $\langle\psi_{nlsjj_z}|V_{so}|\psi_{nlsjj_z}\rangle$ with $j = l \pm 1/2$ for the electron of the hydrogen-like ion in the $nlsj$ orbitals. (Hint: Use the Clebsch–Gordan coefficients that couple orbital angular momentum and spin.)
- Problem 3. Calculate the effect on the atomic spectrum of a Hydrogen-like ion due to interaction between the electron spin and the nuclear spin (hyperfine structure).
- Problem 4. Let us consider a nucleus with A nucleons in its ground state. The proton–proton repulsive Coulomb force tends to destabilize the nucleus. Using the Bethe–Weizsäcker (see Sect. 14.4.2), calculate the maximum number of protons Z_c the nucleus can bear without destabilizing. Comment the result as regards the mass number A and compare with some data from the nuclear mass table.
- Problem 5. Let us describe asymmetric nuclear matter as a degenerate Fermi gas made up of two components: neutron gas with density ρ_n and proton gas with density ρ_p. Using Eq. (16.29), write the expression of the energy per particle of asymmetric nuclear matter as a function of the total density $\rho = \rho_n + \rho_p$ and of the neutron–proton imbalance $\beta = \frac{N-Z}{A}$.

Chapter 15
The Dirac Equation

Quantum Mechanics describes systems moving at a speed much lower than the speed of light c, i.e. $\frac{v^2}{c^2} \ll \frac{v}{c}$. However, in the microscopic world, we often have to deal with high speeds. Hence, we must extend the Schrödinger equation to the relativistic limit. Carrying out such a goal, we shall encounter the spin degree of freedom, which restores the conservation of angular momentum, as soon as the spin is added to the orbital part. Afterwards, we shall derive the spin–orbit interaction, as a relativistic effect on spinful particles like electrons and nucleons, already introduced in the atomic and nuclear spectroscopy (see Chap. 14).

15.1 The Klein–Gordon Equation

The generator of the Lorentz transformations, considered as space–time translations, is the energy-momentum four-vector. In classical mechanics energy and momentum of a free particle are related by the equation $E = \frac{p^2}{2m}$, m being the particle rest mass. In relativistic mechanics the corresponding equation is $\frac{E^2}{c^2} = m^2c^2 + p^2$, such relation is often called the *mass-shell condition*. From the correspondences $E \to \mathrm{i}\hbar\frac{\partial}{\partial t}$ and $\vec{p} \to -\mathrm{i}\hbar\vec{\nabla}$ the time-dependent Schrödinger equation $\mathrm{i}\hbar\frac{\partial\psi}{\partial t} = -\frac{\hbar^2}{2m}\nabla^2\psi(\vec{r}, t)$ transforms into the Klein–Gordon equation

$$\left(\nabla^2 - \frac{\partial^2}{\partial t^2}\right)\psi(x) = m^2\psi(x),$$

Supplementary Information The online version contains supplementary material available at https://doi.org/10.1007/978-981-97-1670-8_15.

U. Lombardo et al., *Quantum Mechanics*, UNITEXT for Physics,
https://doi.org/10.1007/978-981-97-1670-8_15

where $x \equiv (\vec{r}, t)$ and $\hbar = c = 1$. Solutions of the KG equation are the plane waves $\psi(x) = \mathrm{e}^{-\mathrm{i}px}$, with $px = \vec{p} \cdot \vec{r} - Et$, being $E = \pm\sqrt{p^2 + m^2}$ the energy eigenvalues. There are a couple of drawbacks to the two eigenvalues. First, negative energy states should be rejected because free particles have $E \geqslant 0$, even if transitions from positive to negative energy states could be possible in case of interacting particles. A second drawback is related to the probabilistic nature of the wave function. Indeed, we consider the equations obtained by multiplying the Klein–Gordon equation on the left by $-\mathrm{i}\psi^*$, and its complex conjugate on the left times $-\mathrm{i}\psi$. Subtracting the two equations from one another, we get the relativistic version of the continuity equation

$$\partial_t \left[\frac{\mathrm{i}}{2m} (\psi^* \partial_t \psi - \psi \partial_t \psi^*) \right] + \vec{\nabla} \cdot \left[\frac{\mathrm{i}}{2m} (\psi^* \vec{\nabla} \psi - \psi \vec{\nabla} \psi^*) \right] = 0, \tag{15.1}$$

where $\dfrac{\partial}{\partial t} \equiv \partial_t$. The denominator $2m$ has been introduced to define probability density ρ and current j as follows:

$$\rho = \frac{\mathrm{i}}{2m} (\psi^* \partial_t \psi - \psi \partial_t \psi^*),$$
$$\vec{j} = \frac{1}{2\mathrm{i}m} (\psi^* \vec{\nabla} \psi - \psi \vec{\nabla} \psi^*) \cdot$$

The probability density is not necessarily positive, because it depends on the time derivative of the wave function. In particular, for plane waves, $\rho = \dfrac{E}{m} |\psi|^2$ is negative for negative energies. Negative probability is nonsense! This difficulty could be overcome, if we consider charged particles and introduce the charge density $-e\rho$ and the charged current $-ej$, where $e > 0$ is the absolute value of the charge. In such a way, the charge density is always positive if we associate particles with negative charge to $E > 0$ and particles with positive charge to $E < 0$. This consideration suggests the existence of an electron with positive charge. Indeed, it does exist, known as a positron! It was discovered [POS] after Dirac had made such a prediction in the context discussed below. Despite the two drawbacks, there is a case in which the Klein–Gordon equation can hold a physical meaning, i.e. with the massless particles. This happens with photons. As a matter of fact, within the quantum theory of the electromagnetic field, the photon wave function is identified with the four-vector potential $A_\mu(x)$ of the electromagnetic field. Thus, the Klein–Gordon equation

$$\left(\nabla^2 - \frac{\partial^2}{\partial t^2} \right) A_\mu(x) = 0$$

identifies with the classical Maxwell equations in the absence of charges and currents.

15.2 Dirac Equation

In order to overcome the drawbacks of the Klein–Gordon equation, we should require the new equation to be linear in the time derivative and, due to the covariance constraint, it has to be linear also in the space derivative. This implies that the relativistic Hamiltonian must be linear in the momentum variable, that is: $\mathbf{H} = \vec{\alpha} \cdot \vec{p} + m\beta$.[1] The two constants α and β must be independent of space and time because of the space–time symmetries. Thus, the new form of the equation of motion of a free particle is

$$\mathrm{i}\partial_t \psi(x) = \mathrm{H}\psi(x) = (\vec{\alpha} \cdot \vec{p} + m\beta)\psi(x). \tag{15.2}$$

Equation (15.2) is known as the Dirac equation. Recalling the mass-shell condition, and applying $\mathrm{i}\partial_t$ to the Dirac equation, one obtains the identity

$$-\partial_t^2 \psi(x) = (\vec{\alpha} \cdot \vec{p} + m\beta)\mathrm{i}\partial_t \psi(x) = (\vec{\alpha} \cdot \vec{p} + m\beta)^2 \psi(x) = (p^2 + m^2)\psi(x),$$

which enables us to determine the coefficients α and β. One immediately has $m^2\beta^2 = m^2$ and $\vec{\alpha}\beta + \beta\vec{\alpha} = 0$, for which the coefficients cannot be integer numbers but matrices, otherwise $\vec{\alpha} = -\vec{\alpha}$. The other relations are easily obtained

$$\alpha_k^2 = 1, \quad \beta^2 = 1,$$

$$\alpha_k \alpha_l + \alpha_l \alpha_k = 0, \quad k \neq l,$$

$$\alpha_k \beta + \beta \alpha_k = 0.$$

The wave functions $\psi_\sigma(x)$, solutions of the Dirac equation, are N-component vectors. In order to determine such components, let us define the matrices σ_k and τ_k

$$\alpha_k = \tau_1 \cdot \sigma_k, \quad \beta = \tau_3, \quad \tau_2 = \mathrm{i}\tau_1 \cdot \tau_3.$$

The σ and τ matrices have the following properties:

1. The τ matrices are Hermitian, they anti-commute and $\tau_i^2 = 1$;
2. The σ matrices are Hermitian, they anti-commute and $\sigma_i^2 = 1$;
3. The matrices τ and σ commute.

In accordance with property 1, $\tau_3^2 = 1$ so that the eigenvalues are ± 1 and the eigenvectors $|\pm 1\rangle$. The matrices τ_1 and τ_2 transform their eigenvectors into the eigenvectors of the other matrix. It follows that the two-dimensional space described by the $|\pm 1\rangle$ vectors is closed for the τ_k matrices, and these matrices are two-dimensional matrices. The σ_k matrices are two-dimensional matrices as well. Finally, since τ and σ commute (property 3), the space of their common eigenstates is the tensor

---

[1] Henceforth, we do not use the bold symbols for operators unless necessary.

product of their respective states. Thus, the space of their common eigenstates is a four-dimensional space. It follows that since the Hamiltonian has to be a Hermitian operator, the α and β matrices also have to be Hermitian.

15.2.1 Diagonalization of the Hamiltonian

Several representations of the matrices α_k and β exist. The most popular, which is adopted here, is the Dirac–Pauli one

$$\vec{\alpha} = \begin{pmatrix} 0 & \vec{\sigma} \\ \vec{\sigma} & 0 \end{pmatrix}, \quad \beta = \begin{pmatrix} 1 & 0 \\ 0 & -1 \end{pmatrix},$$

where 0, 1, and $\vec{\sigma}$ are 2×2 sub-matrices, in particular, the components of $\vec{\sigma}$: σ_x, σ_y, and σ_z are the so-called Pauli matrices, which are defined as follows:

$$\sigma_x = \begin{pmatrix} 0 & 1 \\ 1 & 0 \end{pmatrix}, \quad \sigma_y = \begin{pmatrix} 0 & -\mathrm{i} \\ \mathrm{i} & 0 \end{pmatrix}, \quad \sigma_z = \begin{pmatrix} 1 & 0 \\ 0 & -1 \end{pmatrix} \cdot \tag{15.3}$$

In this representation, the stationary Dirac equation takes the form

$$\begin{pmatrix} m & \vec{p} \cdot \vec{\sigma} \\ \vec{p} \cdot \vec{\sigma} & -m \end{pmatrix} \begin{pmatrix} u \\ v \end{pmatrix} = E \begin{pmatrix} u \\ v \end{pmatrix},$$

where the eigenvectors are expressed in terms of column vectors whose components are u and v, which are two-component column sub-vectors. The four-dimensional Dirac equation decouples into two coupled two-dimensional equations

$$(m - E)u + \vec{\mathrm{p}} \cdot \vec{\sigma} v = 0,$$
$$(-m - E)v + \vec{\mathrm{p}} \cdot \vec{\sigma} u = 0 \cdot$$

In general, there exist four independent eigenvectors for each momentum vector $\vec{p}$. For $\vec{p} = 0$, the Hamiltonian is diagonal, and its energy eigenvalues are $E_0 = m, m, -m, -m$, while the corresponding eigenvectors are

$$\begin{pmatrix} 1 \\ 0 \\ 0 \\ 0 \end{pmatrix}, \begin{pmatrix} 0 \\ 1 \\ 0 \\ 0 \end{pmatrix}, \begin{pmatrix} 0 \\ 0 \\ 1 \\ 0 \end{pmatrix}, \begin{pmatrix} 0 \\ 0 \\ 0 \\ 1 \end{pmatrix}.$$

For $\vec{p} \neq 0$, the secular equation again provides four energy eigenvalues $E, E, -E, -E$ with $E = \sqrt{p^2 + m^2}$ with the corresponding eigenvectors

$$\begin{pmatrix} 1 \\ 0 \\ \dfrac{\vec{\sigma}\cdot\vec{p}}{E+m} \\ 0 \end{pmatrix}, \begin{pmatrix} 0 \\ 1 \\ 0 \\ \dfrac{\vec{\sigma}\cdot\vec{p}}{E+m} \end{pmatrix}, \begin{pmatrix} \dfrac{\vec{\sigma}\cdot\vec{p}}{E-m} \\ 0 \\ 1 \\ 0 \end{pmatrix}, \begin{pmatrix} 0 \\ \dfrac{\vec{\sigma}\cdot\vec{p}}{E-m} \\ 0 \\ 1 \end{pmatrix}.$$

The eigenvectors are orthogonal, but they should be normalized to unity to be orthonormal. Hence, the Dirac equation admits four eigenstates, two with positive energy and two with negative energy. Besides the drawback of the negative energy, that can only be solved introducing the anti-particles, there is the problem of the double degeneracy of the two eigenvalues.

15.2.2 The Spin Variable

The double degeneracy of the two energy eigenvalues hides one more observable, which can remove such a degeneracy. Let us consider, indeed, the observable $\vec{\Sigma}\cdot\hat{p}$ represented by the 4×4 matrix

$$\vec{\Sigma}\cdot\hat{p} = \begin{pmatrix} \vec{\sigma}\cdot\hat{p} & 0 \\ 0 & \vec{\sigma}\cdot\hat{p} \end{pmatrix},$$

where $\hat{p} = \vec{p}/p$. It is a constant of motion, because it commutes with the Hamiltonian; it also commutes with the momentum, thus $\vec{p}$ and $\vec{\Sigma}\cdot\hat{p}$ form a complete set of observables for the free particle. The eigenvalues of $\vec{\Sigma}\cdot\hat{p}$ are $\lambda = +1$, if $\vec{\Sigma}\cdot\hat{p}$ is parallel to $\vec{p}$, and $\lambda = -1$, if $\vec{\Sigma}\cdot\hat{p}$ is anti-parallel to $\vec{p}$. The $\vec{\Sigma}$, in the $\hbar$ units, is named spin and it can be interpreted as an intrinsic angular momentum of the particle. Such an interpretation is justified by the fact that it has the same commutation rules as the orbital angular momentum $\vec{l}$, even if it is not a true angular momentum (see Sect. 6.6). The observable $\vec{\Sigma}\cdot\hat{p}$ is the projection of the spin in the direction of motion $\hat{p}$ and it is known as *helicity*. Since it takes only two values, the spin modulus is $\dfrac{1}{2}\hbar$ for the electrons.[2] The helicity is the counterpart of the circular (clockwise or counterclockwise) polarization of photons.

The relativistic Hamiltonian does not commute with the orbital angular momentum

$$[\mathbf{H}, \vec{\mathbf{l}}] = -i\vec{\alpha}\times\vec{\mathbf{p}},$$

so that it is no longer a conserved observable for a free particle. However, combining this commutator with that of Hamiltonian with spin

$$[\mathbf{H}, \vec{\Sigma}] = 2i\vec{\alpha}\times\vec{\mathbf{p}},$$

[2] In the more general context of elementary particle physics, the spin can be different from $\dfrac{1}{2}\hbar$.

we get

$$[\mathbf{H}, \vec{\mathbf{J}}\,] = 0,$$

where we have introduced the total angular momentum $\vec{\mathbf{J}} = \vec{\mathbf{l}} + \frac{1}{2}\vec{\Sigma}$. As a result, the conservation of the angular momentum is restored as a conservation of the total angular momentum $\vec{\mathbf{J}}$ in the presence of spin. This property is in keeping with an analogous symmetry of the fundamental inter-particle interactions. In conclusion, the wave function of a free particle is completely defined by momentum $\vec{p}$ and spin projection and the total angular momentum is also a constant of motion.

15.3 Covariant Form of the Dirac Equation

The Dirac equation is form-invariant under Lorentz transformations. Thus, it can be recast in the four-dimensional representation, named also covariant form. To express it in the covariant form, we define

$$\begin{aligned}\gamma^0 &= \beta \ , \ \gamma^k = \beta \alpha^k, \\ 2g^{\nu\mu} &= \gamma^\nu \gamma^\mu + \gamma^\mu \gamma^\nu,\end{aligned}$$

where $g^{\nu\mu}$ is the four-dimensional metric tensor. Multiplying Eq. (15.2) times β, we obtain the covariant form of the Dirac equation

$$[\mathrm{i}\gamma^\mu \partial_\mu - m]\psi = 0. \tag{15.4}$$

To establish the continuity equation, we consider the adjoint of Eq. (15.4)

$$-\,\mathrm{i}\partial_\mu \psi^\dagger \gamma^{\mu^\dagger} - m\psi^\dagger = 0.$$

$\psi^\dagger$ is the row vector whose components are the complex conjugates of the ψ vector components. By exploiting the property $\gamma^{\mu^\dagger} = \gamma^0 \gamma^\mu \gamma^0$, and adopting the definition $\bar{\psi} = \psi^\dagger \gamma^0$, we obtain the conjugate equation

$$-\,\mathrm{i}\partial_\mu \bar{\psi} \gamma^\mu = m\bar{\psi}. \tag{15.5}$$

By multiplying to the left of Eq. (15.4) times $\bar{\psi}$, and to the right of Eq. (15.5) times ψ, and then subtracting term by term the two resulting expressions, we get to the covariant form of the continuity equation:

$$\mathrm{i}\partial_\mu (\bar{\psi} \gamma^\mu \psi) = 0.$$

The four-vector within brackets has to be identified with the density of charge-current four-vector $j^\mu = \bar{\psi}\gamma^\mu\psi$. Precisely, the zeroth component $j^0 = \bar{\psi}\gamma^0\psi = \psi^\dagger\psi = |\psi|^2$

is the probability density, which is devoid of the difficulties carried out by the Klein–Gordon equation. The vector $j^k = \bar{\psi}\gamma^k\psi = \psi^\dagger\gamma^0\gamma^k\psi = \psi^\dagger\alpha^k\psi$ instead represents the probability current vector. Indeed, the α^k vector is just the velocity vector. This can be seen from the equations of motion

$$\frac{\mathrm{d}\vec{r}}{\mathrm{d}t} = -\mathrm{i}[\vec{r}, \mathbf{H}] = \vec{\alpha}.$$

The Dirac equation does not suffer from the drawback of the Klein–Gordon equation with regard to the continuity equation. This latter, in non-covariant form, is proven as follows. Bearing in mind that $\vec{\mathbf{p}} = -\mathrm{i}\vec{\nabla}$, let us multiply the Dirac equation on the right times $\psi^\dagger$, and its Hermitian conjugate on the left times ψ

$$\psi^\dagger \mathrm{i}\partial_t\psi = -\mathrm{i}\psi^\dagger\vec{\alpha}\cdot(\vec{\nabla}\psi) + \psi^\dagger m\beta\psi,$$
$$(-\mathrm{i}\partial_t\psi^\dagger)\psi = \mathrm{i}(\vec{\nabla}\cdot\psi^\dagger)\vec{\alpha}\psi + m\psi^\dagger\beta\psi,$$

and subtract term to term. One obtains

$$\partial_t(\psi^\dagger\psi) + \vec{\nabla}\cdot(\psi^\dagger\vec{\alpha}\psi) = 0.$$

Introducing the relativistic probability density as well as the current

$$\rho = \psi^\dagger\psi,$$
$$\vec{J} = \psi^\dagger\vec{\alpha}\psi,$$

respectively, we recover the continuity equation in the usual form

$$\partial_t\rho + \vec{\nabla}\cdot\vec{J} = 0,$$

where the probability density is never negative at variance with the Klein–Gordon case. However, the drawback of negative energy states is still unsolved. Dirac speculated that these states (Dirac sea) are all occupied by electrons (see Fig. 15.1). Adding a particle with negative charge (electron), which is a fermion, it can only occupy a positive energy state thanks to the Pauli principle. Removing a particle with negative energy, a hole is created in the Dirac sea that can be interpreted as a particle with positive charge (positron) and positive energy. This interpretation was confirmed by the observed decay of a photon of high energy E_γ into an electron–positron couple, as shown in Fig. 15.1. However, the Dirac sea model had so many inconsistencies that it was soon abandoned and later replaced by quantum field theory.

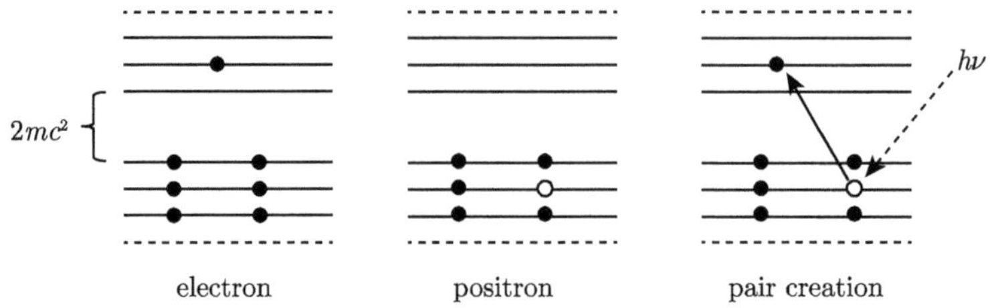

Fig. 15.1 The Dirac sea made of negative and positive energy states separated by the gap of $2mc^2$

15.4 The Spin–Orbit Interaction

The spin–orbit interaction has a relativistic nature, the spin being derived from the Dirac equation. Nevertheless, it can be qualitatively explained with classical concepts. Indeed, an electron orbiting around the nucleus, say the hydrogen nucleus, is equivalent to the nuclear proton orbiting around the electron with the same rotation period T. The proton magnetic field is the same as that of a current loop, $B \sim i$, where in this case the current is $i = e/T$. In turn, the period T is related to the orbital angular momentum of the electron by the relation $l = mvr = 2\pi mr^2/T$, r being the Bohr radius and v the electron speed. As a result, the proton magnetic field turns out to be proportional to $\vec{l}$. Since the electron has a magnetic moment proportional to the spin, a new interaction $V_{so} \sim \vec{\sigma} \cdot \vec{l}$ has to be added to the Coulomb interaction between electron and nucleus. V_{so} is called spin–orbit interaction.

The general form of spin–orbit interaction will be deduced now from the Dirac equation. Let us consider the stationary Dirac equation in an external Coulomb field Φ. It can be split into two two-dimensional coupled equations

$$(\varepsilon - e\Phi)u = \vec{\mathbf{p}} \cdot \vec{\sigma} v, \tag{15.6}$$

$$(\varepsilon - e\Phi + 2m)v = \vec{\mathbf{p}} \cdot \vec{\sigma} u, \tag{15.7}$$

where $\varepsilon = E - m$. With the introduction of the wave function, we get

$$\begin{aligned} \int \mathrm{d}V |\psi|^2 &\equiv \int \mathrm{d}V (|u|^2 + |v|^2) \\ &= \int \mathrm{d}V \left[|u|^2 + \frac{1}{(\varepsilon - e\Phi + 2m)^2} |\vec{\sigma} \cdot \vec{\mathbf{p}} u|^2 \right] \\ &\approx \int \mathrm{d}V \left(|u|^2 + \frac{1}{4m^2} |\vec{\sigma} \cdot \vec{\mathbf{p}} u|^2 \right) . \end{aligned}$$

We have neglected the term $\varepsilon - e\Phi$, because it is small as compared to the electron rest mass m. Integrating by parts the term including $\vec{p}u = -\mathrm{i}\vec{\nabla} u$, one obtains

$$\int \mathrm{d}V|\psi|^2 \equiv \int \mathrm{d}V \left[u^*u + \frac{1}{8m^2}(u^*\mathbf{p}^2 u + u\mathbf{p}^2 u^*)\right],$$

and to the order $1/mc^2$

$$\psi = \left(1 + \frac{\mathbf{p}^2}{8m^2}\right) u, \tag{15.8}$$

$$u = \left(1 - \frac{\mathbf{p}^2}{8m^2}\right) \psi\,. \tag{15.9}$$

After resuming the exact expression of v and applying it in the equation giving u (see Eqs. (15.6) and (15.7)), one gets

$$(\varepsilon - e\Phi)u = (\vec{\sigma}\cdot\vec{\mathbf{p}})\frac{1}{\varepsilon - e\Phi + 2m}(\vec{\sigma}\cdot\vec{\mathbf{p}})u, \tag{15.10}$$

then, expressing u in terms of ψ, after easy calculations (see Sect. 15.5), one recovers the Schrödinger equation in the form $H\psi = \varepsilon\psi$, where the Hamiltonian is given by

$$\mathbf{H} = \frac{\mathbf{p}^2}{2m} + e\Phi - \frac{\mathbf{p}^4}{8m^3} - \frac{e}{4m^2}\vec{\sigma}\cdot(\vec{E}\times\vec{\mathbf{p}}) - \frac{e}{8m^2}\vec{\nabla}\cdot\vec{E},$$

where $\vec{E} = -\vec{\nabla}\Phi$. The term to the order p^4 is the relativistic correction to the kinetic energy, the last term (Darwin term) is non-vanishing only in the regions where the electric charges are located. The fourth term is named spin–orbit interaction. In a spherical symmetric field, it takes the form

$$V_{\mathrm{so}} = -\frac{e}{4m^2 r}\frac{\mathrm{d}\Phi}{\mathrm{d}r}\vec{s}\cdot\vec{l}, \tag{15.11}$$

where the spin $\vec{s} = 2\vec{\sigma}$ has been introduced. The effect of the spin–orbit interaction is the level splitting of, e.g. the p orbital in atomic and nuclear spectroscopy, as discussed in Chap. 14.

15.5 Complement to Sect. 15.4: Semi-classical Hamiltonian

Let us write an expansion in powers of $\frac{1}{mc}$ of Eq. (15.10) up to the $\frac{1}{(mc)^2}$ order (for more details see [LAN]).

$$\begin{aligned}(\varepsilon - e\Phi)u &= (\vec{\sigma}\cdot\vec{\mathbf{p}})\frac{1}{\varepsilon - e\Phi + 2m}(\vec{\sigma}\cdot\vec{\mathbf{p}})u \\ &= \frac{1}{2m}(\vec{\sigma}\cdot\vec{\mathbf{p}})\left[1 - \frac{(\varepsilon - e\Phi)}{2m}\right](\vec{\sigma}\cdot\vec{\mathbf{p}})u.\end{aligned} \tag{15.12}$$

By using Eq. (15.9), we can express u in terms of ψ. Doing so, Eq. (15.12) becomes a homogeneous equation in ψ

$$(\varepsilon - e\Phi)\left(1 - \frac{\mathbf{p}^2}{8m^2}\right)\psi$$
$$= \frac{1}{2m}(\vec{\sigma}\cdot\vec{\mathbf{p}})\left[1 - \frac{(\varepsilon - e\Phi)}{2m}\right](\vec{\sigma}\cdot\vec{\mathbf{p}})\left(1 - \frac{\mathbf{p}^2}{8m^2}\right)\psi. \quad (15.13)$$

By calculating the products, and collecting the several terms obtained according to the form $H\psi = \varepsilon\psi$, Eq. (15.13) turns into

$$\varepsilon\left(1 + \frac{\mathbf{p}^2}{8m^2}\right)\psi = \frac{\mathbf{p}^2}{2m}\psi - \frac{\mathbf{p}^4}{16m^3}\psi + e\Phi\left(1 - \frac{\mathbf{p}^2}{8m^2}\right)\psi$$
$$+\frac{1}{4m^2}(\vec{\sigma}\cdot\vec{\mathbf{p}})e\Phi(\vec{\sigma}\cdot\vec{\mathbf{p}})\psi\,. \quad (15.14)$$

To obtain Eq. (15.14), we have used the property of the Pauli matrices $(\vec{\sigma}\cdot\vec{\mathbf{A}})(\vec{\sigma}\cdot\vec{\mathbf{B}}) = \vec{\mathbf{A}}\cdot\vec{\mathbf{B}} + \mathrm{i}\vec{\sigma}\cdot(\vec{\mathbf{A}}\times\vec{\mathbf{B}})$, so that $(\vec{\sigma}\cdot\vec{\mathbf{p}})(\vec{\sigma}\cdot\vec{\mathbf{p}}) = \mathbf{p}^2$. By multiplying both sides of Eq. (15.14) times $\left(1 - \frac{\mathbf{p}^2}{8m^2}\right)$, and neglecting the higher than $\frac{1}{(mc)^2}$ terms, we get

$$\varepsilon\psi = \frac{\mathbf{p}^2}{2m}\psi - \frac{\mathbf{p}^4}{8m^3}\psi + e\Phi\psi + \frac{e}{8m^2}(\Phi\mathbf{p}^2 - \mathbf{p}^2\Phi)\psi + \frac{e}{4m^2}\mathrm{i}(\vec{\sigma}\cdot\vec{E})(\vec{\sigma}\cdot\vec{p})\psi, \quad (15.15)$$

where $\left(1 - \frac{\mathbf{p}^2}{8m^2}\right)\left(1 + \frac{\mathbf{p}^2}{8m^2}\right) \approx 1$. Since $(\vec{\sigma}\cdot\vec{E})(\vec{\sigma}\cdot\vec{\mathbf{p}}) = \vec{E}\cdot\vec{\mathbf{p}} + \mathrm{i}\vec{\sigma}\cdot(\vec{E}\times\vec{\mathbf{p}})$ and $\Phi\mathbf{p}^2 - \mathbf{p}^2\Phi = \nabla^2\Phi - 2\mathrm{i}\vec{E}\cdot\vec{\mathbf{p}}$, we finally obtain the Hamiltonian

$$\mathbf{H} = \frac{\mathbf{p}^2}{2m} - \frac{\mathbf{p}^4}{8m^3} + e\Phi - \frac{e}{8m^2}\vec{\nabla}\cdot\vec{E} - \frac{e}{4m^2}\vec{\sigma}\cdot(\vec{E}\times\vec{\mathbf{p}})\,.$$

15.6 Notes and References

EMP The four-potential $A_\mu(x) \equiv (\vec{A}, \mathrm{i}\phi)$ is the relativistic version of the electromagnetic field. The electric and magnetic fields are derived by A_μ as follows: $\vec{H} = \mathrm{rot}\vec{A}$ and $\vec{E} = -\nabla\phi - \partial\vec{A}/\partial t$.

DIR Dirac, P.A.M.: Proc. Royal Soc. **117**, 610 (1928).

POS Anderson, C.D.: Phys. Rev. **43**, 491 (1933).

LAN Landau, L.D., Lifshitz, E.M.: Quantum Electrodynamics. Pergamon Press (1982), Sect. 4.33.

15.7 Problems

- Problem 1. Show that the spin observable arises from the breakdown of the rotational symmetry. (Hint: Apply an infinitesimal rotation to the Dirac Hamiltonian).
- Problem 2. Prove the following properties of the Pauli matrices:

  1. $(\vec{\sigma} \cdot \vec{p})(\vec{\sigma} \cdot \vec{q}) = \vec{p} \cdot \vec{q} + \mathrm{i}\vec{\sigma} \cdot (\vec{p} \times \vec{q})$. And for $\vec{p} = \vec{q}$: $(\vec{\sigma} \cdot \vec{p})(\vec{\sigma} \cdot \vec{p}) = p^2$.
  2. $\sigma_j \sigma_i - \sigma_i \sigma_j = 2\mathrm{i} \sum_k \epsilon_{jik} \sigma_k$.

- Problem 3. Determine the normalization factor of the Dirac spinors.
- Problem 4. Determine the kinetic energy of an electron both in the ultra-relativistic ($p >> mc$) and non-relativistic limits ($p << mc$).
- Problem 5. Calculate the relativistic density $\rho = \psi^\dagger \psi$ and current $\vec{J} = \psi^\dagger \vec{\alpha} \psi$ of a free electron and compare with the non-relativistic case.

Chapter 16
Homogeneous Many-Body Systems

In this chapter, we shall deal with systems whose number of particles N is of the order of the Avogadro number. In these systems, i.e. quantum fluids and nuclear matter, the volume component is weakly coupled to the surface, in the case of particles interacting via a short-range force. This justifies the extrapolation to an infinite system of particles through the thermodynamic limit, i.e. a system with N particles, $N \to \infty$, in a volume V, $V \to \infty$ such that the density $\rho = \frac{N}{V}$ is finite. We shall adopt the methods of quantum statistical mechanics with the purpose of establishing a connection with the laws of thermodynamics.

16.1 Gibbs Statistical Approach

In principle, the study of systems of many interacting particles can be carried out by solving a certain number of equations of motion coupled by the interparticle interaction. If the number of particles is quite large, as large as Avogadro's number, as it occurs in atomic or molecular gases, in practice, this is not possible. But, even if it were, it would be a hard task to interpret the results. On the contrary, thermodynamics, which deals with such systems, provides very simple laws able to calculate their global physical properties, i.e. the average behaviour of particles rather than the behaviour of the individual particles.

Suitable methods have been elaborated in statistical mechanics, the most general one being the Gibbs method. It can be applied to classical as well as quantum systems. But, for identical particles, quantum mechanics requires (see Chap. 14) their states to be symmetrized (in the case of bosons), or anti-symmetrized (in case of fermions).

Supplementary Information The online version contains supplementary material available at https://doi.org/10.1007/978-981-97-1670-8_16.

U. Lombardo et al., *Quantum Mechanics*, UNITEXT for Physics,
https://doi.org/10.1007/978-981-97-1670-8_16

In particular, fermions fulfill the Pauli principle, that is two fermions cannot occupy the same state. Its first empirical confirmation was the atomic electron structure of the Mendeleev periodic table.

16.2 Time Average and Statistical Average

Let us assume we can trace the dynamical evolution of each particle belonging to an interacting system of N particles with total energy E, occupying the volume V. Fixed values of N, E and V characterize the boundary conditions of an isolated system, i.e. it does not interact with the external world. As time passes, each particle crosses its own trajectory in the particle phase-space, which is defined as the space of coordinates and momenta of individual particles. At the same time, the system crosses a trajectory in the system phase-space, which consists of the coordinates and momenta of all particles. The trajectory of the whole system is quite cumbersome, resulting from the individual trajectories of a huge number of particles. We can imagine a time interval T so large that such a trajectory crosses all the states of the subspace $\mathcal{H}$ of phase-space compatible with the boundary conditions. Then, we can also imagine dividing 'geometrically' such a subspace into n elementary cells $\mathcal{H}_i$ $(i = 1, \ldots, n)$. Running across its path the system crosses the cell $\mathcal{H}_i$ in a fraction δt_i of the full-time interval T. The ratio $\delta t_i/T$ can be interpreted as the probability P_i of finding the system inside the cell $\mathcal{H}_i$. Let us consider now the state of the system, i.e. coordinates and momenta of all particles inside the cell $\mathcal{H}_i$, where a given dynamical observable takes the value O_i. Thus, we can define the time average of the observable as

$$\langle \mathbf{O} \rangle_t = \sum_{i=1}^{n} P_i O_i = \lim_{T\to\infty} \frac{1}{T} \sum_{i=1}^{n} O_i \delta t_i \, .$$

In practice, the depicted road is not viable, as the time average just defined requires following the system time after time through countless cells of the subspace $\mathcal{H}$ during the time T. Statistical Mechanics offers an alternative road!

Let us consider a sample of $\mathcal{N}$ ideal copies of our system, $\mathcal{N}$ being very large, and distribute them in the subspace $\mathcal{H}$. Primarily, we see that we are changing perspective: we are not considering a single many-body system evolving in time through the countless states of the subspace $\mathcal{H}$, but a number $\mathcal{N}$ of identical many-body systems distributed at the same time among all states of $\mathcal{H}$. The way the $\mathcal{N}$ identical systems are distributed in $\mathcal{H}$ defines a statistical ensemble, where the probability P_i is defined as the ratio between the number $\mathcal{N}_i$ of identical systems dropping inside the ith cell and the total number $\mathcal{N}$.[1] Once the probabilities have been set, one can calculate the statistical average of a physical observable as

---

[1] It is worth noting that this probability has nothing to do with the probabilistic character of quantum mechanics.

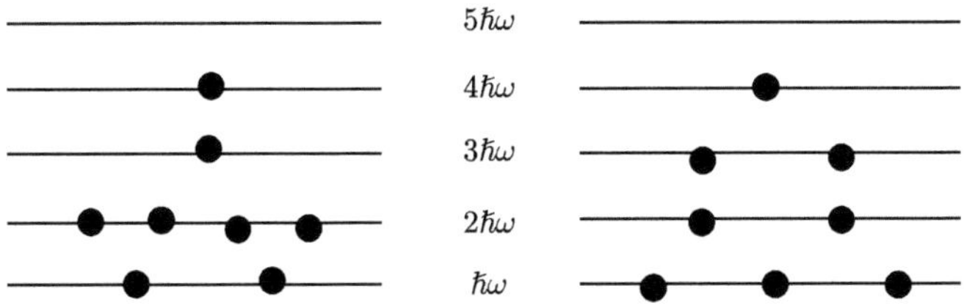

Fig. 16.1 Two possible microstates of an ideal gas of identical particles (bosons) in a harmonic oscillator energy spectrum. They belong to the same macrostate with $N = 8$ and $E = 17\hbar\omega$

$$\langle \mathbf{O} \rangle_s = \lim_{\mathcal{N}\to\infty} \frac{1}{\mathcal{N}} \sum_{i=1,N} \mathcal{N}_i \mathcal{O}_i \, . \tag{16.1}$$

The validity of statistical mechanics relies on the equivalence between time average and statistical average of observables (ergodic hypothesis[2]), that has been demonstrated only in a limited number of cases. The main task of statistical mechanics is to provide the thermodynamics of macroscopic systems with a microscopic interpretation in terms of the structure of their constituents and their interactions.

16.3 Microcanonical Ensemble

The question is: how do we distribute the ideal copies of the system in the Hilbert space to set the statistical ensemble? We will do this by focusing on a non-interacting gas of N particles with total energy E placed in a volume V. Within the thermodynamic limit, there are only two independent parameters: E and $\rho = N/V$. These two parameters define a macroscopic state, also known as *macrostate* of the system. In addition to the macrostate, we define the *microstate* as a given distribution of the N particles among the single-particle levels with quantized energies. The single-particle energy levels are

$$\epsilon_n = \frac{4\pi^2\hbar^2}{2mV^{2/3}} n^2, \tag{16.2}$$

such that the total energy of particles is equal to E. Figure 16.1 depicts two possible microstates of a system in the macrostate with N particles and total energy E.

Introducing the number of particles f_n occupying the state with energy ϵ_n, each microstate λ corresponds to a definite set of occupation numbers $\lambda \equiv [f_1^{(\lambda)}, f_2^{(\lambda)}, f_3^{(\lambda)}, ...]$ with the constraints

$$\rho = \frac{N}{V} \sum_n f_n^{(\lambda)} \, , \tag{16.3}$$

[2] The hypothesis states that over a sufficient amount of time the system, during its evolution, crosses all states of Hilbert's space once.

$$E = \sum_n \epsilon_n f_n^{(\lambda)} . \tag{16.4}$$

The occupation number of each level is limited to 0 or 1 for a Fermi gas, whereas it is limitless for a Bose gas. It is clear that a huge number of microstates corresponds to each macrostate, whereas only one macrostate corresponds to each microstate. Since there is no reason to favor any particular microstate, we assign the same probability to all microstates fulfilling the constraints Eqs. (16.3) and (16.4) *principle of a priori equiprobability*. As a consequence, the probability $\mathcal{P}$ of a macrostate is only specified by the number $\Gamma(E, N, V)$ of the corresponding microstates called *sum of the states*. We can assign a constant value to any microstate, according to the principle of *a priori* equiprobability. The choice of the inverse of $\Gamma(E, N, V)$ ensures the sum of the probabilities of all the microstates will be equal to unity. By including the Dirac delta function into the constraints Eqs. (16.3) and (16.4), the sum of the states is written as

$$\Gamma(E, V, N) = \sum_\lambda \delta(N - \sum f_n^{(\lambda)})\delta(E - \sum \epsilon_n f_n^{(\lambda)}) .$$

Now we can calculate the average value of a physical observable $\boldsymbol{O}$ for a given microstate as follows: $\langle\langle O \rangle\rangle = \sum_\lambda O_\lambda \mathcal{P}_\lambda$. Introducing the density of states operator

$$\rho = \mathrm{Tr}\, \delta(\mathbf{H} - \sum_n \epsilon_n f_n^{(\lambda)})\delta(\mathbf{N} - \sum_n f_n^{(\lambda)}) ,$$

where the trace Tr is the sum over all microstates λ. The average value of $\boldsymbol{O}$ can be written in a compact form[3]

$$\langle\langle \boldsymbol{O} \rangle\rangle = \frac{\mathrm{Tr}(\boldsymbol{O}\rho)}{\mathrm{Tr}(\rho)} .$$

The density operator ρ commutes with the Hamiltonian operator $\mathbf{H}$ as well as the number of particles operator $\mathbf{N}$. Thus, by applying the Heisenberg equations of motion, the density of states and statistical averages turn out to be independent of time. Hence, the microcanonical ensemble describes stationary states of macroscopic systems, namely equilibrium states.

16.4 Connection with Thermodynamics

The microscopic interpretation of thermodynamic quantities such as temperature, entropy and pressure involves the statistical averages of physical observables. The key role is played by the sum of the states $\Gamma(E, N, V)$ from the statistical side and by the entropy $S(E, N, V)$ from the thermodynamic side. Indeed, for an isolated system with parameters E, N and V, the sum of the microstates Γ represents the probability of the macrostate, while the entropy S is the parameter that, in accordance

[3] Henceforth, in this chapter, we will omit one of the double brackets.

with the law of increasing entropy, drives the evolution of the system towards the stationary state. This latter justifies the interpretation of the stationary state of an isolated system as the most probable state among the numerous states with total energy E, and the total number of particles N contained within the fixed volume V. Empirical examples are spontaneous processes like the free diffusion of a gas, the heat transfer from an object with a higher temperature to another one with a lower temperature, and the friction converting work to heat. As a result, there should be a relation between entropy and sum of the states. Boltzmann set the simplest form for this relation

$$S = k_{\mathrm{B}} \ln \Gamma, \tag{16.5}$$

the entropy being an additive quantity (i.e. the total entropy of two independent systems, A and B, is $S = S_A + S_B$) and the sum of the states a multiplicative one (the total probability of two independent events, A and B, is $P = P_A \times P_B$). The constant k_B is called the Boltzmann constant.

The Boltzmann equation enables us to connect the thermodynamic parameters with the statistical averages of the corresponding observables as shown below. From the first and second laws of thermodynamics, we have

$$\mathrm{d}E = T\mathrm{d}S - p\mathrm{d}V + \mu\mathrm{d}N, \tag{16.6}$$

where T, p, and μ are temperature, pressure, and chemical potential, respectively. From Eq. (16.6), we understand the identification of the thermodynamic energy, named the internal energy, with E. Rearranging Eq. (16.6) in the following form:

$$T\mathrm{d}S = \mathrm{d}E + p\mathrm{d}V - \mu\mathrm{d}N.$$

Since S is a function of state and $\mathrm{d}S$ is an exact differential, it follows that

$$\frac{1}{T} = \left(\frac{\partial S}{\partial E}\right)_{V,N} \qquad (V = const, \quad N = const),$$

$$\frac{p}{T} = \left(\frac{\partial S}{\partial V}\right)_{E,N} \qquad (E = const, \quad N = const),$$

$$\frac{\mu}{T} = -\left(\frac{\partial S}{\partial N}\right)_{E,V} \qquad (E = const, \quad V = const).$$

Hence, from the Boltzmann equation, Eq. (16.5), we get

$$\frac{1}{k_{\mathrm{B}}T} = \frac{1}{\Gamma}\left(\frac{\partial \Gamma}{\partial E}\right)_{V,N} \qquad (V = const, \quad N = const), \tag{16.7}$$

$$\frac{p}{k_{\mathrm{B}}T} = \frac{1}{\Gamma}\left(\frac{\partial \Gamma}{\partial V}\right)_{E,N} \qquad (E = const, \quad N = const), \tag{16.8}$$

$$\frac{\mu}{k_B T} = -\frac{1}{\Gamma}\left(\frac{\partial \Gamma}{\partial N}\right)_{E,V} \qquad (V = const, \quad E = const). \tag{16.9}$$

Rounding off this section, we can say that the microcanonical ensemble plays a very important theoretical role in statistical mechanics and establishing its relationship with thermodynamics. Nonetheless, for further developments, it is a better idea to move to the grand canonical ensemble.

16.5 Grand Canonical Ensemble

So far, we have focused our discussion on an isocated system using the Gibbs microcanonical ensemble. There the sum of the states $\Gamma(E, N, V)$ plays the role of probability of a macroscopic state with N particles and total energy E, contained inside a volume V. We are now introducing an open system, i.e. a system interacting with an external environment (or thermostat) which is much larger. The system is in thermal (same temperature T) and chemical (same chemical potential μ) equilibrium with the thermostat. The equilibrium is reached thanks to the heat and particle exchange between the two systems. As a result, thus, the thermodynamic parameters defining the state of an open system are the temperature T instead of the energy, and the chemical potential μ instead of the number of particles. The reason for introducing an open system is twofold: This is experimentally and theoretically easier to be studied. It is difficult, in fact, to prepare a system with precise values of energy and particle number, as required by the microcanonical ensemble. Temperature and chemical potential are, instead, much easier adjustable observables thanks to the thermostat.

The open systems are studied within the grand canonical ensemble to answer the question: for a given temperature and chemical potential, what is the probability that a system has an average energy E and an average number of particles N? Despite being more general,[4] the answer will be given only for a gas of non-interacting particles (ideal gas).

Let us consider an ideal gas A, contained in the volume V, in thermal and chemical equilibrium with the environment. The combination system+environment, known as universe, is supposed to be an isolated system with fixed energy E_0 and fixed number of particles N_0, contained in the volume V_0. The two subsystems A and B do not interact except for the mutual exchange of energy and particles. Hence, we can deal with the three systems within the microcanonical ensemble. Assume A has energy E and particle number N, then the number of microstates of the universe will be the product of the microstates of the two subsystems A and B. But, due to the mutual energy and particles exchange, we must sum over all possible energy and particle distributions between A and B, hence

$$\Gamma_U(E_0, N_0, V_0) = \sum_{E,N} \Gamma_A(E, N, V) \cdot \Gamma_B(E_0 - E, N_0 - N, V_0 - V). \tag{16.10}$$

[4] It is valid also for systems of particles interacting via short-range forces.

In the sequel, the system A will be assumed to have energy, particle number, and volume much smaller than the environment B. Moreover, the two subsystems will be treated in the thermodynamic limit. Under such conditions it will be shown in Sect. 16.9, that the probability that the system, in equilibrium with the environment, has energy E and number of particles N is

$$P \sim \mathrm{e}^{-\beta(E-\mu N)}\Gamma(E, V, N),$$

where $\beta = 1/kT$ and k is the Boltzmann constant. The parameters T and μ are those of the thermostat. The sum of all possible values of energy and number of particles is called the grand partition function

$$\mathcal{Z}(T, \mu, V) = \sum_E \sum_N \mathrm{e}^{-\beta(E-\mu N)}\Gamma(E, N, V) . \tag{16.11}$$

In the grand canonical ensemble, it plays the role of the sum of the states Γ. In other words, it is the overall sum of microscopic configurations of the system in the microcanonical ensemble with energy and number of particles weighted by the Boltzmann factors $\mathrm{e}^{-\beta E}$ and $\mathrm{e}^{\beta\mu N}$, respectively. Applying the definition of Γ in Eq. (16.11), we get

$$\mathcal{Z}(T, \mu, V) = \sum_i \mathrm{e}^{-\beta[\sum_n (\epsilon_n - \mu) f_n^{(i)}]} = \mathrm{Tr}[z^N \mathrm{e}^{-\beta \mathbf{H}}], \tag{16.12}$$

where Tr is the sum over the microscopic states and $z = \exp(\beta\mu)$ is named fugacity. Similarly to the micro-canonical ensemble, we can introduce the operator density of states $\boldsymbol{\rho}$

$$\boldsymbol{\rho} = \frac{1}{\mathcal{Z}} \mathrm{e}^{-\frac{\mathbf{H}-\mu\mathbf{N}}{kT}} .$$

This is used to calculate the statistical average of any observable $\mathbf{O}$ using the relation $\langle \mathbf{O} \rangle = \mathrm{Tr}(\mathbf{O}\rho)$. For instance, assuming $\mathbf{O}$ to be the Hamiltonian

$$\begin{aligned} \langle \mathbf{H} \rangle = \mathrm{Tr}(\mathbf{H}\rho) &= \frac{1}{\mathcal{Z}} \mathrm{Tr}(z^N \mathbf{H} \mathrm{e}^{-\beta \mathbf{H}}) \\ &= -\frac{1}{\mathcal{Z}} \frac{\partial}{\partial \beta} \mathrm{Tr}(z^N \mathrm{e}^{-\beta \mathbf{H}}) = -\frac{1}{\mathcal{Z}} \frac{\partial}{\partial \beta} \mathcal{Z} = -\frac{\partial}{\partial \beta} \ln \mathcal{Z}, \end{aligned} \tag{16.13}$$

or the number of particles

$$\langle \mathbf{N} \rangle = z \frac{\partial}{\partial z} \ln \mathcal{Z}. \tag{16.14}$$

In the following, this formalism will be applied to study ideal systems of fermions and bosons at finite temperature.

16.6 Finite-Temperature Ideal Fermi Gas

In a non-interacting fermion gas, the single-particle levels are classified by the quantized $\epsilon_{\vec{p}}$ given by Eq. (16.2). The grand partition function takes the form

$$\mathcal{Z} = \sum_i e^{-\beta \sum_{\vec{p}} (\epsilon_{\vec{p}} - \mu) f_{\vec{p}}^{(i)}} .$$

According to the Pauli principle, the occupation numbers $f_{\vec{p}}^{(i)}$ can be 0 (unoccupied state) or 1 (occupied state), so that

$$\begin{aligned}\mathcal{Z} &= \sum_{f_{\vec{p}_1}=0,1} e^{-\beta(\epsilon_{\vec{p}_1}-\mu) f_{\vec{p}_1}} \cdot \sum_{f_{\vec{p}_2}=0,1} e^{-\beta(\epsilon_{\vec{p}_2}-\mu) f_{\vec{p}_2}} \cdot \sum_{f_{\vec{p}_3}=0,1} e^{-\beta(\epsilon_{\vec{p}_3}-\mu) f_{\vec{p}_3}} \dots \\ &= \Pi_{\vec{p}}(1 + e^{-\beta(\epsilon_{\vec{p}}-\mu)}) = \Pi_{\vec{p}}(1 + z e^{-\beta\epsilon_{\vec{p}}}) .\end{aligned}$$

Applying Eqs. (16.13) and (16.14), the average number of particles and energy are

$$\langle N \rangle = z \frac{\partial}{\partial z} \ln \mathcal{Z} = \sum_{\vec{p}} \frac{1}{1 + e^{\beta(\epsilon_{\vec{p}}-\mu)}} , \qquad (16.15)$$

$$\langle E \rangle = -\frac{\partial}{\partial \beta} \ln \mathcal{Z} = \sum_{\vec{p}} \frac{\epsilon_{\vec{p}}}{1 + e^{\beta(\epsilon_{\vec{p}}-\mu)}} . \qquad (16.16)$$

At temperature T, we define the average occupation number as

$$\langle f_{\vec{p}}(T) \rangle = [1 + e^{\beta(\epsilon_{\vec{p}}-\mu)}]^{-1} , \qquad (16.17)$$

and its values range in the interval $0 < \langle f_{\vec{p}}(T) \rangle < 1$, as shown in Fig. 16.2. At the limit $T \to 0$, the exponential tends to 0 for energies smaller than $\mu(T = 0)$ and then $\langle f_{\vec{p}} \rangle \to 1$; it diverges to infinity for energies bigger than $\mu(T = 0)$ and then $\langle f_{\vec{p}} \rangle \to 0$. The zero temperature chemical potential coincides with the Fermi energy

Fig. 16.2 Distribution of the occupation numbers of a low-temperature Fermi gas

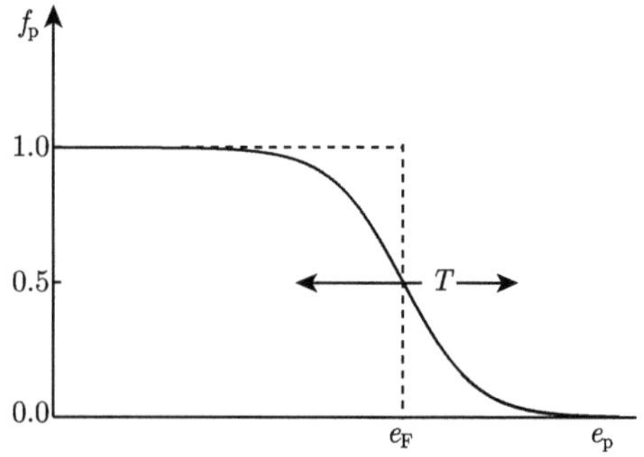

ϵ_F, i.e. the energy of the last occupied energy level with momentum p_F. Hence, each microscopic stare is occupied ($\epsilon_p \leq \epsilon_F$) or unoccupied ($\epsilon_p > \epsilon_F$), as shown by the dashed step function in Fig. 16.2. This is true not only at zero temperature but also at finite temperature. However, the macroscopic state of the system is a mixture of microscopic states, and then a microscopic state can be partially occupied and partially unoccupied, as shown in the region around μ. This justifies the definition of $\langle f_{\vec{p}}(T)\rangle$ as average occupation number. At the $T = 0$ limit, the macroscopic state coincides with a single microscopic state, where all microscopic states with $p \leq p_F$ are occupied and the others are empty. In other words, the macroscopic state at $T = 0$ is achieved by one and only one microscopic state: the sum over states turns out to be $\Gamma = 1$ and so the entropy vanishes according to the Boltzmann equation. This interesting property is in keeping with the third principle of thermodynamics.

Heating weakly the system from $T = 0$ to a low temperature ($T \ll \varepsilon_F$), particles occupying states close to the Fermi surface, i.e. $\varepsilon = \varepsilon_F$, are excited to unoccupied states, depleting the discontinuity of the function $\langle f_{\vec{p}}\rangle$ at the Fermi energy. Conversely, particles sitting deeper are blocked by the Pauli principle. The Pauli blocking has remarkable effects on the system's thermodynamics, e.g. the heat capacity c_V of a Fermi gas. In a classical gas, all particles contribute to c_V, which thus turns out to be temperature independent (Dulong–Petit law), whereas in a quantum Fermi gas the Pauli-blocked particles do not contribute. Since their number increases when lowering T, c_V will be reduced until it vanishes at $T = 0$, where all particles are frozen in the ground state by the Pauli principle.

At high temperature, most particles go to occupy higher energy levels $\varepsilon_p \gg \varepsilon_F$. As a consequence, the Pauli principle loses effectiveness and the gas behaves like a classical system with constant heat capacity.

Quantum Fermi systems, such as compact astrophysical objects, can be found in states of such high density and low temperature as to behave like ideal systems. These systems are named *degenerate Fermi systems*. In such conditions, the chemical potential can be approximated [SOM] as

$$\mu = \varepsilon_F \left[1 - \frac{\pi^2}{12}\left(\frac{k_B T}{\varepsilon_F}\right)^2\right], \tag{16.18}$$

and, since $k_B T \ll \varepsilon_F$, the occupation numbers are well approximated by step functions. As a consequence, the behaviour of an ideal Fermi system at $T = 0$ is worth discussing to a greater extent. In particular, its equation of state can be easily derived. The numbers of particles and energy, Eq. (16.16), take an analytic form, so that the integration over momentum can be easily performed. Applying Eq. (4.37), we write down

$$N = g\frac{V}{(2\pi\hbar)^3}4\pi\int_0^{p_F} p^2 \mathrm{d}p = g\frac{V}{(2\pi\hbar)^3}\frac{4\pi}{3}p_F^3, \tag{16.19}$$

$$E = g\,\frac{V}{(2\pi\hbar)^3}4\pi\int_0^{p_F} p^2 \mathrm{d}p\frac{p^2}{2m} = \frac{V}{(2\pi\hbar)^3}\frac{4\pi}{10m}p_F^5 = \frac{3}{5}N\varepsilon_F, \tag{16.20}$$

where V is the volume of the fluid and $\varepsilon_{\mathrm{F}} = p_{\mathrm{F}}^2/2m$. The factor g is the degeneracy of each momentum state: 2 for fermions of spin 1/2, e.g. electrons, and 4 for fermions of spin and isospin 1/2, e.g. symmetric nuclear matter. To derive the equation of state, we resort to the first principle of thermodynamics $\mathrm{d}E = T\mathrm{d}S - P\mathrm{d}V + \mu \mathrm{d}N$, where S is the entropy and P the pressure. For a homogeneous system at $T = 0$, it simplifies as $E = -PV + \varepsilon_{\mathrm{F}} N$. With E and N given by Eq. (16.19) and (16.20), respectively, the first principle yields[5]

$$PV = \frac{2}{5}\varepsilon_{\mathrm{F}} N \, . \tag{16.21}$$

It is noteworthy that the pressure does not vanish, even though the temperature is zero. Indeed, due to the Pauli principle, the particles distribute in all states of momentum $0 \leq p \leq p_{\mathrm{F}}$, giving rise to the pressure.

The equation of state in relativistic regime is often requested for astrophysical applications. An analytic expression can be obtained at the ultra-relativistic limit, $p_{\mathrm{F}} \gg mc$, where the single-particle energy is $\varepsilon = p\,c$. Following the same procedure as in the non-relativistic case, we find

$$PV = \frac{1}{4} N \varepsilon_{\mathrm{F}}^{ur} , \tag{16.22}$$

but $\varepsilon_{\mathrm{F}}^{ur} = p_{\mathrm{F}}\, c$. Thus the ultra-relativistic limit predicts P to be linear in p_{F}, whereas P is quadratic in p_{F} at the non-relativistic limit according to Eq. (16.20). This behaviour could have significant consequences for massive systems, where the density changes during their evolution (see following section).

16.7 Fermi Systems in Astrophysics

16.7.1 Degenerate Electron Gas in White Dwarf Stars

Degenerate Fermi systems are encountered in astrophysics, e.g. the electron gas in equilibrium at low temperature with ionized atoms in white dwarf stars. In white dwarfs with low mass, say a fraction of the solar mass $M_{\odot}$, the thermal energy is not enough to trigger the fusion of nuclei heavier than the ^{4}He atom (nucleosynthesis). Thus, we may suppose that these stars are made mostly of ionized helium atoms mixed with their electrons. The electron pressure is able to compensate the ion gravitational pressure, preventing the star from collapsing. Let us calculate the two pressures assuming that the electrons can be treated as a degenerate Fermi gas and the ^{4}He ions as a self-gravitating system. In the early stage of the star evolution, its mass is so low that the non-relativistic equation of state, Eq. (16.21), can be applied to the electron gas. Since the Fermi energy is $\varepsilon_{\mathrm{F}} = p_{\mathrm{F}}^2/2m$ and the Fermi momentum

[5] Henceforth, in this section, the pressure will be denoted with P and the momentum with p.

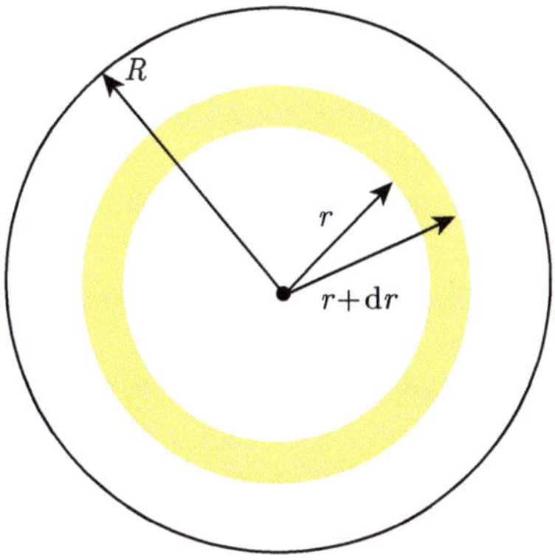

Fig. 16.3 Profile of a self-gravitating star

is proportional to the density $(N/V)^{\frac{2}{3}}$ we can express the pressure in terms of the density alone. To within constant factors, it is

$$P_{\text{electron}} \sim \left(\frac{N}{V}\right)^{\frac{5}{3}} \sim \frac{M^{5/3}}{R^5}, \tag{16.23}$$

the number of electrons N being the ratio between the total star mass M and two times the nucleon mass, and the star volume V proportional to the cubic radius R^3.

The pressure is that of a self-gravitating system. It will be calculated assuming the electrons do not contribute, because their mass is three orders of magnitude smaller than the ion mass.

As shown in Fig. 16.3, the pressure gradient on a layer of thickness dr at distance r from the center of the star must be equal to the force *per* unitary volume

$$\frac{\mathrm{d}p(r)}{\mathrm{d}r} = -G\frac{m(r)\rho(r)}{r^2}, \tag{16.24}$$

due to the gravitational attraction between the mass $m(r)$ contained in the sphere of radius r and the mass

$$\mathrm{d}m(r) = \rho(r)\mathrm{d}V.$$

The two equations can be integrated with the boundary conditions that the density vanishes at the star surface, i.e. $\rho(R) = 0$ where R is the radius, and it takes a constant value at the star center, i.e. $\rho(0) = \rho_c$. The details are reported in the Complement to this section, where it is shown $P \sim \frac{M^2}{R^4}$, according to Eq. (16.42). Comparing with the electron pressure, an equilibrium configuration is found with $M^{1/3}R = const.$ In other words, the white dwarf follows the slow gravitational compression reducing volume through quasi-stationary states. But, increasing the density, the electron gas gets into the relativistic regime with equation of state intermediate between Eqs. (16.21) and (16.22). Using this latter, the electron pressure is $P \sim \frac{M^{4/3}}{R^4}$, which can no longer adjust its volume to the mass increase in an equilibrium configuration. In fact, a mass limit $M_c \approx 1.4 M_\odot$ has been estimated, beyond which no physical solution exists for the white dwarfs [HUA]. M_c is known as Chandrasekhar mass.

To summarize, we can imagine the following scenario for a white dwarf star with a small mass $M < M_\odot$. In the early stage, the thermal energy is not sufficiently large to fuse nuclei heavier than ^{4}He atoms, so that only ^{4}He ions and electrons, ripped from the ^{4}He atoms, are produced. Since the electrons are non-relativistic, their pressure can equilibrate the ion pressure, such that $MR^3 \sim const$, preventing the gravitational collapse. Increasing the ^{4}He mass, the volume shrinks and the electron density increases, reaching the relativistic regime. Our calculation, with electrons in the ultra-relativistic regime, shows that no stable configuration is possible in this case. The mass limit is the Chandrasekhar mass. As a matter of fact, this endpoint of the low-mass white dwarfs has never been observed, supporting the expectation that reaching this point may take longer than the age of the Universe.

16.7.2 Neutron Stars

A supernova explosion is the last phase of a massive star, when the nuclear fuel runs out and a dense core of the synthesized elements is formed. In a few seconds, the star collapses as an effect of its own strong gravity and then, bouncing off the core, explodes. The neutron star is the remnant of the explosion. The observations say that neutron stars have mass $M \sim (1.5–2.5)\ M_\odot$, where $M_\odot$ is the solar mass. This mass is confined to a volume with radius $R \sim 10$ km, which is significantly smaller than the sun radius of 696×10^3 km. Due to the strong gravitational field, the neutron-star matter could be in a nucleonic phase in equilibrium with leptons, mainly electrons (Fig. 16.4). In this model, the average distance between two nucleons is less than 10^{-15} meters, i.e. it is in the range where the nuclear force becomes repulsive. Thus, the gravitational attraction and the nuclear repulsion would ensure the neutron-star stability configuration with fixed mass and radius. Accordingly Eq. (16.24) is written as

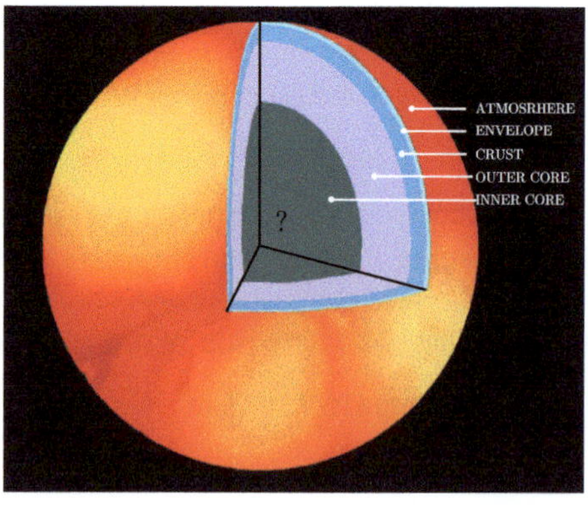

Fig. 16.4 Neutron-star structure

$$\frac{\mathrm{d}p}{\mathrm{d}\rho}\frac{\mathrm{d}\rho}{\mathrm{d}r} = -G\frac{m(r)\rho(r)}{r^2}, \tag{16.25}$$

where $\frac{\mathrm{d}p}{\mathrm{d}\rho}$ depends on the composition of the neutron-star matter. This latter is determined by the chemical equilibrium between nucleons and electrons

$$p + e^- \to n + \nu_e \quad \text{(electron capture)}, \tag{16.26}$$

$$n \to p + e^- + \bar{\nu}_e \quad (\beta^- \text{decay}), \tag{16.27}$$

where ν_e and $\bar{\nu}_e$ are electronic neutrino and antineutrino, respectively. We assume the nuclear matter temperature to be $T = 0$ and that the total charge conservation $Z = N_{el}$ (charge neutrality) is also locally valid. The chemical equilibrium among neutrons, protons, and electrons is achieved when the respective chemical potentials fulfill the condition

$$\mu_n - \mu_p = \mu_{el}, \tag{16.28}$$

where neutrinos and anti-neutrinos are neglected. From Eq. (16.28), we obtain the proton fraction $x = Z/A$ at equilibrium. For this purpose, we first calculate the neutron-to-proton chemical potential shift from the energy per nucleon, Eq. (14.18), in isospin-asymmetric ($N \neq Z$) nuclear matter

$$B_A(\rho, x) = a_A(\rho) + a_I(\rho)(1 - 2x)^2, \tag{16.29}$$

where $x = (1 - \beta)/2$. It is easily found that

$$\mu_n - \mu_p = \frac{\partial B_A}{\partial \rho_n} + \frac{\partial B_A}{\partial \rho_p} = 4\beta a_I \cdot \tag{16.30}$$

We can then calculate the electron chemical potential (equal to the Fermi energy at $T = 0$) from the ultra-relativistic limit of the free Fermi gas model, combining Eqs. (16.29) and (16.30)

$$N_{el} = \frac{1}{3\pi^2}\left(\frac{\mu_{ele}}{\hbar c}\right)^3, \tag{16.31}$$

where $N_{el} = Z$.

Turning back to neutron-star hydrostatic equilibrium from the balance between gravitational and nuclear pressures, we now have all the ingredients to integrate equilibrium differential equation Eq. (16.25) with the boundary conditions: $\rho(0) = \rho_c$ and $\rho(R) = 0$. The central density ρ_c is not known, so it plays the role of a parameter, but its value is not expected to be less than $1\,\mathrm{fm}^{-3}$, unless there are signals of a transition to the quark-gluon plasma phase. The calculation is reported in Fig. 16.5. In the interval $0.2 < \rho_c < 0.4\mathrm{fm}^{-3}$, the star mass M ranges from 1 to 2 units of solar mass (see dashed line in the right panel) with a radius R of the order of 12 Km (see dashed line in the left panel) in agreement with observations. The most severe drawback is the absence of a mass limit, expected by the observations. In fact, the gravitational mass is so large as to enter the domain of general relativity.

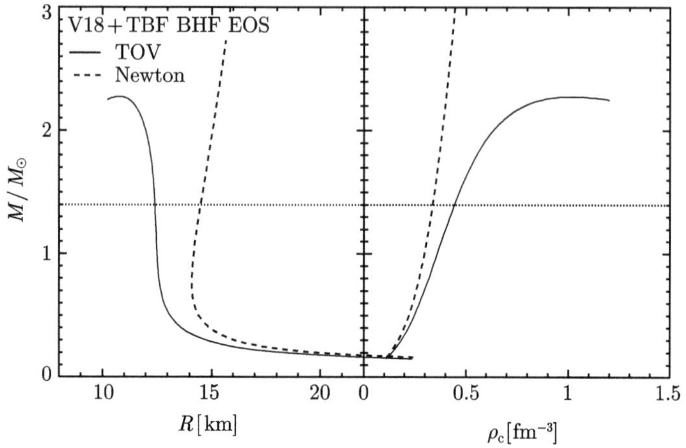

Fig. 16.5 Left panel: mass–radius plot in the non-relativistic (dashed line) and relativistic (solid line) limits. Right panel: neutron-star mass Versus central density [LOM]

Indeed, applying the hydrostatic equilibrium equation, based on general relativity, a mass limit is found, as shown in Fig. 16.5, in good agreement with observations. Nevertheless, the main achievement of this calculation is that the nuclear matter equation of state seems to successfully justify the neutron-star stability and to predict an inner structure consistent with observed properties.

16.8 Finite-Temperature Ideal Bose Gas: Black-Body Radiation

16.8.1 Spectral Decomposition of the Electromagnetic Field

At each point $\vec{r}$ of a cavity the electric and magnetic fields take the values $\vec{E}(\vec{r}, t)$ and $\vec{H}(\vec{r}, t)$, but they vanish at the inner walls. Being $V = L^3$ the cavity volume, one can develop the functions $\vec{E}$ and $\vec{H}$ in Fourier series

$$\vec{E}(\vec{r}, t) = \sum_{\vec{k}} \vec{E}_{\vec{k}}(t) \mathrm{e}^{\mathrm{i}\vec{k}\cdot\vec{r}}, \qquad \vec{H}(\vec{r}, t) = \sum_{\vec{k}} \vec{H}_{\vec{k}}(t) \mathrm{e}^{\mathrm{i}\vec{k}\cdot\vec{r}},$$

where $k_x = 2\pi\frac{n_x}{L}$, $k_y = 2\pi\frac{n_y}{L}$, $k_z = 2\pi\frac{n_z}{L}$ $(n_x, n_y, n_z = \pm 1, \pm 2, ...)$. Being solutions of the Maxwell equations, the respective Fourier coefficients $\vec{E}_{\vec{k}}$ and $\vec{H}_{\vec{k}}$ form with the wave number $\vec{k}$ three orthogonal axes. Accordingly, $\vec{k} \times \vec{E}_{\vec{k}}(t) = \frac{\omega_{\vec{k}}}{c} \vec{H}_{\vec{k}}(t)$ with $k = \omega_k / c$ and c is the speed of light.

As a result, one can express the magnetic field in terms of the electric one and obtain

$$\vec{H}(\vec{r},t) = \sum_{\vec{k}} \frac{c}{\omega_{\vec{k}}}[\vec{k} \times \vec{E}_{\vec{k}}(t) \mathrm{e}^{\mathrm{i}\vec{k}\cdot\vec{r}}].$$

The advantage of the Fourier decomposition is that the description of the electromagnetic (e.m.) field in terms of continuous variables, $E(\vec{r},t)$ and $\vec{H}(\vec{r},t)$, is replaced by a description in terms of discrete variables, $\vec{E}_{\vec{k}}(t)$ and $\vec{E}^*_{\vec{k}}(t)$, thanks to the boundary condition of vanishing at the walls. This change has important theoretical implications, because now the Hamiltonian of the e.m. field $\mathcal{H} = \frac{1}{8\pi}\int \mathrm{d}V\, (E^2 + H^2)$ in terms of the new variables takes the form

$$\begin{aligned} \mathcal{H} &= \frac{V}{8\pi}\sum \left[\vec{E}_{\vec{k}}(t)\cdot \vec{E}_{\vec{k}}(t) + \frac{c^2}{\omega^2_{\vec{k}}}(\vec{k}\times \vec{E}_{\vec{k}}(t))\cdot(\vec{k}\times\vec{E}_{\vec{k}}(t))\right] \\ &= \frac{V}{4\pi}\sum |E_{\vec{k}}|^2 = \frac{V}{4\pi}\sum_{\sigma}\sum_{\vec{k}} |E_{\vec{k},\sigma}|^2 . \end{aligned} \tag{16.32}$$

In order to obtain Eq. (16.32), we have used the property of plane waves to form an orthonormal set

$$\frac{1}{V}\int \mathrm{d}V \mathrm{e}^{\mathrm{i}(\vec{k}-\vec{k}')\cdot\vec{r}} = \delta_{\vec{k},\vec{k}'}$$

and the condition on the field to be a real quantity $\vec{E}_{-\vec{k}} = \vec{E}^*_{\vec{k}}$. The time dependence disappears because $\vec{E}_{\vec{k}}(t) = \vec{E}_{\vec{k}}\mathrm{e}^{-\mathrm{i}\omega t}$. Thus $\vec{E}_{\vec{k}}$ has only two independent components which are specified by the letter σ in Eq. (16.32). Applying the variable transformation

$$\begin{aligned} q_k &= \sqrt{\frac{V}{4\pi c^2}}(E_{k,\sigma} + E^*_{k,\sigma}), \\ p_k &= -\mathrm{i}\sqrt{\frac{V\omega^2_k}{4\pi c^2}}(E_{k,\sigma} - E^*_{k,\sigma}), \end{aligned}$$

the Hamiltonian takes the form of a system of harmonic oscillators

$$\mathcal{H} = \sum_k \sum_{\sigma=1,2} (p^2_{k,\sigma} + \omega^2_k q^2_{k,\sigma}) . \tag{16.33}$$

In conclusion, we have found that the Hamiltonian of the electromagnetic field in a cavity is equivalent to the Hamiltonian of an infinite series of independent harmonic oscillators.

16.8.2 Quantization of the Electromagnetic Field: Photon Gas

The energy spectrum of each harmonic oscillator of the e.m. field is

$$\epsilon_{\vec{k}}^{(n)} = \hbar\omega_{\vec{k}}\left(n_{\vec{k}} + \frac{3}{2}\right), \qquad n_k = 0, 1, 2, \ldots$$

The excitation energy between two nearest levels $\hbar\omega_{\vec{k}}$ is the elementary excitation, called *quantum*, that is a constant because the levels are equally spaced. If it is assumed for a while that the latter is a particle, the excitation states of each harmonic oscillator can be defined by the number $n_{\vec{k}}$ of such particles. We call $n_{\vec{k}}$ the occupation number of the harmonic oscillator with wave number $\vec{k}$, the same as in the case of a Fermi gas. Thus, a microstate of the full electromagnetic field can be characterized by the set of occupation numbers $n_{\vec{k}}$ of all harmonic oscillators. This description of the state of a many-body system is named *second quantization*. The above-mentioned assumption was proved to be real by the Compton scattering of electrons off the e.m. field. This particle, known as a photon (Einstein 1905), is a boson with two spin states corresponding to the two independent polarization states of a transverse wave. Hence the electromagnetic field in a cavity amounts to a non-interacting photon gas (the harmonic oscillators are independent of each other). An important difference from the ideal particle gas is that the excitation energy depends linearly on the wave number, $\epsilon_k = \hbar\omega_{\vec{k}} = \hbar ck$, whereas it is quadratic, $\epsilon_k = \dfrac{\hbar^2 k^2}{2m}$ for particles.

The photon gas in equilibrium with the cavity walls at the temperature T is an open system, because it exchanges energy and particles with its walls. Thus it is to be statistically treated within the grand canonical ensemble. Despite the particle exchange, the number of particles in the gas remains indefinite, so that the chemical potential is equal to zero. In addition, there are no restrictions on the occupation number $f_{\vec{k}}^{(i)}$, because photons are bosons, so that the grand partition function, given by Eq. (16.12), is

$$\begin{aligned}\mathcal{Z} &= \sum_i \mathrm{e}^{-\beta\sum_{\vec{k}}\hbar\omega_{\vec{k}} f_{\vec{k}}^{(i)}} = \prod_{\vec{k}}\sum_{f_{\vec{k}}=1}^{\infty} \mathrm{e}^{-\beta(\hbar\omega_{\vec{k}})f_{\vec{k}}} \\ &= \prod_{\vec{k}}(1-\mathrm{e}^{-\beta\hbar\omega_{\vec{k}}})^{-1},\end{aligned} \tag{16.34}$$

after summing up a geometric series. Using this grand partition function we can calculate the average energy

$$\langle E\rangle = -\frac{\partial}{\partial\beta}\ln\mathcal{Z} = 2\sum_{\vec{k}}\frac{h\nu_{\vec{k}}}{\mathrm{e}^{\beta h\nu_{\vec{k}}}-1}, \tag{16.35}$$

where the pre-factor 2 takes into account the two polarization states of photons.

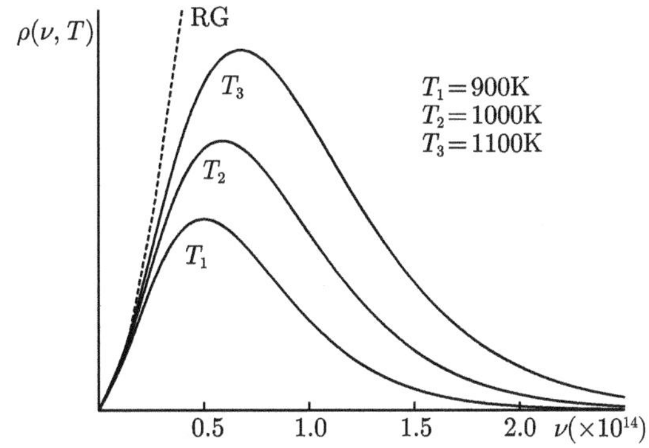

Fig. 16.6 Quantum (solid line) and classical (dashed line) energy spectral density

The spectral energy density can be obtained from the average energy of the individual oscillators times the number of oscillators *per* unitary volume and unitary frequency. The result is the famous Plank formula of the black-body radiation

$$\rho_\nu(T) = \frac{8\pi\nu^2}{c^3} \cdot \frac{h\nu}{e^{\beta h\nu} - 1},$$

which agrees quite well with the experimental data. It is plotted in Fig. 16.6 for several temperatures. The classical Rayleigh–Jeans law, is also plotted (dashed curve): it can be deduced by the Plank law in the limit of low frequency $h\nu \ll 1$, in which the denominator can be approximated by $\frac{h\nu}{k_B T}$. At high frequency the average energy is driven by the Boltzmann factor $e^{-\frac{h\nu}{k_B T}}$, which avoids the occurrence of the so-called ultraviolet catastrophe of the spectral energy density. The integration of the Plank formula over all frequencies yields $\int d\nu \rho_\nu(T) \sim const \cdot T^4$. This is the Stefan–Boltzmann law used to estimate the star temperature, approximating the star to a black body.

16.8.3 Black-Body Radiation from the Classical Point of View

One of the main achievements of classical statistical mechanics is the theorem of equipartition of energy. It states that each form of energy of a system contributes $\frac{1}{2}k_B T$ to the average energy, times the number f of degrees of freedom. For instance, the thermal energy of a gas of N non-interacting particles is $\frac{3}{2}Nk_B T$, considering that the particles move in three dimensions. This theorem is valid for a large class of systems, including a gas of non-interacting particles and harmonic oscillators. If the Hamiltonian of the system has kinetic and potential terms the average energy for each degree of freedom is $k_B T$. For an ideal gas (no potential) with N particles the total number of degrees of freedom is $3N$ and the total average energy is $3Nk_B T/2$. For the black body, that is equivalent to an infinite set of harmonic oscillators, the total average energy *per* unitary volume is determined by the number of harmonic oscillators *per* unitary volume. This can be calculated assuming that the wave numbers can be considered as continuous variables, if the volume is quite large. The wave numbers contained in the interval between k and $k + \mathrm{d}k$ are

$$4\pi n^2 \mathrm{d}n = 4\pi \frac{V}{(2\pi)^3} k^2 \mathrm{d}k = 4\pi V \frac{\nu^2}{c^3} \mathrm{d}\nu,$$

applying the relation between wave number and frequency $k = 2\pi\nu/c$. As a result, taking into account the fact that we have only two independent polarization states, the number of degrees of freedom of the e.m. field in the black-body cavity × unitary volume and frequency is $\frac{4\pi n^2 \mathrm{d}n}{V \mathrm{d}\nu} = \frac{8\pi\nu^2}{c^3}$ and the total average energy × unitary volume and frequency, called energy spectral density, is

$$\rho_\nu(T) = \frac{8\pi\nu^2}{c^3} k_B T.$$

This relation, which is the Rayleigh–Jeans law, is in agreement with the experimental data at low frequency, but it diverges at high frequency, leading to the so-called ultraviolet catastrophe. In fact, due the parabolic frequency dependence, ν^2, the frequency integration diverges:

$$\int_0^\infty \rho_\nu \mathrm{d}\nu = \infty.$$

This result has been of historic relevance in the transition from classical to quantum mechanics. We will see now how quantum mechanics reconciles theory with experimental black-body radiation.

16.9 Complement to Sect. 16.5: Grand Canonical Probability

Let us turn back to Eq. (16.12) for the system A, exchanging energy and particles with the environment B to preserve the thermodynamic equilibrium, that is

$$\Gamma_U = \sum_N \sum_E \Gamma_A(E, N, V) \cdot \Gamma_B(E_0 - E, N_0 - N, V_0 - V),$$

where E, N and V refer to the system A and E_0, N_0 and V_0 to the system U. Assume that the energy, number of particles, and volume of universe are much larger than those of subsystem A.[6] This approximation amounts to the assumption that that the number of energy and particle configurations (microstates) in A is very small compared to that in B. Hence, the series expansion of Γ_{B}, truncated at the first order of energy, number of particles, and volume, is justified. Since we want to introduce the thermodynamic parameters, which are connected to the logarithm of the sum over states, we expand the logarithm of Γ_{B} instead of Γ_{B} itself, obtaining

$$\ln\Gamma_B \approx \ln\Gamma_U - E\cdot\left(\frac{1}{\Gamma}\frac{\partial\Gamma}{\partial E}\right)_U - N\cdot\left(\frac{1}{\Gamma}\frac{\partial\Gamma}{\partial N}\right)_U - V\cdot\left(\frac{1}{\Gamma}\frac{\partial\Gamma}{\partial V}\right)_U .$$

Applying Eqs. (16.7) and (16.8) and (16.9), we get

$$\Gamma_{\mathcal{U}} = \Gamma_{\mathcal{U}}\sum_E\sum_N \mathrm{e}^{-\frac{E}{k_{\mathrm{B}}T}-\frac{pV}{k_{\mathrm{B}}T}+\frac{\mu N}{k_{\mathrm{B}}T}}\Gamma(E,N,V),$$

and finally

$$1 = \mathrm{e}^{-\frac{pV}{k_{\mathrm{B}}T}}\sum_E\sum_N \mathrm{e}^{-\frac{E}{k_{\mathrm{B}}T}+\frac{\mu N}{k_{\mathrm{B}}T}}\Gamma(E,N,V),$$

where T and μ are temperature and chemical potential of the universe, respectively. This result is interpreted as follows: the probability $P(E,N,V)$ of the system A, in equilibrium with B, being in the state with energy E and number of particles N is

$$P(E,N,V) = const\cdot\mathrm{e}^{-\frac{E-\mu N}{k_{\mathrm{B}}T}}\Gamma(E,N,V),$$

and the sum is, accordingly, equal to one. $P(E,N,V)$ is called *grand canonical Gibbs probability distribution*. The number of energy and particle configurations contributing to Γ_U is quite small, because the product $\Gamma_{\mathrm{A}}\cdot\Gamma_{\mathrm{B}}$ is sharply peaked around the energy and particle values at which the derivatives of $\Gamma_A\,\Gamma_B$ with respect to E and N vanish, that is

$$1/\Gamma_{\mathrm{A}}\mathrm{d}\Gamma_{\mathrm{A}}/\mathrm{d}E = 1/(k_{\mathrm{B}}T)$$

$$1/\Gamma_{\mathrm{A}}\mathrm{d}\Gamma_{\mathrm{A}}/\mathrm{d}N = -\mu/(k_{\mathrm{B}}T),$$

where T and μ are temperature and chemical potential of the universe, respectively. Now, according to Eqs. (16.7) and (16.9), we conclude that $T_{\mathrm{A}} = T$ and $\mu_{\mathrm{A}} = \mu$, consistently with the assumption that the system is in thermal and chemical equilibrium with the environment.

[6] In this case, B plays the role of the so-called *thermostat* in thermodynamics.

16.10 Complement to Sect. 16.7: Newtonian Hydrostatic Equilibrium

Let us consider a self-gravitating system, assumed to be spherical. The hydrostatic equilibrium equation, Eq. (16.24), and the auxiliary equation for the radial mass are

$$\frac{\mathrm{d}P(r)}{\mathrm{d}r} = -G\,\frac{m(r)\rho(r)}{r^2}, \tag{16.36}$$

$$\frac{\mathrm{d}m(r)}{\mathrm{d}r} = 4\pi\rho(r)r^2, \tag{16.37}$$

respectively. They are to be solved simultaneously with the boundary conditions $\rho(0) = \rho_c$ and $\rho(R) = 0$, where ρ_c is the central density. After isolating $-G\,m(r)$ to the second member, we take the derivative of Eq. (16.36), obtaining

$$\frac{\mathrm{d}}{\mathrm{d}r}\left[\frac{r^2}{\rho(r)}\frac{\mathrm{d}P(r)}{\mathrm{d}r}\right] = -G\frac{\mathrm{d}m(r)}{\mathrm{d}r}. \tag{16.38}$$

So, we integrate over the interval $0 \le r \le R$. The integral in the right hand side is just the total mass M of the system, according to Eq. (16.37). Thus, we are left with

$$\left|\frac{r^2}{\rho(r)}\frac{\mathrm{d}P(r)}{\mathrm{d}r}\right|_{r=R} = -GM. \tag{16.39}$$

The next integration can be carried out only numerically. But, for our purpose, we don't need the integration result. Indeed, with the variable changes $s = r/R$ and $u(r) = \rho(r)/\rho_c$ Eq. (16.39) and the integral of Eq. (16.37) are written

$$\frac{R}{\rho_c}\left|\frac{s}{u(s)}\frac{\mathrm{d}P(s)}{\mathrm{d}s}\right|_{s=1} = -GM, \tag{16.40}$$

$$M = R^3\rho_c 4\pi\int_0^1 s^2\mathrm{d}s\,u(s), \tag{16.41}$$

respectively. Combining the two equations, we get

$$P \sim \frac{M\rho_c}{R} \sim \frac{M^2}{R^4}, \tag{16.42}$$

unless constants that are irrelevant in the present context.

16.11 Notes and References

HUA K. Huang, Statistical Mechanics, John Wiley and Sons, 1963, Chap. 11.2.

SOM For calculations of degenerate Fermi systems, a good approximation is the Sommerfeld formula

$$\int_0^{\infty} \mathrm{d}\varepsilon\, \phi(\varepsilon)\, f(\varepsilon) = \int_0^{\mu} \mathrm{d}\varepsilon\, \phi(\varepsilon) + \frac{\pi^2}{6}(k_{\mathrm{B}}T)^2 \left(\frac{\mathrm{d}\phi}{\mathrm{d}\varepsilon}\right)_{\varepsilon=\mu}, \tag{16.43}$$

where $f(\varepsilon)$ is the Fermi function (or occupation number). For the demonstration see, e.g. Yu.B. Rumer and M.Sh. Ryvkin, *Thermodynamics, Statistical Physics and Kinetics*, MIR (1980), Appendix X.

CHA S. Chandrasekhar, Stellar Structure, Dover (N.Y.) 1957, Chap. 11.

LOM These calculations are based on the equation of state from a meson-exchange model of the nuclear interaction (see Z.H. Li, U. Lombardo, H.J. Schulze, and W. Zuo, Phys. Rev. C **77** (2008) 034316).

16.12 Problems

- Problem 1. Calculate the energy and specific heat of a photon gas inside a cavity at thermal equilibrium.
- Problem 2. Derive Eq. (16.18) from the Sommerfeld approximation [SOM].
- Problem 3. During the evolution of a white dwarf, the speed of electrons changes from $v/c \ll 1$ to $v/c \sim 1$. Prove that the star evolves from quasi-equilibrium to non-equilibrium states, correspondingly.[Remind that $\rho \sim k_F^3$, k_F is the wave number].
- Problem 4. Let us consider a fluid made of A nucleons in equilibrium with the electrons inside a volume V at temperature $T = 0$. The electrons are emitted by β-decay from neutrons and created by e-capture from protons. Neglecting neutrinos and antineutrinos, demonstrate the equilibrium condition expressed in terms of the chemical potentials of the reaction $n \rightleftarrows p + e^-$ from the first law of thermodynamics.
- Problem 5. Within the quantum grand canonical ensemble, derive the average kinetic energy $\langle K \rangle$ of an ideal classical gas at temperature T.
- Problem 6. Calculate the energy and pressure of a free Fermi gas at low and zero temperatures.

Chapter 17
Semi-classical Limit

In this chapter, we present a semi-classical approach based on the Wigner transform of the quantum density matrix [WIG]. We prefer this approach to the more famous WKB approximation [WKB] for two main reasons. Firstly, it is more suitable for studying the relationship between classical and quantum mechanics, in particular the correspondence between Poisson brackets and the commutation brackets, which are the basic tools of conservation theorems and the equations of motion. Secondly, the Wigner transform of the density is the natural extension of the particle distribution in the phase space and so it is suitable for the semi-classical treatment of quantum fluid dynamics.

17.1 Wigner and Weyl Transforms

Let us consider a state $|\psi\rangle$ (normalized to one) describing the one-dimensional motion of a system in the potential $V(\mathbf{q})$. Instead of the state vector, we can consider the density operator $\boldsymbol{\rho} = |\psi\rangle\langle\psi|$, which is the projection operator projecting any state onto $|\psi\rangle$. The formalism of quantum mechanics can be recast in terms of the density operator, e.g. the expectation value of a quantum variable $\mathbf{O}$ can be written in the diagonal-coordinates representation

$$\langle\psi|\mathbf{O}|\psi\rangle = \int \mathrm{d}q\mathrm{d}q'\psi^*(q)O(q,q')\psi(q') = \mathrm{Tr}(\mathbf{O}\rho),$$

where the expression in terms of the trace symbol Tr on the right-hand side emphasizes that it is independent of the representation. The time-dependent Schrödinger equation can also be written in terms of the time evolution of the density operator

---

Supplementary Information The online version contains supplementary material available at https://doi.org/10.1007/978-981-97-1670-8_17.

U. Lombardo et al., *Quantum Mechanics*, UNITEXT for Physics,
https://doi.org/10.1007/978-981-97-1670-8_17

$$\mathrm{i}\hbar\frac{\partial\rho(t)}{\partial t} = \mathrm{i}\hbar\,|\partial_t\psi(t)\rangle\langle\psi(t)| + \mathrm{i}\hbar|\psi(t)\rangle\langle\partial_t\psi(t)| = [\mathbf{H}, \rho(t)], \tag{17.1}$$

where we have applied Eq. (7.7) and its dual form. We should note that the order of the operators in the commutator is inverted with respect to the Heisenberg equations of motion.

The physical meaning of the density matrix is apparent. The diagonal matrix element $\psi(q_1)\psi^*(q_1)$ is the probability density for a particle to be detected in the position q_1, and the matrix element $\psi(q_2)\psi^*(q_2)$ is the probability density for a particle to be detected in the position q_2. The off-diagonal matrix element $\psi(q_1)\psi^*(q_2)$, being in general non-vanishing, means that the two events are not exclusive. In order to establish a connection with the language of classical mechanics, we introduce the real function, called Wigner transform[1]

$$f_W(q, p) = \frac{1}{(2\pi\hbar)}\int \mathrm{d}q_{12}\,\mathrm{e}^{\frac{\mathrm{i}}{\hbar}pq_{12}}\langle q_1|\rho|q_2\rangle, \tag{17.2}$$

which replaces the dependence on the variable $q_{12} = q_1 - q_2$ with that of the conjugate momentum p (henceforth, we omit the lower case in f_W, unless needed). The Wigner transform plays the role of a probability density in phase space, but it is not always a definite positive number. For instance, in the case of the harmonic oscillator, the Wigner transform of the ground-state is a definite positive number, whereas that of excited states is not (see Problem 1).

The representation of a dynamic variable in phase space is the Weyl transform [WEY] defined as follows:

$$O_W(q, p) = \int \mathrm{d}q_{12}\,\mathrm{e}^{\frac{\mathrm{i}}{\hbar}pq_{12}}\langle q_1|\mathbf{O}|q_2\rangle. \tag{17.3}$$

Let us calculate, for example, the Weyl transform of the Hamiltonian of a particle of mass m in a scalar potential $U(\mathbf{q})$, in the one-dimensional case. The kinetic energy K_w and potential energy U_w are

$$K_w = \int \mathrm{d}q_{12}\mathrm{e}^{\frac{\mathrm{i}}{\hbar}p\,q_{12}}\left\langle q_1\left|\frac{\mathbf{p}^2}{2m}\right|q_2\right\rangle = \int \mathrm{d}p'\,\delta(p - p')\frac{p'^2}{2m} = \frac{p^2}{2m},$$
$$U_w = \int \mathrm{d}q_{12}\mathrm{e}^{\frac{\mathrm{i}}{\hbar}p\,q_{12}}\langle q_1|U(\mathbf{q})|q_2\rangle = \int \mathrm{d}q_{12}\mathrm{e}^{\frac{\mathrm{i}}{\hbar}p\,q_{12}}U(q)\delta(q_1 - q_2) = U(q),$$

respectively. Notice that, to obtain K_w, we made use of the momentum representation. In the case of non-local potentials, the Weyl transform is

$$U_w = \int \mathrm{d}q_{12}\,\mathrm{e}^{\frac{\mathrm{i}}{\hbar}p\,q_{12}}\langle q_1|U(\mathbf{q})|q_2\rangle = \int \mathrm{d}q_{12}\,\mathrm{e}^{\frac{\mathrm{i}}{\hbar}p\,q_{12}}U(q_1, q_2) = U(q, p),$$

[1] The generalization to many degrees of freedom is straightforward.

showing that a non-local potential is equivalent to a momentum-dependent potential. The average value of a dynamical variable **O** with respect to the state $|\psi\rangle$, expressed in terms of the Weyl transform, is obtained inverting Eqs. (17.2) and (17.3)

$$\langle\psi|\mathbf{O}|\psi\rangle = \mathrm{Tr}(\mathbf{O}\rho) = \int \mathrm{d}q\,\mathrm{d}p\, O_w(q,p)\, f_W(q,p) \qquad (17.4)$$

One main reason for introducing the Wigner transform is the connection of quantum mechanics with classical mechanics at the limit $\hbar \to 0$. This can be significantly illustrated by the connection between the quantum commutation relations and the classical Poisson brackets. In the Complement to this section, the Wigner transform of the commutator between two observables is derived, Eq. (17.4). In the $\vec{r}$-space, the classical limit is the Poisson bracket, i.e.

$$[\mathbf{A},\mathbf{B}]_W(\vec{r},\vec{p}) \Rightarrow \{A_W(\vec{r},\vec{p}),\, B_W(\vec{r},\vec{p})\}. \qquad (17.5)$$

Suppose **A** is the Hamiltonian of the system and **O** is a dynamical variable. In Sect. 10.3, we learned that **O** is a constant of motion, as long as it commutes with the Hamiltonian. Equation (17.5) shows that this is also true in the classical limit with O_W and H_W, but exchanging the role of commutators with that of the Poisson brackets.

17.2 Ehrenfest Theorem Revisited

In order to get familiar ourselves with the new formalism, we reconsider the Ehrenfest theorem of Sect. 10.4 using the Wigner transforms. We start with the derivation of the equations of motion of a particle in a one-dimensional potential $U(q_1, q_2) = U(q_1)\delta(q_1 - q_2)$. The Wigner transform version of the equation of motion Eq. (17.1) can be obtained from Eq. (17.4), in which the Wigner transform of the commutator between two observables is expressed in terms of the Wigner transforms of the two observables. The classical limit is just the Poisson bracket of the two Wigner transforms as shown in Eq. (17.5). Thus, we obtain (the arguments q and p are understood)

$$\frac{\mathrm{d}f}{\mathrm{d}t} - \{H_W, f\} = 0 \qquad (17.6)$$

where $H_W = p^2/2m + U(q)$. Analogously to Sect. 10.4, we consider the equations of motion for the average values of position and momentum. Applying Eq. (17.4) for the average values, it is easily proved that the Wigner transforms of the commutators are equal to the Poisson brackets with the Hamiltonian H_w, as in the semi-classical limit. As a consequence, we get the same quantum equations of motion of coordinate and momentum as in Sect. 10.4

$$\frac{\mathrm{d}\langle q\rangle}{\mathrm{d}t} = \int \mathrm{d}q\,\mathrm{d}p\, q\, \{H_W(q,p), f(q,p)\} = \frac{\langle p\rangle}{m},$$

$$\frac{\mathrm{d}\langle p\rangle}{\mathrm{d}t} = \int \mathrm{d}q\,\mathrm{d}p\; p\;\{H_W(q,p), f(q,p)\} = -\left\langle \frac{\mathrm{d}V}{\mathrm{d}q} \right\rangle,$$

after performing part integrations with f vanishing at infinity. As we learned in Sect. 10.4, in this case, the only quantum corrections can arise from the potential gradient. Equation (17.3) furnishes the $\hbar$ expansion from which to calculate all quantum corrections to the equations of motion. The first non-vanishing correction to the time derivative of $\langle q(t)\rangle$ is of the third-order in $\hbar$, that is

$$\left[\frac{\mathrm{d}\langle q\rangle}{\mathrm{d}t}\right]_{\mathrm{corr}} = \frac{-\hbar^2}{24} \int \mathrm{d}q\,\mathrm{d}p\, q\;\, H_W(q,p)[\overleftarrow{\partial}_q \cdot \overrightarrow{\partial}_p - \overleftarrow{\partial}_p \cdot \overrightarrow{\partial}_q]^3 f(q,p).$$

The derivatives with left (right) arrow act on the left (right) function. It is worthwhile stressing that there are no first-order $\hbar$ corrections.

17.3 Semi-classical Limit of the HF Approximation

The definition of the one-body density matrix, introduced in Sect. 17.1, can be extended to N particle systems as it follows:

$$\rho_N = |\psi^N\rangle\langle\psi^N|,$$

where $|\psi^N\rangle$ are orthonormalized wave functions of N particles $\psi_N(\vec{r}_1, \vec{r}_2, \cdots, \vec{r}_N)$ (the spins of particles are understood). From ρ_N, one can extract the n-body ($n \leq N-1$) density matrix, describing n-body dynamics of the system, integrating over particles $n+1, \ldots, N$ of the N-body density matrix with $\vec{r}_{n+1} = \vec{r}_{n+1}{}', \ldots, \vec{r}_N = \vec{r}_N{}'$. The first two density matrices are

$$\rho_1(\vec{r}_1; \vec{r}_1{}') = \int \mathrm{d}\vec{r}_2 ... \mathrm{d}\vec{r}_N \; \rho_N(\vec{r}_1, \vec{r}_2, \ldots, \vec{r}_N; \vec{r}_1{}', \vec{r}_2, \ldots, \vec{r}_N),$$

$$\rho_2(\vec{r}_1, \vec{r}_2; \vec{r}_1{}', \vec{r}_2{}') = \int \mathrm{d}\vec{r}_3 ... \mathrm{d}\vec{r}_N \; \rho_N(\vec{r}_1, \vec{r}_2, \vec{r}_3, \ldots, \vec{r}_N; \vec{r}_1{}', \vec{r}_2{}', \vec{r}_3, \ldots, \vec{r}_N).$$

In the case of N-body systems with independent particles, there are no correlations among the particles, so that the N-body density matrix factorizes into the product of N one-body density matrices. Since we want to restrict our investigation to independent particle systems in the Hartree–Fock approximation (see Sect. 14.2), we only consider the one-body density matrix ρ_1 (henceforth, we omit the subscript 1). In the Hartree–Fock approximation, the density matrix is (see Sect. 14.2)

$$\rho(\vec{r}, \vec{r}\,') = \sum_k \phi_k(\vec{r}, t)\phi_k^*(\vec{r}\,', t),$$

where ϕ_k are the single particle (s.p.) wave functions. The time-dependent Hartree–Fock equation of motion in operator form is written as

$$\mathrm{i}\hbar\frac{\partial\rho}{\partial t} = [\mathbf{H}, \rho].$$

The Hamiltonian contains a local mean field and a non-local exchange term. Afterwards, the exchange term will be neglected.

Moving on to the Wigner transform, the semi-classical limit can be easily obtained by applying Eq. (17.6). The resulting equation, known as Vlasov equation, is formally written as

$$-\frac{\mathrm{d}f}{\mathrm{d}t} = \{h_{\mathrm{HF}}, f\} = \vec{\nabla}U_{\mathrm{HF}} \cdot \vec{\nabla}_p f - \vec{p}/m \cdot \vec{\nabla} f. \tag{17.7}$$

It keeps the main quantum peculiarities of the Hartree–Fock model, namely the self-consistent density dependence of the potential $U = U_{\mathrm{HF}}(\rho)$, as described in Chap. 14, and the Pauli blocking effect in the momentum distribution of $f(\vec{r}, \vec{p})$. The Vlasov equation looks like the collisionless equation of motion of a fluid in phase space. According to the fluid dynamics theory, it can be recast in the form of a chain of coupled equations for the collective variables of a fluid. Integrating upon momentum space, one obtains the continuity equation

$$\frac{\partial\rho(\vec{r}, t)}{\partial t} + \vec{\nabla} \cdot \vec{J}(\vec{r}, t) = 0.$$

It simply tells us that, in the absence of collisions, at any time t, the number of particles inside any elementary space cell $d^3\vec{r}$, i.e. $\rho(\vec{r}, t)d^3\vec{r}$, must be equal to the incoming particles minus the outgoing particles, i.e. the *current* $\vec{J}(\vec{r}, t)$. Density and current are defined, respectively, as follows:

$$\rho(\vec{r}, t) = \int \mathrm{d}\vec{p}\; f(\vec{r}, \vec{p}, t),$$
$$\vec{J}(\vec{r}, t) = \frac{1}{m}\int \mathrm{d}\vec{p}\vec{p}\; f(\vec{r}, \vec{p}, t).$$

In order to determine $\rho(\vec{r}, t)$ from the continuity equation we need to know the current $\vec{J}(\vec{r}, t)$. The Vlasov equation, first multiplied by $\vec{p}$ and the integrated over $\vec{p}$, yields the equation of motion for the current

$$\frac{\partial\vec{J}(\vec{r}, t)}{\partial t} + \frac{1}{m}\nabla \cdot \tau(\vec{r}, t) + \frac{1}{m}\nabla U(\vec{r}) = 0,$$

called as Euler equation. This equation tells us that the time variation of the collective flow is driven by the interplay between the mean-field gradient $\vec{\nabla}U$ and the *pressure* divergence $\sum_i \nabla_j\tau_{ij}$, the pressure tensor being defined as

$$\tau_{ij}(\vec{r},t) = \frac{1}{m}\int \mathrm{d}\vec{p}\, p_i\, p_j\, f(\vec{r},\vec{p},t) \quad (17.8)$$

Again, in order to determine the $\vec{J}(\vec{r},t)$ from the Euler equation, we need to know the equation of motion for the pressure and so on and so forth, in the next steps. This is the procedure to generate and then solve the chain of fluid dynamical equations. In general, we do not need to solve the whole chain, because, depending on the adopted model, it truncates at some step. In the forthcoming applications, it truncates at the step of the Euler equation.

17.4 Complement to Sect. 17.2

In this complement, we derive the $\hbar$ power expansion of the commutators between dynamical variables transcribed in the Wigner transform formalism. This enables us to make contact with the classical and semi-classical limit of quantum dynamics. In particular, we consider the commutator $[\mathbf{H},\rho] = [\mathbf{K}+\mathbf{U},\rho]$, where $\mathbf{K} = \mathbf{p}^2/2m$ is the kinetic energy and $U(\mathbf{q})$ the potential energy. We restrict the formalism to the one-dimensional case, the generalization to the multi-dimensional case being straightforward. Firstly, we calculate the Wigner transform of the product $(\mathbf{U}\rho)$:

$$\begin{aligned}(U\rho)_W(q,p) &= \int \mathrm{d}q_{12}\;\; \mathrm{e}^{\mathrm{i}pq_{12}/\hbar} U(q_1)\rho(q_1,q_2)\\ &= \int \mathrm{d}q_{12}\mathrm{e}^{\mathrm{i}pq_{12}/\hbar}\int \mathrm{d}p'\mathrm{e}^{-\mathrm{i}p'q_{12}/\hbar} f_W(q,p')\mathrm{e}^{\frac{1}{2}q_{12}\partial_q}U(q)\\ &= \int \mathrm{d}q_{12}\mathrm{e}^{\mathrm{i}pq_{12}/\hbar}\int \mathrm{d}p'\mathrm{e}^{-\mathrm{i}p'q_{12}/\hbar}[\mathrm{e}^{(p'-p)\partial_p} f_W(q,p)]\\ &\quad\times[\mathrm{e}^{\frac{1}{2}q_{12}\partial_q}U(q)], \end{aligned} \quad (17.9)$$

where $q = (q_1+q_2)/2$ and $q_{12} = q_1 - q_2$. In the first step, the locality of the potential $\langle q_1|\mathbf{U}|q'\rangle = U(q_1)\delta(q_1-q')$ was applied; in the second step, the Wigner transform of the density was introduced and the potential $U(q_1) = U(q+q_{12}/2)$ expanded in Taylor's series around q; in the last step, $f_W(p') = f_W(q, p'-p+p)$ was developed in Taylor's series around p. The last expression allows us to perform two integrations:

$$(U\rho)_W(q,p) = \int \mathrm{d}q_{12}\delta(p-p'+\frac{\hbar}{2i})\partial_q f_W(q,p)\mathrm{e}^{\frac{1}{2}q_{12}\partial_q}U(q) \quad (17.10)$$

$$= f_W(q,p)\mathrm{e}^{(1/2)\overleftarrow{\partial_p}\overrightarrow{\partial_q}}U(q) \quad (17.11)$$

The Wigner transform of the product ($\mathbf{K}\rho$) can easily be calculated with the same procedure by replacing the coordinate representation with the equivalent diagonal-momentum representation, i.e. exchanging in coordinate with momentum in the differential operators. The result is

$$(K\rho)_w(q,p) = f_W(q,p)\mathrm{e}^{(1/2)\frac{\hbar}{i}\overleftarrow{\partial}_q\overrightarrow{\partial}_p}K_W, \tag{17.12}$$

where $K_W = p^2/2m$ (notice that here p^2 is not an operator but a c-number). Combining the two results, Eqs. (17.11) and (17.12), we obtain the Wigner transform of the commutator

$$[H,\rho]_W(q,p) = 2i\frac{2}{\hbar}H_W(q,p)\sin\frac{\hbar}{2}\left(\overleftarrow{\partial}_q\overrightarrow{\partial}_p - \overleftarrow{\partial}_p\overrightarrow{\partial}_q\right)f_W(q,p). \tag{17.13}$$

This equation can be proved to be valid for the commutator of any couple of operators. In three-dimensional space, it is written as

$$[A,B]_W(\vec{r},\vec{p}) = A(\vec{r},\vec{p})\frac{2}{\hbar}\sin\frac{\hbar}{2}\left(\overleftarrow{\partial}_{\vec{r}}\cdot\overrightarrow{\partial}_{\vec{p}} - \overleftarrow{\partial}_{\vec{p}}\cdot\overrightarrow{\partial}_{\vec{r}}\right)B(\vec{r},\vec{p}). \tag{17.14}$$

The main property of Eq. (17.3) is that the classical limit of the quantum commutator is the Poisson bracket, as it results in the first order of the series expansion of the sin function,

$$[A,B]_w(\vec{q},\vec{p}) \sim \{A(\vec{q},\vec{p}),B(\vec{q},\vec{p})\},$$

and the $\hbar$ higher-orders can be interpreted as quantum fluctuations.

17.5 Notes and References

WIG Wigner, E.: On the quantum correction for thermodynamic equilibrium. Phys. Rev **40** (1932) 749.

WKB The WKB approximation assumes the wave function of a stationary state to be written in the form $\psi(\vec{r}\,) = A(\vec{r}\,)\exp\left(\frac{\mathrm{i}}{\hbar}S(\{\vec{r}\})\right)$. This form is suitable for distinguishing the semi-classical limit from the quantum fluctuations.

WEY Weyl, H.: The Theory of Groups and Quantum Mechanics, p. 275. Dover, New York (1931).

17.6 Problems

- Problem 1. Calculate the Wigner transform of the harmonic oscillator ground state and the first excited state wave functions. Comment the results in connection with their interpretation in terms of probability density.
- Problem 2. Calculate the Weyl transform of the one-dimensional harmonic oscillator Hamiltonian and its average value in the ground state. Discuss the result.
- Problem 3. Calculate the exact Weyl transform of the angular momentum $\tilde{\mathbf{L}}$ and of the squared angular momentum $\mathbf{L}^2$ and discuss the result Versus the semi-classical approximation.
- Problem 4. Using the Wigner–Weyl approach, derive and solve the equations of motion of the linear rigid rotor with Hamiltonian $\mathbf{H} = \frac{\mathbf{l}_z^2}{2\mathcal{I}}$ and compare the solution with the full quantum–mechanical one.
- Problem 5. Derive the Vlasov equation of a many-body system interacting through a self-consistent potential $\mathbf{U}$ in a rotating frame with constant angular velocity ω.

Chapter 18
Collective Modes in Atomic and Nuclear Systems

In this presentation we only consider collective modes of Fermi systems, including the nuclei, within the semiclassical apprximation of quantum fluid dynamics of diluted systems. The central role of this approximatio is played by the Vlasov equation We consider only two applications: the sound propagation in quantum homogeneous systems, and the giant resonances in nuclear systems.

18.1 Collective Modes in Fermi Systems

At variance with classical fluids, which exhibit only the ordinary sound (henceforth called first sound), quantum systems also exhibit the so-called zero sound [LAN]. Its existence is supported directly by the experimental sound attenuation curves and indirectly by the models devised to interpret higher multipolarity giant resonances in nuclei.

In classical mechanics, a fluid is described by the particle distribution $f(\vec{r}, \vec{p}, t)$ in the phase space. In quantum mechanics, its analogue is the Wigner transform $f_W(\vec{r}, \vec{p}, t)$ of the one-body density matrix. There is a strong resemblance between the two functions, except that the latter is constrained by the Pauli principle. In both theories, the local equilibrium of the fluid is characterized by an isotropic momentum distribution. The resemblance extends to the equations of motion. In the classical fluid dynamics, this is the transport Boltzmann equation [HUA]

$$\frac{\partial f}{\partial t} = \{K + U, f\} + I_{\text{coll}} \tag{18.1}$$

The original version of the chapter has been revised: A correction to this chapter can be found at https://doi.org/10.1007/978-981-97-1670-8_19.

Supplementary Information The online version contains supplementary material available at https://doi.org/10.1007/978-981-97-1670-8_18.

U. Lombardo et al., *Quantum Mechanics*, UNITEXT for Physics,
https://doi.org/10.1007/978-981-97-1670-8_18

and in the quantum fluid dynamics, this is the Vlasov equation (see Chap. 16)

$$\frac{\partial f_W}{\partial t} = \{K + U_W, f_W\}_. \tag{18.2}$$

The potential U in Eq. (18.1) is an external field and the interaction manifests in terms of collisions between particles in the collision integral I_{coll} [HUA], whereas, in Eq. (18.2), the only effect of the interaction is the Hartree–Fock potential U_W (henceforth the subscript W will be understood). The collision term I_{coll} drives the system towards equilibrium and disappears in the case of local equilibrium, whereas the self-consistent potential U_W cannot do that, so equilibrium must be enforced through approximations, such as the adiabatic limit, in the distribution function f itself.

As discussed in Chap. 11, in atomic fluids the collective modes are strictly characterized by the interplay between the inter-particle collision time τ and the oscillation period $T = 2\pi/\omega$. The treatment can be simplified into two extreme regimes. If $\tau/T \ll 1$ (hydrodynamical regime), then the collisions have enough time to keep local equilibrium with an isotropic distribution of the particle momenta. This happens with first sound either in classical or quantum ideal homogenous fluids. During the propagation, the density oscillations (collective variables) are accompanied by the expansion of momentum distribution, which, however, maintains an isotropic shape as an effect of rapid collision rates. Conversely, if $\tau/T \gg 1$ (collisionless regime), then the collisions have no time to destroy the collective character of the disturbance that generates the anisotropies in momentum space. In nuclear systems, the decision on which of the two regimes prevails depends on the interplay between the oscillation time and single-particle transition time.

18.2 Quantum Fermi Fluids

For the sake of illustration of the two aforementioned regimes of collective motion, we consider the sound propagation in a quantum Fermi fluid. This propagation exhibits a quite different behaviour according to whether a collisional regime (first sound) or a collisionless regime (zero sound) is established in the system.

18.2.1 Quantum First Sound

Let us consider collective oscillations propagating in a diluted Fermi fluid at low temperature ($kT \ll \varepsilon_F$), in particular ^{3}He. Making the assumption that the system is in a local equilibrium state, the distribution function $f(\vec{r}, \vec{p}, t)$ is written as

$$f(\vec{r}, \vec{p}, t) = f_0(\vec{r}, |\vec{p} - \vec{p}_0|), \tag{18.3}$$

where $f_0(p)$ is the ideal Fermi gas distribution (see Fig. 16.2). The variable $\vec{p}_0/m = \vec{u}(\vec{r}, t)$ is the propagation speed of the collective oscillation. This approximation entails the medium undergoing an adiabatic process, as it results from the calculation of the heat transfer ΔQ between the collective motion and the medium

$$\Delta Q \sim \int d^3\vec{p}(\vec{p}-\vec{p}_0)|\vec{p}-\vec{p}_0|^2 f_0(\vec{r},|\vec{p}-\vec{p}_0|)$$
$$= \int d^3\vec{p}\,\vec{p}\,p^2 f_0(\vec{r},p,t) = 0. \quad (18.4)$$

Assuming that the collective variables: density variation $\delta\rho(\vec{r},t)$, current $\vec{J}(\vec{r},t)$, and pressure gradient $\vec{\nabla}P(\vec{r},t)$ are small, we linearize the fluid dynamical equations Eqs. (17.7) and (17.8) to get

$$\frac{d\delta\rho}{dt} + \frac{1}{m}\rho_0\vec{\nabla}\cdot\vec{p}_0 = 0, \quad (18.5)$$
$$\rho_0\frac{d\vec{p}_0}{dt} + m\vec{\nabla}P = 0. \quad (18.6)$$

Combining the time derivative of the first equation and the space divergence of the second one, we obtain

$$\frac{d^2\delta\rho}{dt^2} - \nabla^2 P = 0. \quad (18.7)$$

Since we are dealing with an adiabatic process, the pressure gradient can be expressed in terms of the adiabatic compressibility κ

$$\kappa = \rho\frac{\partial P}{\partial\rho}, \quad (18.8)$$

that enables us to convert Eq. (18.7) into the equation of the sound propagation in terms of density fluctuations

$$\frac{\partial^2\delta\rho}{\partial^2 t} - v^2\nabla^2\delta\rho = 0, \quad (18.9)$$

where $v = \sqrt{\kappa/\rho}$. In order to simplify matters, we consider the zero temperature limit[1], in which the pressure P is generated by the particles with momentum $p \le p_F$. Thus, the adiabatic compressibility can be calculated from the $T = 0$ equation of state: $P = \frac{2}{5}\rho_0\varepsilon_F$ (see Problem 16.2). Then the sound velocity turns out to be

$$v = \hbar k_F/(m\sqrt{3}) = v_F/\sqrt{3}. \quad (18.10)$$

No comparison at all is possible with the classical fluid, where the pressure is sustained only by the thermal motion, and the speed of sound is $v = \sqrt{5kT/3m}$ [HUA].

[1] In reality, at zero temperature, only the zero sound can occur.

18.2.2 Zero Sound

In a quantum Fermi fluid, another kind of sound has been predicted, first theoretically [LAN] and then observed in ^{3}He Fermi gas. This is a purely quantum–mechanical effect occurring at very low temperature. At $T = 10^{-3}\,^0$K the collision rate becomes very low, as low as T^2 because most atoms are blocked below the Fermi energy and only very few are above. The collision time τ is increasing like T^{-2} so that the collisionless regime $\omega\tau \gg 1$ is fulfilled in a large range of the oscillation angular frequency. The collective motion will be sustained by the strong coupling with the self-consistent mean field of the atomic medium.

We study the zero sound in the semi-classical approximation, linearizing the Vlasov equation Eq. (17.7)

$$\begin{aligned}\frac{\mathrm{d}}{\mathrm{d}t}\delta f(\vec{r},\vec{p},t) &= \{\delta h, f_0\} + \{h_0, \delta f\} = \vec{\nabla}\delta U \cdot \vec{\nabla}_p f_0 - \vec{\nabla}_p h_0 \cdot \vec{\nabla}\delta f \\ &= \kappa\vec{\nabla}\delta\rho \cdot \vec{\nabla}_p f_0 - \vec{p}/m \cdot \vec{\nabla}\delta f,\end{aligned} \tag{18.11}$$

where $\delta f = f(\vec{r},\vec{p},t) - f_0(p^2)$ is the first-order expansion of the perturbed momentum distribution around the solution of the static Hartree–Fock equation, $\{h_0, f_0\} = 0$. The quantity $\kappa = (\delta U/\delta\rho)_0$ is the compressibility of the atomic fluid. In contrast with the quasi-static distribution leading to the first sound, we want to look for solutions where the momentum distribution departs from the isotropic shape. For this purpose, we expand δf in plane-wave series (assume $\hbar = 1$)

$$\delta f = \sum_k \delta f_k(\vec{p}) \mathrm{e}^{\mathrm{i}(\vec{k}\cdot\vec{r} - \omega_k t)} . \tag{18.12}$$

Introducing this expansion into Eq. (18.11), the linearized Vlasov equation decomposes into a series of partial-wave equations

$$\left(\omega_k - \frac{\vec{p}\cdot\vec{k}}{m}\right)\delta f_k(p) = \kappa\delta\rho_k \vec{k}\cdot\vec{\nabla}_p f_0, \tag{18.13}$$

where $\delta\rho_k = \int \mathrm{d}\{\vec{p}\}\,\delta f_k(\vec{p})$. After integrating over momentum, we obtain

$$-1 - \frac{1}{4\pi\kappa v_F} = \frac{\omega}{2k v_F}\ln\left|\frac{\omega + k v_F}{\omega - k v_F}\right| . \tag{18.14}$$

The integration has been performed in the free Fermi-gas approximation for the ground state $f_0 = \theta(p_F - p)$, the first derivative of the step function being $\vec{\nabla}_p f_0 = \frac{\vec{p}}{p}\delta(p - p_F)$. In the limit $k \to 0$, we can assume $\omega_k = v\,k$, where v has the meaning of sound velocity. Applying this relation, Eq. (18.14) becomes

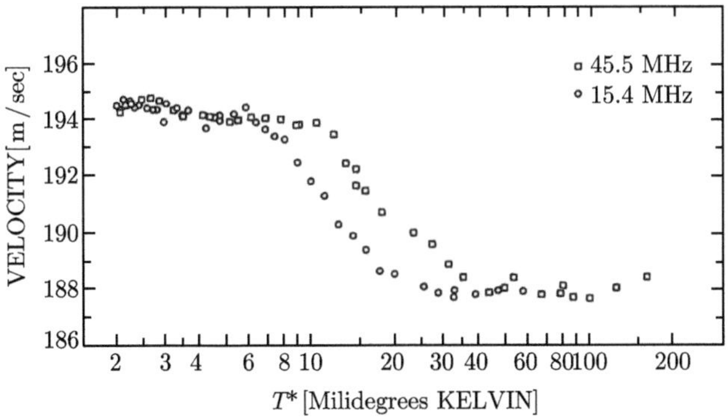

Fig. 18.1 Zero sound attenuation curves. Sound propagation velocity as a function of magnetic temperature in pure liquid ^{3}He at 0.32 atmospheres and for frequencies of 15.4 and 45.5 MHz (ABE). Courtesy of American Physical Soc

$$-1 - \frac{1}{4\pi\kappa\, v_{\mathrm{F}}} = \frac{v}{2v_{\mathrm{F}}}\ln\left|\frac{1+v/v_{\mathrm{F}}}{1-v/v_{\mathrm{F}}}\right|. \tag{18.15}$$

In the case of $\kappa > 0$, the right side of Eq. (18.15) is resonant for $v \to v_{\mathrm{F}}$. The existence of the zero sound mode in atomic Fermi systems is supported by the measurements of sound attenuation in ^{3}He. In Fig. 18.1, the transition from the first sound to the zero sound vs. temperature is shown. The fact that the zero sound could also occur in nuclear matter is confirmed by the success of theoretical models of giant resonances in nuclei (see next section). In the case of $\kappa < 0$, no solution of Eq. (18.15) can exist at $T = 0$. This is interpreted as due to the onset of a superfluid phase at very low temperature, as in the case of ^{3}He [LEG].

18.3 Nuclear Collective Modes

In addition to the single-nucleon excitations, nuclei exhibit collective excitations, in which nucleons perform coherent excitations. Since their excitation energy is much larger than single-particle energy, these modes, called giant resoances, are expected to occur in a collisionless regime ($\tau_{\mathrm{coll}} \ll \tau_{s.p.}$), similarly to the zero sound. They are classified according to whether neutrons and protons oscillate in-phase or out-of-phase. In the first case we speak of isoscalar modes, in the second case of isovector modes. The most important giant resonances we are going to discuss are the isoscalar monopole (ISGMR) and quadrupole (ISGQR) resonances and the isovector dipole (IVGDR) resonance. Their profiles are sketched in Fig. 18.2.

18.3.1 Scaling Model

Giant resonances are induced by external perturbations on the nuclear ground-state[2]. If the perturbations are weak enough, the giant resonances can be studied as small amplitude oscillations about the ground-state, so that the quantum equations of motion can be linearized, as described in Chap. 8. In general, the exact procedure involves cumbersome calculations, which make the interpretation of the experimental data quite difficult. For this reason semiclassical approaches have been devised that give a simpler, even if incomplete, description of their properties. Since the nucleus is a many-body system, the semiclassical Hartree-Fock approximation is suitable for describing the dynamics of giant resonances. We start with linearizing the Vlasov equation, Eq. (18.2), as follows

$$\frac{\partial}{\partial t}\delta f(\vec{r}, \vec{p}, t) + \{h_0, \delta f\} + \{\delta h, f_0\} = 0, \tag{18.16}$$

where $f_0(\vec{r}, \vec{p})$ is the Wigner transforms of the Hartree-Fock nuclear ground-state. Its momentum distribution is isotropic about the Fermi momentum sphere, f_0 being solution of the static Vlasov equation, i.e. $\{h_0, f_0\} = 0$. The Wigner transform of the Hamiltonian is $h = p^2/2m + U_{HF}(\rho)$, where U_{HF} is a simplified version of the HF potential, depending only on the nuclear density. Henceforth , we use the notation $U(\rho) = U_{HF}(\rho)$ and $U_0 = U(\rho_0)$, ρ_0 being the stationary nuclear density.

A simple model for describing the giant resonances is the *scaling model* [DITO], according to which the solution $f(\vec{r}, \vec{p}, t)$ of the Vlasov equation has the form of the stationary solution f_0 with space and momentum displaced by a suitable collective variable $\chi_s(\vec{r}, \vec{p}, t)$, named scaling parameter, i.e.,

$$f(\vec{r}, \vec{p}, t) = f_0(\vec{r} + \vec{\nabla}_p\chi_s, \vec{p} - \vec{\nabla}\chi_s) \tag{18.17}$$

For small oscillations χ_s is small, and then the series expansion of scaled solution can be truncated at the first order

$$\delta f(\vec{r}, \vec{p}, t) = \vec{\nabla}_p\chi_s \cdot \vec{\nabla} f_0 - \nabla\chi_s \cdot \vec{\nabla}_p f_0 = \{\chi_s, f_0(\vec{r}, \vec{p})\},$$

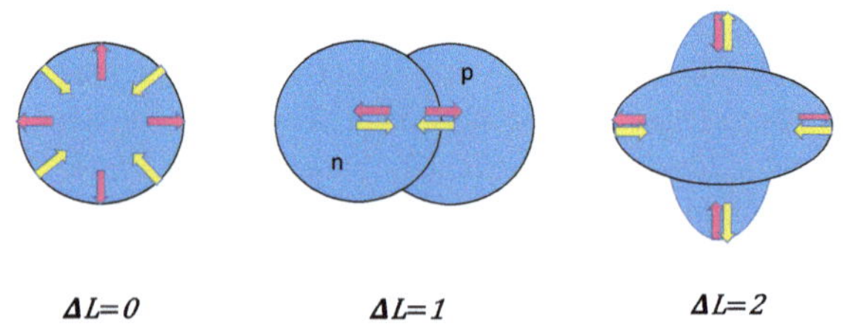

Fig. 18.2 Sketch of the main GRs: $\Delta L = 0$ (monopole), $\Delta L = 1$ (dipole), $\Delta L = 2$ (quadrupole)

[2] We do not consider thermally excited nuclei.

showing that the scaling parameter couples space coordinate with momentum. Thus, the scaling model, in principle, postulates that the space oscillations are sustained by momentum distortions, the same as the zero sound. Following the criterion of maximum collectivity, we truncate to the first-order the series expansion in momentum powers

$$\chi_s(\vec{r}, \vec{p}, t) = \chi(\vec{r}, t) - \vec{p} \cdot \vec{s}(\vec{r}, t), \tag{18.18}$$

where $\vec{s} = \vec{\nabla}_p \chi_s$ is the collective space displacement according to Eq. (18.17).

In the scaling model the transition density $\delta\rho$, the current $\vec{J}$ and the pressure tensor variation $\delta\tau = \tau(\vec{r}, t) - \tau^o(\vec{r})$, due to the oscillations, are

$$\delta\rho(\vec{r}, t) = \int d\vec{p}\{\chi_s, f_0\} = -\nabla \cdot (\rho_0 \vec{s}) \tag{18.19}$$

$$\vec{J}(\vec{r}, t) = \frac{1}{m} \int \vec{p}\, d\vec{p}\{\chi_s, f_0\} = \frac{1}{m} \rho_0 \vec{\nabla} \chi, \tag{18.20}$$

$$\delta\tau_{ij} = \frac{1}{m} \int p_i p_j d\vec{p}\{\chi_s, f_0\} = \nabla_k (s_k \tau^o_{ij}) + \tau^o_{kj} \nabla_i s_k + \tau^o_{ik} \nabla_j s_k$$

respectively. The time derivative of the first equation coincides with the linearized continuity equation if we identify $\rho_0 \dot{\vec{s}}$ with the current $\vec{J}$. By doing so, we get a relationship between the scaling parameters

$$\dot{\vec{s}} = \frac{1}{m} \vec{\nabla} \chi, \tag{18.21}$$

which enables us to figure out the physical meaning of the parameter $\vec{u} = \vec{\nabla}\chi/m$, that is the collective velocity $\vec{u} = \dot{\vec{s}}$. In the language of fluid dynamics the scaling model implies the approximation of irrotational flow, in that the collective velocity $\vec{u}$ is generated by the gradient of a collective field, $\chi(\vec{r})$ in our case, and then no current vortices are produced, that is $rot \vec{\nabla} \chi = 0$.

18.3.2 Giant Resonances

In the study of the giant modes of various multipolarity L the following choice (Tassie model) is frequently adopted

$$\chi(\vec{r}) = \begin{cases} r^2 & (L = 0) \\ r^L Y_{L0} & (L > 0), \end{cases} \tag{18.22}$$

In this model the flow is irrotational for all three modes considered, whereas it is also incompressible ($\vec{\nabla} \cdot \vec{u} = 0$) for the $L > 0$ modes only. The linearized Vlasov equation Eq. (18.16) with the boundary conditions at the nuclear surface gives the energy spectrum of the giant modes, which is experimentally detected. However, a simplification of calculations and a better insight of the inherent physics is obtained if we focus on the average energy or energy centroid).

As a first step, we integrate Eq. (18.16) over momentum, obtaining the linearized continuity equation

$$\delta\dot{\rho} = -\int d\vec{p}\{\delta h, f_0\} + \{h_0, \delta f\} = -\vec{\nabla}\cdot\vec{J},$$

where, in the scaling approximation, the current is $\vec{J} = \rho_0\vec{u}$. This equation links the transition density $\delta\rho$ to the current $\vec{J}$. Then, we multiply by p_k/m the Vlasov equation and integrate over momentum, obtaining the linearized Euler equation

$$\frac{\partial J_k}{\partial t} = -\frac{1}{m}\int d\vec{p}\; p_k[\{\delta h, f_0\} + \{h_0, \delta f\}] = -\frac{1}{m}\sum_i \nabla_i \delta\tau_{ik} - \delta(\rho\nabla_k U),$$

where $\delta(\rho\nabla_k U) = \rho_0\nabla_k\delta U + \delta\rho\nabla_k U_0$. This equation links the current to the variation of pressure tensor $\delta\tau_{ik}$. Multiplying it by $p_i p_j/m$ and integrating, we obtain the equation linking τ to the heat flow Q; iterating N times this procedure we obtain the full fluid dynamical chain, that is a system of N coupled equations, N being the number of particles. In the scaling approximation we do not need to go beyond the Euler equation, as shown below.

As a second step, we multiply the Euler equation by u_k, sum over k, and integrate over space, obtaining

$$\int \mathrm{d}\vec{r}\;\rho_0\ddot{\vec{s}}\cdot\vec{u} + \frac{1}{m}\int \mathrm{d}\vec{r}\sum_k u_k[\sum_i \delta\tau_{ik} + \delta(\rho\nabla_k U)] = 0. \tag{18.23}$$

Since the pressure tensor $\delta\tau$ and the potential δU are linear functions of $\delta\rho$, which, in turn, is a linear function of the space displacement $\vec{s}$, the latter equation is a second-order differential equation in $\vec{s}$. Then, assuming that $\vec{s}(\vec{r}, t) = \alpha(t)\vec{s}(\vec{r})$, it takes the form of wave equation $M\ddot{\alpha}(t) + C\alpha(t) = 0$, where

$$M = m\int \mathrm{d}\vec{r}\;\rho_0\ddot{\vec{s}}\cdot\vec{u} = m\int \mathrm{d}\vec{r}\;\rho_0 u^2(\vec{r}) \tag{18.24}$$

$$C = \int \mathrm{d}\vec{r}\sum_k u_k[\sum_i \nabla_i\delta\tau_{ik} + \delta(\rho\nabla_k U)]. \tag{18.25}$$

are the inertial and stiffness parameters, respectively. These parameters will be calculated for each giant resonance we are considering, according to the boundary conditions imposed by the nuclear structure, and then the average energy $E = \hbar\omega$, ω being the angular frequency of the collective mode, $\omega^2 = C/M$.

- ISGMR ($\Delta L = 0$)
The monopole resonance, or *breathing mode*, is a radial compression-expansion oscillation of in-phase neutrons and protons around the equilibrium configuration. Accordingly, the collective velocity, given by the Tassie model, Eq. (18.22), is $\vec{u} \sim \vec{r}$ and, hence, the monopole resonance is not an incompressible motion, even if an irrotational one. As such, it is sensitive to the compression modulus of the nucleus, as we will see below. The stiffness parameter is given by

$$C = \int \mathrm{d}\vec{r} \sum_k u_k [\sum_i \nabla_i \delta\tau_{ik} + \rho_0 \nabla_k \delta U + \delta\rho \nabla_k U_0]$$
$$= -\int \mathrm{d}\vec{r}\, \delta\rho [(\frac{d\tau}{d\rho})_0 + \rho_0 (\frac{dU}{d\rho})_0] \vec{\nabla} \cdot \vec{u} \tag{18.26}$$

having carried out integration by parts of the first two terms and the cancellation of the two surface contributions. The two terms inside square bracket define the compression modulus. Neglecting, for a while, the contribution from the nuclear potential, we are left with oscillations substained only by the kinetic compression modulus, similar to the sound waves in atomic gas, except for the finite size effects. Assuming a spherical nucleus with density $\rho_0 = constant$ and radius R, the stiffness and inertial parameters, calculated with Eq. (18.25), are

$$C = \frac{4\pi R^3}{3} \rho_0 \varepsilon_F \tag{18.27}$$
$$M = \frac{4\pi R^5}{5} 4m\rho_0, \tag{18.28}$$

respectively, and the IGMR average energy

$$E = \hbar\omega = \hbar\sqrt{\frac{15\varepsilon_F}{m}} R^{-1} \sim 73.8 A^{-1/3} MeV \tag{18.29}$$

where the static pressure tensor has been approximated by the Fermi-gas model value $\tau_0 = \rho_0 \varepsilon_F$. In the last step the empirical law $R = 1.2A^{1/3} Fm$ and ε_F was calculated from the nuclear-matter saturation density (see Sect. 14.4.2). The $A^{-1/3}$ dependence can be easily explained by means of geometrical argument: C is proportional to the volume V, whereas M is proportional to $V^{5/3}$ and the ratio to the radius R^{-1}, then $E \sim A^{-1/2}$.
Despite the approximations made, this estimate of the ISGMR excitation energy reasonably matches the systematics of the experimental data, as shown in Fig. 18.3. In particular, this proves that the contribution of nuclear potential to the compressibility of nuclei is much lower than that of kinetic energy. In recent years the problem of the relation of the compressibility in nuclei and in nuclear matter has been raised [BLA]. The latter, in fact, is of particular concern in the study of

astrophysical compact objects and composite nuclear systems formed in head-on nuclear collisions.

- ISGQR ($\Delta L = 2$)
  The nuclear isoscalar quadrupole modes ($\Delta L = 2$) are in-phase oscillations of protons and neutrons. According to Eq. (18.22) the quadrupole scaling parameter is $\chi(\vec{r}) \sim r^2 Y_{20}$. The GQR is substained by distortions of the Fermi surface (see Problem 4) , similarly to what happens with the zero sound.
  Since in the scaling model, the collective flow is incompressible, $\nabla^2(r^2 Y_{20}) = 0$, the calculation of the oscillation parameters C and M turns out to be significantly simplified. In particular, the contribution of the potential energy to the stiffness parameter vanishes, as shown in the following calculation

$$\int \mathrm{d}\vec{r} \sum_k u_k[\delta(\rho \nabla_k U)] = \int \mathrm{d}\vec{r}\, \delta\rho\, (\frac{dU}{\rho})_0 \vec{u} \cdot \vec{\nabla}\rho_0 - \int \mathrm{d}\vec{r}\, \delta\rho (\frac{dU}{d\rho})_0 \vec{u} \cdot \vec{\nabla}\rho_0 = 0$$

  after integrating by parts the second integral and dropping the contribution due to the differential $\sum_k \nabla_k u_k d\vec{r}... = 0$, owing to the incompressibility.
  As a result, the ISGQR dynamics is only driven by the kinetic term. Its calculation turns out to be also simplified by the above mentioned property of the field and the boundary conditions. The stiffness parameter is

$$C = \int \mathrm{d}\vec{r} \int \mathrm{d}\vec{p} \sum_k u_k \sum_i \nabla_i \delta\tau_{ik} = \frac{4}{3} \int \mathrm{d}\vec{r} \tau_0(\vec{r}) \sum_{ik} (\nabla_i u_k)(\nabla_k u_i),$$

  where the pressure tensor, in the static limit, is $\tau_{ij}^0 = \frac{1}{3}\tau_0 \delta_{ij}$. With $\tau_0 = \dfrac{3}{5}\varepsilon_F(\vec{r})\rho_0$ from the Fermi-gas approximation (see Chap.16.6) and constant density ρ_0, the integration over space of M and C can easily be performed as well as the estimate of the oscillation angular frequency $\omega = \sqrt{\frac{C}{M}}$. Considering a spherical nucleus of radius R, estimated from the empirical law $R = 1.2A^{1/3} Fm$, the energy centroid results (see Problem 18.3)

$$E = \frac{2\hbar}{R}\sqrt{\varepsilon_F/m} =\sim 64.07 A^{-1/3}\mathrm{MeV}, \qquad (18.30)$$

  which reproduces quite well the experimental trend, as shown in Fig.18.4. Also in this case, a geometrical argument can easily explain the A dependence of the IVGDR energy. In fact, with the Tassie model the space integral in C is proportional to R^3 whereas in M is proportional to R^5 so that $E \sim R^{-1} \sim 1/A^{1/3}$.

- IVGDR ($\Delta L = 1$)
  In the IVGDR the two nuclear fluids, proton fluid with Z protons and neutron fluid with N neutrons, oscillate out-of-phase so that $\delta f_p = -\delta f_n$. In order to calculate the oscillation frequency, it is sufficient to consider only one of the two, e.g. the neutron fluid. In the Tassie model the scaling parameter is $\chi(\vec{r}) \sim rY_{10}$ and the collective velocity $\vec{u} = \vec{\nabla}\chi/m$, with cartesian components (0, 0, 1). Hence, the

field is incompressible, $\vec{\nabla} \cdot \vec{u} = 0$, with the Fermi surface oscillating around the p_z-axis. In this case, the stiffness parameter C is contributed only by the potential energy, whereas the kinetic term is easily found to be zero due to incompressibility. Moreover, since the out-of-phase oscillations generate locally pieces of asymmetric nuclear matter, i.e. $\rho_n(r,t) \neq \rho_p(r,t)$, the potential variations in any point of the nucleus are affected separately by both neutron and proton density variations. As a result, the potential variation δU^n is

$$\delta U^n = \frac{\partial U^n}{\partial \rho^n}\delta\rho^n + \frac{\partial U^n}{\partial \rho^p}\delta\rho^p = [\frac{\partial U^n}{\partial \rho^n} - \frac{\partial U^n}{\partial \rho^p}]\delta\rho^n,$$

being $\delta\rho^p = -\delta\rho^n$. Extending the parabolyc law, Eq. (14.17), from nuclear matter to any point $\vec{r}$ of the nucleus (local density approximation), we can write $\epsilon(\vec{r}, \rho^n, \rho^p) = \epsilon_0(\vec{r}, \rho) + \beta^2 a_I^U$, where a_I^U is the contribution of the potential to the symmetry energy. Hence

$$\frac{\partial U^n}{\partial \rho^n} = 2\beta a_I^U/\rho \qquad \frac{\partial U^n}{\partial \rho^p} = -2\beta a_I^U/\rho. \tag{18.31}$$

We see that the giant dipole frequencies are sensitive to the nuclear symmetry energy. Therefore, from Eq. (18.25) the stiffness parameter turns out to be

$$\begin{aligned} C &= \int d\vec{r} \sum_k u_k[\delta(\rho\nabla_k U)] \\ &= \frac{A}{N}\int d\vec{r} \sum_{ki} u_k u_i \nabla_k \rho_0^n \nabla_i \rho_0^n (\frac{\partial U^n}{\partial \rho^p})_0 \end{aligned} \tag{18.32}$$

We see that C is also sensitive to the nuclear surface, owing to the presence of density gradients. Since there is only one oscillation axis, namely the z-axis, the latter equation simplifies as follows

$$C = \frac{N}{A}\int d\vec{r}\, u_z u_z \nabla_z \rho_0 \nabla_z \rho_0 (\frac{\partial U^n}{\partial \rho^p})_0$$

To take properly into account the surface effects, we take for the density a Wood-Saxon profile $\rho_0 = \rho_{00}/(1 + exp[(r - R)/a]$, where R is the radius and a is the diffuseness, assuming the nucleus to be spherical. In the case of $N = Z$ the oscillation frequency is then given by

$$\omega^2 = \frac{C}{M} = \frac{\frac{1}{2}\int d\vec{r}\, (u_z)^2 (\frac{\partial \rho_0}{\partial r})^2 (\frac{\partial U^n}{\partial \rho^p})_0}{\frac{1}{2}m\int d\vec{r}\, (u_z)^2 \rho_0)} \tag{18.33}$$

Employing the Sommerfeld approximation in the Wood-Saxon profile, after a tedious procedure we obtain $((\frac{\partial \rho_0}{\partial r})^2 \sim \frac{\rho_{00}^2}{6a}\delta(r - R)$, which permits us to easily integrate the numerator. We finally obtain With the symmetry parameter calculated from the Skyrme potential of Sect. 14.4.2, we get the mass law

$$E = \hbar\omega \sim 35A^{-1/6}, \tag{18.34}$$

which well agrees with the experimental systematics of the IVGDR energy centroids. Again, a geometrical argument can easily explain the A dependence of the IVGDR energy. In fact, the space integral in C is proportional to R^2 whereas in M is proportional to R^3 so that $E \sim R^{-1/2} \sim 1/A^{1/6}$, because the dipole resonance is a surface effect, as we said above.

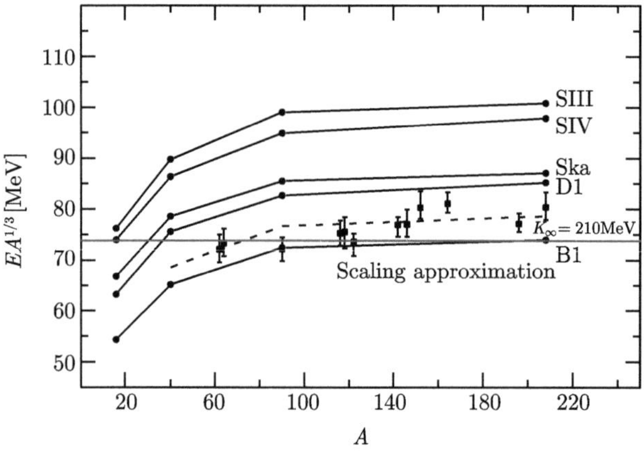

Fig. 18.3 Systematics of ISGMR energy centroids. The symbols represent the experimental data, and the solid lines are some theoretical predictions with Skyrme forces. The red line shows the results with the scaling model.

In conclusion, we have shown that the semiclassical approach to nuclear GRs with the scaling model provides a simple interpretation of these collective modes as well as good estimates of the GR centroids of the excitation energies. Finally, significant constraints to the equation of state of nuclear matter come from GRs: a constraint on the compression modulus comes from GMR as breathing mode and another constraint on the symmetry energy comes from GDR as proton-to-neutron out-of-phase oscillation.

18.4 Notes and References

DITO M. Di Toro, U. Lombardo, and G. Russo, *Il Nuovo Cimento*, 87A (1985) 174.
LAN L.D. Landau,*Oscillations in a Fermi liquid*, Sov. Phys. JETP, **5** (1957) 101.
HUA K. Huang, *Statistical Mechanics*, Wiley and Sons, 1963, Sect. 5.3.

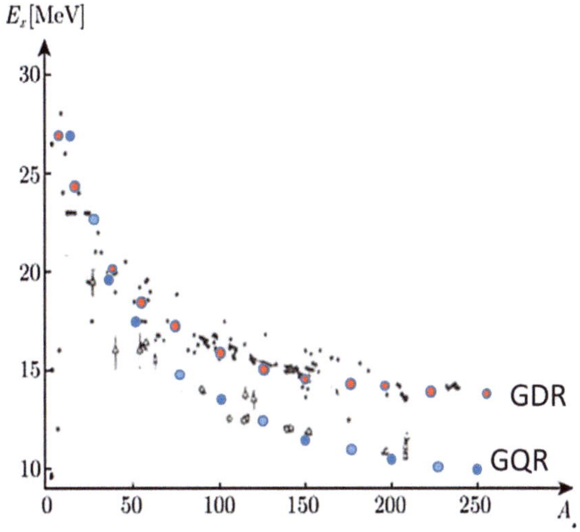

Fig. 18.4 Systematics of the experimental results of the energy centroids of IVGDR and ISGQR and theoretical mass laws (circles)[RSC]

ABE W.R.Abel, A.C.Anderson, and J. C. Wheatley, Physical Review Letters **17** (1966) 74.

BLA J.P.Blaizot, *Nuclear Compressibilities*, Physics Reports **64** (1980) 171.

LEG For the superfluidity in ^{3}He see: A.J. Leggett, *A theoretical description of the new phases of liquid* ^{3}*He*, Rev. Mod. Phys. **69** (1997) 645.

RSC P. Ring and P. Schuck, *The Nuclear Many-Body Problem*, Springer, 2000, Chap. 13 .

18.5 Problems

- Problem 1. Estimate the classical speed of sound at a temperature of 25 °C in air assuming it is made only of nitrogen (^2N) molecules.
- Problem 2. Show that sound propagation in a homogeneous system is an adiabatic process under a local thermal equilibrium [HUA].
- Problem 3. Calculate the energy centroid of ISGQR using the displacement parameter $\chi = \alpha(t)r^2Y_{20}(\hat{r})$.
- Problem 4. A nucleus does harmonic oscillations with the collective velocity field $\phi(\vec{r}, t) = \alpha(t)\phi(\vec{r})$. Using the continuity equation, determine the current $\vec{J}$, at the first-order approximation of momentum expansion of the scaling model. In the GQR case, determine the distortions of the Fermi surface in the scaling approximation.
- Problem 5. Determine within the scaling model the semi-classical equations of motion of giant resonances in a rotating nucleus. Discuss the effects of the rotation.

Correction to: Quantum Mechanics

Correction to:
U. Lombardo et al., *Quantum Mechanics*, UNITEXT for Physics, https://doi.org/10.1007/978-981-97-1670-8

The original version of this book was inadvertently published with incorrect details in Chapter 14 and few additional changed in Chapter 18, which are listed below:

• In Chapter 14, Page 164, <HF|H|HF> is replaced with $\Phi|\mathbf{H}|\Phi$ in the last two equations.

• In Chapter 18, the last paragraph of Section 18.3 "In conclusion, …… non-adiabatic process" has been removed from page 243.

• In Chapter 18, a new reference, "DITOM. Di Toro, U. Lombardo and G. Russo, Il Nuovo Cimento 87A (1985) 174" has been included.

The book has been updated with the changes.

The updated versions of these chapters can be found at
https://doi.org/10.1007/978-981-97-1670-8_14
https://doi.org/10.1007/978-981-97-1670-8_18

U. Lombardo et al., *Quantum Mechanics*, UNITEXT for Physics,
https://doi.org/10.1007/978-981-97-1670-8_19

Appendix A
Vectors and Operators

A.1 Multidimensional Vector Spaces

Classical mechanics employs the vectors of a three-dimensional vector space to describe several observables, including position, force and electromagnetic field. In quantum mechanics, the notion of state requires the extension of the three-dimensional vector space to a multidimensional vector space. The conditions for a non-empty set V_D to be a D-dimensional linear vector space are a straightforward extension of those for the three-dimensional case. Any pair of elements $|f\rangle, |g\rangle \in V_D$[1] and any scalar α (real or complex) must obey the following properties:

1. the sum $|h\rangle = |f\rangle + |g\rangle$ has to belong to V_D, i.e. $|h\rangle \in V_D$;
2. $|f\rangle + |g\rangle = |g\rangle + |f\rangle$, (commutative property of the sum);
3. $|f\rangle + |g\rangle + |h\rangle = |f\rangle + (|g\rangle + |h\rangle) = |g\rangle + (|f\rangle + |h\rangle) = |h\rangle + (|f\rangle + |g\rangle)$ (distributive property of the sum);
4. there is a unique element $|0\rangle \in V_D$, which is called the null element, such that $|f\rangle + |0\rangle = |f\rangle, \forall |f\rangle \in V_D$;
5. for any $|f\rangle \in V_D$, there is a unique element $-|f\rangle \in V_D$ such that $|f\rangle + (-|f\rangle) = |0\rangle$;
6. $\alpha(|f\rangle + |g\rangle) = \alpha|f\rangle + \alpha|g\rangle$ (distributive property of the product with a scalar);
7. $(\alpha\beta)|f\rangle = \alpha(\beta|f\rangle)$, $\alpha, \beta \in \mathbb{C}$ (commutative property of the product with a scalar),

where α and β are real or complex numbers, that is $\alpha, \beta \in C$.

[1] We adopt here the Dirac notation, where a vector is denoted with the symbol $|f\rangle$(*ket*) and its dual with the symbol $\langle f|$ (*bra*).

U. Lombardo et al., *Quantum Mechanics*, UNITEXT for Physics,
https://doi.org/10.1007/978-981-97-1670-8

The set of linearly independent vectors $|f_1\rangle, |f_2\rangle, ..., |f_D\rangle$ forms a basis (or complete set), if any vector $|h\rangle \in V_D$ can be written as a linear combination

$$|h\rangle = \sum_{j=1}^{D} \alpha_j |f_j\rangle,$$

where the α_j coefficients are complex numbers.

In quantum mechanics, the dimensionality of the vector spaces V_D ranges from $D = 2$, as in the case of the spin states, to $D = \infty$ (numerable), as in the case of the hydrogen-bound states. Since the space is complex, before defining the inner product, we must introduce the space V_D^* of vectors $\langle f|$, called *dual* space of V_D. We assume that there exists a one-to-one (anti-linear) correspondence between vectors of the space V_D and vectors of the dual space V_D^* such that, if

$$|v\rangle = \lambda_1|1\rangle + \lambda_2|2\rangle,$$

then

$$\langle v| = \lambda_1^*\langle 1| + \lambda_2^*\langle 2|.$$

The correspondence between ket and bra is analogous to the correspondence between the wave functions ψ and their complex conjugates ψ^*. The inner product of two vectors $|f\rangle$ and $|g\rangle \in V_D$ is defined as the complex function $\langle f|g\rangle$ with the following properties:

1. $\langle f|f\rangle \geq 0$ and $\langle f|f\rangle = 0$, only if $|f\rangle = |0\rangle$;
2. $\langle f|(|g\rangle + |h\rangle) = \langle f|g\rangle + \langle f|h\rangle$;
3. $\langle f|g\rangle = \langle g|f\rangle^*$;
4. $(\alpha\langle f|)|g\rangle = \alpha\langle f|g\rangle$,

where α is a complex number. If $\alpha\langle f|g\rangle = 0$, the two vectors are said to be orthogonal. The quantity $||f|| \equiv \langle f|f\rangle^{\frac{1}{2}}$ is called *norm* of the vector $|f\rangle$. If $||f|| = 1$, the vector $|f\rangle$ is said to be normalized to 1.

A.2 Hilbert Space $\mathcal{H}$

The Hilbert space $\mathcal{H}$ is a linear vector space with an uncountable infinite dimension. It is introduced in quantum mechanics to describe states of dynamical variables which take non-quantized values. Important examples are r-space states of a particle or scattering states, the position in the former case and the energy in the latter case take continuous values. We define the inner product between two elements $|\psi\rangle$ and $|\phi\rangle$, both belonging to $\mathcal{H}$, as follows:

$$\langle\psi|\phi\rangle = \int_{-\infty}^{\infty} \psi^*(x)\phi(x)\mathrm{d}x,$$

and define the norm $||\psi||$ as follows:

$$||\psi||^2 = \int_{-\infty}^{\infty} \psi^*(x)\psi(x)\mathrm{d}x.$$

The two integrals are supposed to be finite. A function having a finite norm is called a square-integrable function.

A.3 Operators

The physical observables are represented by operators of the Hilbert space. Let us consider the transformation

$$\mathbf{A}|u\rangle = |v\rangle,$$

that is, the ket $|v\rangle$ results from the action of the operator $\mathbf{A}$ on the ket $|u\rangle$. The operator $\mathbf{A}$ is said to be linear if the following relation holds

$$\mathbf{A}(\lambda_1|1\rangle + \lambda_2|2\rangle) = \lambda_1\mathbf{A}|1\rangle + \lambda_2\mathbf{A}|2\rangle,$$

where λ_1 and λ_2 are complex numbers. If the ket $|v\rangle = 0$ for any $|u\rangle$, then $\mathbf{A} = 0$.

A.3.1 Properties of Operators

- Two operators $\mathbf{A}$ and $\mathbf{B}$ are equal if and only if $\mathbf{A}|u\rangle = \mathbf{B}|u\rangle$ for any vector $|u\rangle$.
- The transformation $\mathbf{I}|u\rangle = |u\rangle$ defines the Identity Operator $\mathbf{I} = \mathbf{1}$.
- The operators $\mathbf{A}$ and $\mathbf{B}$ are each other's inverse if they satisfy the equations

$$\mathbf{AB} = \mathbf{BA} = \mathbf{1}. \tag{A.1}$$

  We denote the inverse of the $\mathbf{A}$ operator as $\mathbf{B} = \mathbf{A}^{-1}$. It follows the inverse of the operatorial product (A.1)

$$(\mathbf{AB})^{-1} = \mathbf{B}^{-1}\mathbf{A}^{-1}.$$

- Sum of operators: $\mathbf{S} = \mathbf{A} + \mathbf{B}$,

$$\mathbf{S}|u\rangle = \mathbf{A}|u\rangle + \mathbf{B}|u\rangle.$$

- Product of operators: $\mathbf{P} = \mathbf{AB} \neq \mathbf{BA}$.

- Multiplication by a complex constant c:

$$c\mathbf{A}|u\rangle = c(\mathbf{A}|u\rangle).$$

- The Hermitian conjugate of the operator $\mathbf{A}$, also called the adjoint of the operator $\mathbf{A}$, denoted by $\mathbf{A}^\dagger$, is defined as

$$\langle v|\mathbf{A}^\dagger|u\rangle = \langle u|\mathbf{A}|v\rangle^*.$$

Some properties of the Hermitian conjugation:

$$\begin{aligned}(\mathbf{A}^\dagger)^\dagger &= \mathbf{A},\\ (c\mathbf{A})^\dagger &= c^*\mathbf{A}^\dagger,\\ (\mathbf{A}+\mathbf{B})^\dagger &= \mathbf{A}^\dagger + \mathbf{B}^\dagger,\\ (\mathbf{A}\mathbf{B})^\dagger &= \mathbf{B}^\dagger\mathbf{A}^\dagger.\end{aligned}$$

The hermitian conjugate of the operator $|u\rangle\langle v|$ is $(|u\rangle\langle v|)^\dagger = |v\rangle\langle u|$.
- A linear operator $\mathbf{H}$ is Hermitian (or self-adjoint) if $\mathbf{H}^\dagger = \mathbf{H}$.
- The Hermitian operator $\mathbf{H}$ is positive definite if $\langle u|\mathbf{H}|u\rangle \geq 0$ for any vector $|u\rangle$. The operator $|a\rangle\langle a|$ is a positive definite Hermitian operator.
- An operator $\mathbf{U}$ is unitary if it is the inverse of its own adjoint: $\mathbf{U}\mathbf{U}^\dagger = \mathbf{U}^\dagger\mathbf{U} = \mathbf{1}$. The commutator of two operators $\mathbf{A}$ and $\mathbf{B}$ is defined by the operation: $[\mathbf{A}, \mathbf{B}] = \mathbf{A}\mathbf{B} - \mathbf{B}\mathbf{A}$. It has the following properties:

$$[\mathbf{A}, \mathbf{B}] = -[\mathbf{B}, \mathbf{A}],$$

$$[\mathbf{A}, \mathbf{A}] = 0,$$

$$[\mathbf{A}, c] = 0,$$

$$[\mathbf{A}, \mathbf{B}+\mathbf{C}] = [\mathbf{A}, \mathbf{B}] + [\mathbf{A}, \mathbf{C}],$$

$$[\mathbf{A}, \mathbf{B}\mathbf{C}] = \mathbf{B}[\mathbf{A}, \mathbf{C}] + [\mathbf{A}, \mathbf{B}]\mathbf{C},$$

$$[\mathbf{A}\mathbf{B}, \mathbf{C}] = \mathbf{A}[\mathbf{B}, \mathbf{C}] + [\mathbf{A}, \mathbf{C}]\mathbf{B},$$

$$[\mathbf{A}, [\mathbf{B}, \mathbf{C}]] + [\mathbf{B}, [\mathbf{C}, \mathbf{A}]] + [\mathbf{C}, [\mathbf{A}, \mathbf{B}]] = 0.$$

A.3.2 Projection Operators

Let T be a subspace of the Hilbert space $\mathcal{H}$, and $T^{\times}$ its complementary subspace. The ket $|u\rangle$ has a projection $|u_s\rangle$ in T and a projection $|u_s^{\times}\rangle$ in $T^{\times}$, both of which are uniquely defined. The following relation holds true

$$|u\rangle = |u_s\rangle + |u_s^{\times}\rangle.$$

Each ket $|u\rangle$ corresponds to only one ket $|u_s\rangle$, this correspondence being linear. This correspondence defines a linear operator $\mathbf{P}_s$ called the projector (or projection operator) on the subspace T.

$$\mathbf{P}_s|u\rangle = |u_s\rangle.$$

The projection operator is Hermitian and idempotent, that is, $\mathbf{P}_s^2 = \mathbf{P}_s$. Hence, its eigenvalues are 0 and 1. As an example of projection operator, we consider a set of orthonormal vectors $|1\rangle, |2\rangle, \ldots, |N\rangle$, with $\langle m|n\rangle = \delta_{mn}$. These vectors span a subspace E_1 with N dimensions. The operator $\mathbf{P}_1 = \sum_{m=1}^{N}|m\rangle\langle m|$ is the projector operator upon E_1.

A.4 Eigenvalue Problem

- Eigenvalue equation

  Let $\mathbf{A}$ be a linear operator: if there exists a ket $|u\rangle$ (different from the null vector) such that

  $$\mathbf{A}|u\rangle = a|u\rangle,$$

  where a is a number. The ket $|u\rangle$ is called eigenvector of $\mathbf{A}$, belonging to the eigenvalue a. The above equation is known as the eigenvalue equation. If $\mathbf{A}$ is Hermitian, its eigenvalues are real numbers and the eigenvectors, belonging to eigenvalues different from each other, are orthogonal.
- Eigenvalue subspace and degeneracy

  If there exist several linearly independent eigenvectors belonging to the same eigenvalue a, any linear combination of them is still an eigenvector of $\mathbf{A}$ belonging to the eigenvalue a. The set of the eigenvectors of $\mathbf{A}$, belonging to a given eigenvalue a, forms a subspace of Hilbert space. If this subspace has only one dimension, the eigenvalue is said to be non-degenerate; otherwise, it is said to be degenerate. The order of degeneracy is the number of dimensions of this subspace.

A.5 Representation of Vectors and Operators

A.5.1 Matrix Representation

- A matrix A of the type $M \times N$ is a set of MN quantities $A_{mn}(m = 1, 2, \ldots, M,$ $n = 1, 2, \ldots, N)$, which one usually arranges in a rectangular array with M rows and N columns

$$\begin{pmatrix} A_{11} & A_{12} & \ldots & A_{1N} \\ A_{21} & A_{22} & \ldots & A_{2N} \\ \vdots & \vdots & \vdots & \vdots \\ A_{M1} & A_{M2} & \ldots & A_{MN} \end{pmatrix}. \tag{A.2}$$

  A_{mn} is the matrix element of the m_{th} row and the n_{th} column. If $M = N$, it is called a square matrix. The number of its rows and its columns is the number of dimensions or order of the matrix. We call column vector a matrix with one column ($N = 1$), and row vector a matrix with one row ($M = 1$). A scalar is a special matrix for which $M = N = 1$.
- Certain conjugation operations on matrix A:

  (i) complex conjugate matrix A^*: $(A^*)_{kl} = A^*_{kl}$;
  (ii) transposed matrix $\tilde{A}$: $\tilde{A}_{kl} = A_{lk}$;
  (iii) hermitian conjugate matrix $A^\dagger$: $(A^\dagger)_{kl} = A^*_{lk}$.
- Algebraic operations with matrices:

  (i) multiplication of a matrix A by a constant c: $(cA)_{mn} = cA_{mn}$;
  (ii) sum $S = A + B$: $S_{mn} = A_{mn} + B_{mn}$;
  (iii) product $P = AB$: $P_{mn} = \sum_k A_{mk} B_{kn}$;
  (iv) tensor product $A^{(12)} \equiv A^{(1)} \otimes A^{(2)}$: $A^{(12)}_{m_1 m_2; n_1 n_2} = A^{(1)}_{m_1 n_1} A^{(2)}_{m_2 n_2}$.
- A matrix O is orthogonal if $O\tilde{O} = I$, where I is a unit matrix $I_{mn} = \delta_{mn}$.
- A matrix U is unitary if $UU^\dagger = I$.

A.5.2 Matrix Representation of Vectors and Operators

Consider a vector space E and choose as the basis in this space the set of eigenvectors of a given observable $\mathbf{Q}$, i.e.

$$\mathbf{Q}|n\rangle = q_n|n\rangle.$$

They form a complete set of orthonormal vectors

$$\langle m|n\rangle = \delta_{mn}, \tag{A.3}$$

$$P_Q \equiv \sum_n |n\rangle\langle n| = 1. \tag{A.4}$$

For any ket $|u\rangle$,

$$|u\rangle = \mathbf{P}_Q|u\rangle = \sum_n |n\rangle\langle n|u\rangle. \tag{A.5}$$

The quantities $u_n = \langle n|u\rangle$ are the matrix elements of a column vector, whose index is n. This column vector represents $|u\rangle$ in the $\{Q\}$ representation. For any bra $\langle v|$,

$$\langle v| = \langle v|\mathbf{P}_Q = \sum_v \langle v|n\rangle\langle n|. \tag{A.6}$$

The quantities $\langle v|n\rangle$ are the complex conjugates of the components v_n. The elements v_n form a row vector in the $\{Q\}$ representation. In this representation, any operator $\mathbf{A}$ is written as

$$\mathbf{A} = \mathbf{P}_Q\mathbf{A}\mathbf{P}_Q = \sum_{mn} |m\rangle\langle m|\mathbf{A}|n\rangle\langle n|.$$

The coefficients of the expansion $A_{mn} = \langle m|\mathbf{A}|n\rangle$ uniquely define $\mathbf{A}$ in the $\{Q\}$ representation.

A.5.3 Eigenvalue Problem in Matrix Form

The eigenvalue equation

$$\mathbf{A}|u\rangle = a|u\rangle,$$

in the Q representation is written as

$$\sum_m A_{nm}u_m = au_n. \tag{A.7}$$

The solution of this system of homogeneous equations goes through the condition

$$\det(A - aI) = 0, \tag{A.8}$$

which determines the eigenvalues of $\mathbf{A}$. It is known as characteristic (or secular) equation.

A.6 Continuous Spectrum and Dirac δ-function

In quantum mechanics, some physical observables, such as position and momentum, take continuous instead of discrete values. The eigenfunctions do not have a finite norm. To address this drawback, Dirac introduces a δ-function to describe the orthonormality relation, e.g. between two position eigenstates

$$\langle q|q'\rangle = \delta(q - q'), \tag{A.9}$$

where

$$\delta(x) = \begin{cases} 0 & (x \neq 0) \\ \infty & (x = 0) \end{cases}, \tag{A.10}$$

and

$$\int_{-\infty}^{\infty} \delta(x) = 1. \tag{A.11}$$

The main properties of the δ-function are

- $\delta(q - q_0) = \dfrac{1}{2\pi}\displaystyle\int_{-\infty}^{\infty} \mathrm{d}k e^{\mathrm{i}k(q-q_0)}$,
- $\psi(x) = \displaystyle\int \mathrm{d}x'\psi(x')\delta(x - x')$,
- $x\delta(x - x') = x'\delta(x - x')$,
- $\delta(f(x)) = \sum_k \dfrac{\delta(x - x_k)}{|f'(x_k)|}$,

where the x_k are the roots of the equation $f(x) = 0$. Before closing this section, we consider two realizations of the δ-function to clarify that, as a matter of fact, it is not a true function, but a functional. The first example is the step function

$$\delta(x) = \lim_{L\to\infty} \frac{1}{L}\theta(|x| \leq L/2), \tag{A.12}$$

which is plotted in the r.h.s. of Fig. A.1. It is easily shown that this form satisfies Eqs. (A.10) and (A.11). The second realization is

$$\delta(x) = \lim_{t\to\infty} \frac{1}{\pi}\frac{\sin^2(xt)}{x^2 t}, \tag{A.13}$$

which is plotted in the l.h.s. of Fig. A.1. The behavior of this form was discussed in Sect. 11.2: for t getting bigger and bigger, the secondary peaks disappear, whereas the central peak increases indefinitely, in agreement with Eq. (A.10). The agreement with Eq. (A.11) is verified by integrating over x.

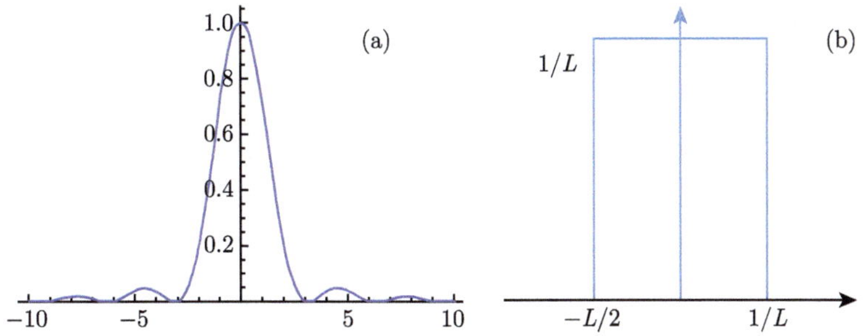

Fig. A.1 Two realizations of the Dirac δ functional

Both examples are not functions, but the asymptotic limits of well defined functions $f_p(x)$, depending on the p parameter, for $p \to \infty$.

Appendix B
Special Functions in Quantum Mechanics

In this appendix, we discuss a few special functions encountered in the study of the exact solutions of the Schrödinger equation. Table B.1 provides a synoptic view of this subject.[2]

Table B.1 Special functions in quantum mechanics

Angular momentum	Spherical harmonics	Sect. 6.4
Free particle	Spherical Bessel functions	Sect. 12.1
1D harmonic oscillator	Hermite polynomials	Sect. 8.1
Hydrogen atom	Laguerre polynomials	Sect. 12.2
3D harmonic oscillator	Laguerre polynomials	Sect. 14.5

B.1 Orbital Angular Momentum in Polar Coordinates

The relation between Cartesian and polar coordinates is

$$x = r \sin\theta \cos\phi, \tag{B.1}$$

$$y = r \sin\theta \sin\phi, \tag{B.2}$$

$$z = r \cos\theta, \tag{B.3}$$

where (x, y, z) are the Cartesian coordinates and (r, θ, ϕ) the polar coordinates, θ and ϕ being the polar and azimuthal angle, respectively.

In the polar coordinate system, the components of orbital angular momentum are

[2] For a comprehensive summary see *Handbook of Mathematical Functions*, Eds. M. Abramowitz and I.A. Stegun, Dover (N.Y.) 1964.

U. Lombardo et al., *Quantum Mechanics*, UNITEXT for Physics,
https://doi.org/10.1007/978-981-97-1670-8

$$\mathbf{l}_z = \hbar \mathrm{i} \frac{\partial}{\partial \phi}$$
$$\mathbf{l}_\pm \equiv \mathbf{l}_x \pm \mathrm{i}\mathbf{l}_y = \hbar \mathrm{e}^{\pm \mathrm{i}\phi} \left(\pm \frac{\partial}{\partial \theta} + \mathrm{i} \cot \theta \frac{\partial}{\partial \phi} \right) \tag{B.4}$$

$$\mathbf{l}^2 \equiv -\hbar^2 \left[\frac{1}{\sin\theta} \frac{\partial}{\partial \theta} \left(\sin\theta \frac{\partial}{\partial \theta} \right) + \frac{1}{\sin^2\theta} \frac{\partial^2}{\partial \phi^2} \right]. \tag{B.5}$$

B.2 Spherical Harmonics and Legendre Polynomials

In Sect. 6.4, we learned that spherical harmonics $Y_{lm}(\theta, \phi)$ are simultaneous eigenfunctions of the $\mathbf{l}^2$ and $\mathbf{l}_z$ operators, that is

$$\mathbf{l}^2 Y_{lm}(\theta, \phi) = \hbar^2 l(l+1) Y_{lm}(\theta, \phi), \tag{B.6}$$
$$\mathbf{l}_z Y_{lm}(\theta, \phi) = \hbar m\, Y_{lm}(\theta, \phi), \tag{B.7}$$

where $0 \le l \le \infty$ and $-l \le m \le l$. Henceforth, we set $\hbar = 1$. The spherical harmonics form a complete set of orthonormal functions, i.e.

$$\int Y^*_{lm}(\Omega) Y_{l'm'}(\Omega) \mathrm{d}\Omega = \delta_{ll'} \delta_{mm'} \qquad \text{(orthonormality)}, \tag{B.8}$$
$$\sum_{lm} Y^*_{lm}(\Omega) Y_{l'm'}(\Omega') = \delta(\Omega - \Omega') \qquad \text{(completeness)}, \tag{B.9}$$

where $\mathrm{d}\Omega = \sin\theta \mathrm{d}\theta \mathrm{d}\phi$ and $\delta(\Omega - \Omega') = \delta(\cos\theta - \cos\theta')\delta(\phi - \phi')$. In the sum of Eq. (B.9), $0 \le l \le \infty$ and $-l \le m \le +l$.

Two properties of the spherical harmonics, which are frequently applied in quantum mechanics, are

$$Y_{lm}(\pi - \theta, \phi + \pi) = (-1)^l Y_{lm}(\theta, \phi) \qquad \text{(reflection)}, \tag{B.10}$$
$$Y^*_{lm}(\theta, \phi) = (-1)^m Y_{l,-m}(\theta, \phi) \qquad \text{(complex conjugation)}. \tag{B.11}$$

In Sect. 6.4, it was shown that for $m = l$ the angular momentum eigenfunctions factorize in the azimuthal part, given by $\mathrm{e}^{\mathrm{i}m\phi}$, and in the polar part, given by $(\sin\theta)^l$. Hence,

$$Y_{ll} = c_l (\sin\theta)^l \mathrm{e}^{\mathrm{i}l\phi}, \tag{B.12}$$

The constant c_l is fixed by the normalization $\int \mathrm{d}\Omega |Y_{ll}|^2 = 1$. This turns out to be

$$c_l = (-1)^l \left[\frac{(2l+1)(2l)!}{4\pi 2^{2l} (l!)^2} \right]^{1/2}. \tag{B.13}$$

For $m < l$, the eigenfunctions are obtained by repeatedly applying to the eigenfunction with $m = l$, the operator $\mathbf{l}_-$, which lowers by one the m value at each application. Thus, according with Eq. (B.4), we get

$$Y_{lm-1} = [l(l+1) - m(m-1)]^{-\frac{1}{2}} l_- Y_{l,m} \tag{B.14}$$

$$= [l(l+1) - m(m-1)]^{-\frac{1}{2}} \mathrm{e}^{-\mathrm{i}\phi} \left[-\frac{\partial}{\partial\theta} + \mathrm{i}\cot\theta \frac{\partial}{\partial\phi} \right] Y_{lm}(\theta, \phi), \tag{B.15}$$

where the normalization coefficient is given by Eq. (B.13), for each l value. The spherical harmonics, for the lowest values of l and m, are

$$Y_{00} = \frac{1}{\sqrt{4\pi}},$$

$$Y_{10} = \sqrt{\frac{3}{4\pi}} \cos\theta, \qquad Y_{1,\pm1} = \mp\sqrt{\frac{3}{8\pi}} \mathrm{e}^{\pm\mathrm{i}\phi} \sin\theta,$$

$$Y_{20} = \sqrt{\frac{5}{16\pi}} (3\cos^2\theta - 1), \qquad Y_{2,\pm1} = \mp\sqrt{\frac{15}{8\pi}} \sin\theta\cos\theta \mathrm{e}^{\pm\mathrm{i}\phi},$$

$$Y_{2,\pm2} = \sqrt{\frac{15}{32\pi}} \sin^2\theta \mathrm{e}^{\pm2\mathrm{i}\phi}.$$

Spherical harmonics are closely related to the Legendre polynomials $P_l(u)$, which are the solutions of the differential equation

$$\frac{\mathrm{d}}{\mathrm{d}u} \left[(1-u^2) \frac{\mathrm{d}}{\mathrm{d}u} \right] P_l(u) + l(l+1) P_l(u) = 0, \tag{B.16}$$

where the u variable runs over the range $-1 \leq u \leq 1$ and $l = 0, 1, 2, \ldots$. Their explicit expression is

$$P_l(u) = \frac{1}{2^l l!} \frac{\mathrm{d}^l}{\mathrm{d}u^l} (u^2 - 1)^l. \tag{B.17}$$

The Legendre polynomials form a complete set of orthonormal functions, that is

$$\int_{-1}^{+1} \mathrm{d}u \, P_l(u) P_{l'}(u) = \delta_{ll'} \frac{2}{2l+1}, \tag{B.18}$$

$$f(u) = \frac{2l+1}{2} \sum_l f_l P_l(u), \tag{B.19}$$

where $f(u)$ can be any single-valued function, finite in the interval $|u| \leq 1$, which exhibits, at most, a finite number of singularities.

Particular values of the Legendre polynomials are

$$
\begin{aligned}
P_l(1) &= 1, \\
P_l(-1) &= (-1)^l, \\
P_l(0) &= (-1)^p \frac{(2p)!}{2^l (p!)^2} \qquad (l = 2p), \\
P_l(0) &= 0 \qquad (l = 1 + 2p).
\end{aligned}
$$

Two useful relations with the spherical harmonics are

$$
Y_{l0}(\theta, \phi) = \sqrt{\frac{2l+1}{4\pi}} P_l(\cos\theta), \tag{B.20}
$$

$$
\frac{2l+1}{4\pi} P_l(\cos\theta) = \sum_{m=-l}^{+l} Y^*_{lm}(\theta', \phi') Y_{lm}(\theta'', \phi'') \qquad \text{(composition law)}. \tag{B.21}
$$

where θ is the angle between the two directions $\hat{u}' = (\theta', \phi')$ and $\hat{u}'' = (\theta'', \phi'')$.

B.3 Spherical Bessel Functions

The radial Schrödinger equation for a free particle, Eq. (12.2) with $V = 0$, takes the form of the so-called Bessel's equation

$$
\left[\frac{\mathrm{d}^2}{\mathrm{d}\rho^2} + \frac{2}{\rho}\frac{\mathrm{d}}{\mathrm{d}\rho} - \frac{l(l+1)}{\rho^2} + 1 \right] y_l = 0, \tag{B.22}
$$

where $\rho = kr$ and $k = \sqrt{2mE/\hbar^2}$. Among the solutions of Bessel's equation, only the spherical Bessel functions $j_l(\rho)$ are physically acceptable, as they are regular at the origin $r = 0$. For $l = 0$ the solution of Eq. (B.22) is $j_0(\rho) = \dfrac{\sin(\rho)}{\rho}$. The solutions for $l > 0$ are obtained by subsequent ρ-derivatives of j_0 according to the formula

$$
j_l(\rho) = \left[\rho^l \left(-\frac{1}{\rho}\frac{\mathrm{d}}{\mathrm{d}\rho} \right)^l \right] j_0(\rho). \tag{B.23}
$$

The asymptotic form is

$$
j_l \underset{\rho\to\infty}{\sim} \frac{\sin\left(\rho - \frac{l\pi}{2}\right)}{\rho}. \tag{B.24}
$$

According to the discussion of Sect. 12.1, the plane wave can be expanded as a sum over the spherical harmonics $Y_{lm}(\theta, \phi)$ as for the angular part and the Bessel functions $j_l(kr)$ as for the radial part, that is

$$e^{i\vec{k}\cdot\vec{r}} = \sum_{lm} c_{lm} j_l(kr) Y_{lm}(\theta, \phi), \tag{B.25}$$

The coefficients c_{lm} can be calculated as follows. If we choose the wave number $\vec{k}$ to lie along the z axis, then $\vec{k} \cdot \vec{r} = kr \cos\theta$. Hence, this equation can be rewritten as

$$e^{ikr\cos\theta} = \sum_{l} c_{lm} j_l(kr) Y_{lm}(\theta, \phi). \tag{B.26}$$

The left- and the right-hand sides are both independent of the azimuthal angle, hence only the spherical harmonics Y_{l0} with $m = 0$ can contribute to the sum as well as the coefficients $c_{l0} \equiv c_l$. As a result, the plane wave is written as

$$e^{ikr\cos\theta} = \sum_{l} \sqrt{\frac{2l+1}{4\pi}} c_l j_l(kr) P_l(\cos\theta), \tag{B.27}$$

after applying Eq. (B.20). In order to calculate the c_l coefficients, we multiply both sides of this equation by $P_{l'}(\cos\theta)$ and integrate over $\cos\theta$. By applying the orthogonality property of the Legendre polynomials, Eq. (B.18), the expansion is reduced to a single partial wave

$$\int d(\cos\theta) e^{kr\cos\theta} P_{l'}(\cos\theta) = \sqrt{\frac{2l'+1}{4\pi}} \frac{2}{2l'+1} c_{l'} j_{l'}(kr). \tag{B.28}$$

The next step would be a full integration of the left-hand side of Eq. (B.28), but we can simplify things based on the asymptotic behavior ($r \to \infty$) of both sides, as the coefficients c_l are independent of the space coordinates r. After integrating parts on the left-hand side, the result can be split into an asymptotically dominant term and a negligible one. The asymptotic Bessel function, Eq. (B.24), can be split in the same way, on the right-hand side. We end up with

$$\begin{aligned} &\frac{1}{ikr}[e^{ikr} - e^{-ikr} e^{i\pi l'}] + \mathcal{O}\left(\frac{1}{r^2}\right) \\ &= \frac{1}{\sqrt{\pi(2l'+1)}} c_{l'} \left[\frac{1}{kr} \sin(kr - l'\pi/2)\right] + \mathcal{O}\left(\frac{1}{r^2}\right). \end{aligned} \tag{B.29}$$

Finally, the coefficients turn out to be

$$c_{l'} = i^{l'} \sqrt{4\pi(2l'+1)}, \tag{B.30}$$

and the partial wave expansion of the plane wave in terms of Legendre polynomials take the following form:

$$e^{ikr\cos\theta} = \sum_l i^l (2l+1) j_l(kr) P_l(\cos\theta) . \tag{B.31}$$

Plugging Eq. (B.21) into the latter, we end up with the spherical-harmonics expansion of the plane wave in any polar coordinate system

$$e^{i\vec{k}\cdot\vec{r}} = 4\pi \sum_{lm} i^l j_l(kr) Y^*_{lm}(\hat{k}) Y_{lm}(\hat{r}) . \tag{B.32}$$

B.4 Hermite Polynomials

As it has been proven in Sect. 8.1, the eigenfunctions of the one-dimensional harmonic oscillator can be written in a compact form

$$\psi_n(q) = (2^n n! \sqrt{\pi})^{-\frac{1}{2}} \left(\frac{m\omega}{2\hbar}\right)^{\frac{n}{2}} \left(q - \frac{\hbar}{m\omega}\frac{d}{dq}\right)^n \psi_0(q), \tag{B.33}$$

where $\psi_0(q)$ is the ground-state wave function, given by Eq. (8.2). The eigenfunctions $\psi_n(q)$ are usually expressed in terms of the Hermite polynomials $H_n(\xi)$ by the relation

$$\psi_n(q) = (2^n n! \sqrt{\pi})^{-\frac{1}{2}} e^{-\frac{\xi^2}{2}} H_n(\xi), \tag{B.34}$$

$$H_n(\xi) = e^{\xi^2/2} \left(\xi - \frac{d}{d\xi}\right)^n e^{-\xi^2/2}, \tag{B.35}$$

where $\xi = \left(\frac{m\omega}{\hbar}\right)^{\frac{1}{2}} q$. The first Hermite polynomials are

$$H_0 = 1, \qquad H_1 = 2\xi, \tag{B.36}$$

$$H_2 = 4\xi^2 - 2, \qquad H_3 = 8\xi^3 - 12\xi. \tag{B.37}$$

The eigenfunctions of the three-dimensional isotropic harmonic oscillator, discussed in Sect. 14.5, are the product of three eigenfunctions

$$\psi_{\vec{n}}(x, y, z) = \psi_{n_x}(x)\psi_{n_y}(y)\psi_{n_z}(z), \tag{B.38}$$

corresponding to the three independent harmonic oscillators along the Cartesian axes. In the system of polar coordinates, the eigenfunctions can be obtained from

$\psi_{\vec{n}}(x, y, z)$ by an orthogonal transformation. Otherwise, they can be directly derived from the Schrödinger equation in polar coordinates, by using the generalized Laguerre polynomials, which will be discussed in the next section.

B.5 Laguerre Polynomials

The Laguerre polynomials are associated to the radial solution of the Schrödinger equation for the hydrogen atom and the three-dimension spherical harmonic oscillator. In fact, the true interest is in the generalized Laguerre polynomials. In closed form, these can be defined by means of the Rodrigues formula

$$L_n^k(x) = \frac{e^x x^{-k}}{n!} \frac{d^n}{dx^n}(e^{-x} x^{n+k}), \tag{B.39}$$

with $k > -1$. The L_n^0 polynomials are the ordinary Laguerre polynomials.

In the case of the hydrogen atom the iteration procedure to determine the energy eigenfunctions led to the first few Laguerre polynomials, with k integer, as discussed in Sect. 12.2. In general, the entire spectrum of the radial eigenfunctions is given by

$$\psi_{nl}(r) = C_{nl}(2k_n r)^l e^{-k_n r}(-1)^{(2l+1)} L_{n+l}^{2l+1}(2k_n r) \qquad (n = 1, 2, \ldots), \tag{B.40}$$

where $k_n = \sqrt{-2mE_n/\hbar^2}$, E_n being the energy eigenvalues. The normalization factor is given by

$$C_{nl} = -a^{-3/2} \frac{2^{l+1}}{2/n^2} \sqrt{\frac{(n-l-1)!}{[(n+l)!]^3}}, \tag{B.41}$$

where a is the Bohr radius. The lowest energy eigenfunctions can be built up by using the following table of generalized Laguerre polynomials with $n \leq 2$:

$$L_0^k(x) = 1,$$
$$L_1^k(x) = k + 1 - x,$$
$$L_2^k(x) = \frac{(k+1)(k+2)}{2} - (k+2)x + \frac{x^2}{2} \quad (k = 2l + 1).$$

In the case of the three-dimensional spherical harmonic oscillator, a similar iterative procedure also leads to the Laguerre polynomials with k half-integer. The radial eigenfunctions with energy $E_n = \left(n + \frac{3}{2}\right)\hbar\omega$ are

$$\psi_{nl}(r) = D_{kl} r^l e^{-\frac{1}{2}\nu r^2} L_k^{l+\frac{1}{2}}(\nu r^2), \tag{B.42}$$

where $k = (n - l)/2$ is an even integer, $\nu = m\omega/\hbar$ and

$$D_{kl} = \left(\frac{\nu^3}{4\pi}\right)^{\frac{1}{4}} \left[\frac{2^{k+l+3} k! \nu^l}{(2k + 2l + 1)!!}\right]^{\frac{1}{2}} . \tag{B.43}$$

The lowest energy eigenfunctions can be built up using the same table with half-integer k.

Appendix C
Coupling of Angular Momenta

Consider the sum of two angular momenta $\vec{\mathbf{L}} = \vec{\mathbf{l}}_1 + \vec{\mathbf{l}}_2$, with $\mathbf{L}_z = \mathbf{l}_{1z} + \mathbf{l}_{2z}$. The four operators $\mathbf{l}_1^2$, $\mathbf{l}_2^2$, $\mathbf{l}_{1z}$, and $\mathbf{l}_{2z}$ commute and hence they have a complete set of simultaneous eigenstates $|l_1 m_1 l_2 m_2\rangle$, fulfilling the eigenvalue equations (here and henceforth, we set $\hbar = 1$):

$$
\begin{aligned}
\mathbf{l}_1^2|l_1 m_1 l_2 m_2\rangle &= |l_1 m_1 l_2 m_2\rangle l_1(l_1+1),\\
\mathbf{l}_2^2|l_1 m_1 l_2 m_2\rangle &= |l_1 m_1 l_2 m_2\rangle l_2(l_2+1),\\
\mathbf{l}_{1z}|l_1 m_1 l_2 m_2\rangle &= |l_1 m_1 l_2 m_2\rangle m_1,\\
\mathbf{l}_{2z}|l_1 m_1 l_2 m_2\rangle &= |l_1 m_1 l_2 m_2\rangle m_2.
\end{aligned}
$$

The operators $\mathbf{L}^2$, $\mathbf{l}_1^2$ and $\mathbf{l}_2^2$ commute as well, hence they also have a complete set of simultaneous eigenstates, $|(l_1 l_2)LM\rangle$, fulfilling the eigenvalue equations:

$$
\begin{aligned}
\mathbf{L}^2|(l_1 l_2)LM\rangle &= |(l_1 l_2)LM\rangle L(L+1),\\
\mathbf{L}_z|(l_1 l_2)LM\rangle &= |(l_1 l_2)LM\rangle M,\\
\mathbf{l}_1^2|(l_1 l_2)LM\rangle &= |(l_1 l_2)LM\rangle l_1(l_1+1),\\
\mathbf{l}_2^2|(l_1 l_2)LM\rangle &= |(l_1 l_2)LM\rangle l_2(l_2+1).
\end{aligned}
\tag{C.1}
$$

As shown in Chap. 6, for each angular momentum eigenvalue l, there are $2l+1$ eigenstates, hence for each couple (l_1, l_2) there are $N = (2l_1+1)(2l_2+1)$ eigenstates. This number must be equal to the number of eigenstates of the eigenvalue L related to the vector $\vec{\mathbf{L}}$, because the two subspaces must have the same dimension. It can be shown that the range of L values is the same as the classical one, namely $|l_1 - l_2| \le L \le l_1 + l_2$, but the values are quantized. Moreover, since $\mathbf{L}_z = \mathbf{l}_{1z} + \mathbf{l}_{2z}$, then $M = m_1 + m_2$.

U. Lombardo et al., *Quantum Mechanics*, UNITEXT for Physics,
https://doi.org/10.1007/978-981-97-1670-8

C.1 Clebsch–Gordan Coefficients

Since the two above-mentioned sets of eigenstates realize different representations of the same space, their eigenvectors must be connected by a unitary transformation $\mathbf{U}$

$$|L(l_1l_2)M\rangle = \sum_{m_1m_2} C(l_1m_1, l_2m_2|LM)\,|l_1m_1, l_2m_2\rangle,$$
$$C(l_1m_1, l_2m_2|LM) = \langle l_1m_1, l_2m_2|L(l_1l_2)M\rangle,$$

where $C(l_1m_1, l_2m_2|LM)$, called Clebsch–Gordan (C-G) coefficients, are the matrix elements of the transformation $\mathbf{U}$.[3] The C-G coefficients are vanishing except for $M = m_1 + m_2$ and $|l_1 - l_2| \leq L \leq l_1 + l_2$. To calculate the non-vanishing coefficients, there have been devised recurrence relations, which will be derived with the help of the rising and lowering operators $\mathbf{l}_\pm$, introduced in Sect. 6.3. Their main property is

$$\mathbf{l}_\pm|lm\rangle = C^\pm_{lm}|lm \pm 1\rangle. \tag{C.2}$$

The normalization factors $C^\pm_{lm}$ are determined as follows:

$$[C^\pm_{lm}]^2 \equiv \langle lm|\mathbf{l}_\mp\mathbf{l}_\pm|lm\rangle = \langle lm|(\mathbf{l}^2 \mp \mathbf{l}_z - \mathbf{l}_z^2)|lm\rangle = l(l+1) - m(m \pm 1). \tag{C.3}$$

The imaginary part of the $C^\pm_{lm}$ factors is not physically significant and will be neglected hereafter. Consider now the sum of two angular momenta $\vec{\mathbf{L}} = \vec{\mathbf{l}}_1 + \vec{\mathbf{l}}_2$ and calculate the matrix element of L± between two angular momentum states with the two coupled and uncoupled angular momenta. By applying Eq. (C.2), it follows:

$$\langle l_1m_1l_2m_2|\mathbf{L}_\pm|(l_1l_2)LM\rangle = C^\pm_{LM}\langle l_1m_1l_2m_2|(l_1l_2)LM \pm 1\rangle, \tag{C.4}$$

$$\langle l_1m_1l_2m_2|(\mathbf{l}_{1\pm} + \mathbf{l}_{2\pm})|(l_1l_2)LM\rangle = C^\mp_{l_1m_1}\langle l_1m_1 \mp 1$$
$$l_2m_2|(l_1l_2)LM\rangle + C^\mp_{l_2m_2}\langle l_1m_1l_2m_2 \mp 1|LM\rangle, \tag{C.5}$$

After equating the right-hand sides of Eqs. (C.4) and (C.5), we obtain the recurrence equations

$$C^\pm_{LM}\langle l_1m_1l_2m_2|(l_1l_2)LM \pm 1\rangle = C^\mp_{l_1m_1}\langle l_1m_1 \mp 1 l_2m_2|(l_1l_2)LM\rangle$$
$$+ C^\mp_{l_2m_2}\langle l_1m_1l_2m_2 \mp 1|(l_1l_2)l_1l_2LM\rangle, \tag{C.6}$$

which enable us to calculate the C-G coefficients. Since the coupling transformation is orthogonal, the C-G coefficients fulfill the orthogonality and completeness relations

[3] The unitary transformation $\mathbf{U}$ can be assumed to be orthogonal and hence its matrix elements, the (C-G) coefficients, are real.

$$\sum_{LM} \langle l_1 m_1 l_2 m_2 | (l_1 l_2) LM \rangle \langle l_1 m_{1'} l_2 m_{2'} | (l_1 l_2) LM \rangle = \delta_{m_1 m_{1'}} \delta_{m_2 m_{2'}}, \tag{C.7}$$

$$\sum_{m_1 m_2} \langle l_1 m_1 l_2 m_2 | (l_1 l_2) LM \rangle \langle l_1 m_1 l_2 m_2 | (l_1 l_2) L' M' \rangle = \delta_{LL'} \delta_{MM'}, \tag{C.8}$$

respectively.

C.2 Tensor Operators: Wigner–Eckart Theorem

The operators can be classified according to their transformation property under space rotations. In quantum mechanics, operators involved as physical variables are rank zero tensor (or scalar) operators, such as Hamiltonian, rank one tensor (or vector) operators, such as space coordinates and momentum, and rank two tensor operators, such as the electromagnetic field. An operator $\mathbf{O}$ is a scalar operator if it is constant under space rotations, so that, as was shown in Chap. 6, a scalar operator commutes with angular momentum: $[\mathbf{L}_i, \mathbf{O}] = \mathbf{0}$. A main property of scalar operators is that they do not change the angular momentum of any state. Indeed, one has

$$\mathbf{L}^2(\mathbf{O}|\tau, LM\rangle) = [\mathbf{L}^2, \mathbf{O}]|\tau, LM\rangle + \mathbf{O}\mathbf{L}^2|\tau, LM\rangle = L(L+1)(\mathbf{O}|\tau, LM\rangle),$$
$$\mathbf{L}_z(\mathbf{O}|\tau, LM\rangle) = [\mathbf{L}_z, \mathbf{O}]|\tau, LM\rangle + \mathbf{O}\mathbf{L}_z|\tau, LM\rangle = M(\mathbf{O}|\tau, LM\rangle).$$

The quantum number τ incorporates the state quantum numbers other than angular momentum. A noticeable property of $\mathbf{O}$ is that its matrix elements between two eigenstates of the angular momentum $\vec{\mathbf{L}}$ do not depend on its z-projection $\mathbf{L_z}$. From the simple algebra, it follows:

$$\begin{aligned}
\langle \tau', LM|\mathbf{O}|\tau, LM\rangle &= [L(L+1) - M(M-1)]^{-\frac{1}{2}} \langle \tau', LM|\mathbf{O}\mathbf{L}_+|\tau, LM-1\rangle \\
&= [L(L+1) - M(M-1)]^{-\frac{1}{2}} \langle \tau', LM|\mathbf{L}_+\mathbf{O}|\tau, LM-1\rangle \\
&= \langle \tau', LM-1|\mathbf{O}|\tau, LM-1\rangle.
\end{aligned}$$

The last two properties tell us that the matrix elements of a scalar operator are diagonal in L and M, and independent of M, i.e.

$$\langle \tau', L'M'|\mathbf{O}|\tau, LM\rangle = \delta_{LL'}\delta_{MM'}\langle \tau'||\mathbf{O}_L||\tau\rangle. \tag{C.9}$$

This is the **Wigner-Eckart** theorem for scalar operators.
A vector operator $\vec{\mathbf{O}}$ is a set of three operators $\mathbf{O}_i$ which transform under rotation according to

$$\mathbf{U}^\dagger \mathbf{O}_i \mathbf{U} = \sum_j \mathcal{R}_{ij} \mathbf{O}_j,$$

where $\mathcal{R}$ is the rotation matrix corresponding to the rotation operator $\mathbf{U}$. Analogously to the scalar operators, the vector operators $\mathbf{O}_i$ fulfill commutation rules deriving from the infinitesimal rotations. To prove it, let us assume the z-axis to be the rotation axis and $\delta\alpha$ the infinitesimal rotation angle, by applying the corresponding unitary transformation

$$\vec{\mathbf{O}} - \mathrm{i}\delta\alpha\,[\mathbf{L_z}, \vec{\mathbf{O}}] = \begin{pmatrix} 1 & -\delta\alpha & 0 \\ \delta\alpha & 1 & 0 \\ 0 & 0 & 1 \end{pmatrix} \vec{\mathbf{O}}\,. \tag{C.10}$$

Remember that $\vec{\mathbf{O}}$ is a column vector, the preceding equation amounts to a set of three equations. After developing this equation, one easily obtains

$$[\mathbf{L_z}, \mathbf{O}_i] = \mathrm{i}\epsilon_{zij}\mathbf{O}_j, \tag{C.11}$$

that can be extended to any cartesian component

$$[\mathbf{L_k}, \mathbf{O}_i] = \mathrm{i}\epsilon_{kij}\mathbf{O}_j. \tag{C.12}$$

With the three cartesian components of the operator $\vec{\mathbf{O}}$ one can build the vector operator: $\mathbf{O}_1 = -1/\sqrt{2}[\mathbf{O}_x + \mathrm{i}\mathbf{O}_y], \mathbf{O}_2 = 1/\sqrt{2}[\mathbf{O}_x - \mathrm{i}\mathbf{O}_y], \mathbf{O}_3 = \mathbf{O}_z$. As an example, let us consider the position operator $\vec{\mathbf{r}}/r \equiv (\mathbf{x}/r, \mathbf{y}/r, \mathbf{z}/r)$, which, written in polar coordinates, can be expressed in terms of the spherical harmonics: $Y_1^{(1)}(\vec{\mathbf{r}}/r), Y_{-1}^{(1)}(\vec{\mathbf{r}}/r), Y_0^{(1)}(\vec{\mathbf{r}}/r)$, with $l = 1$ angular momentum quantum number. These latter, as function of operators, form a class of the so-called irreducible spherical tensors. In particular, the preceding vector is a three-dimensional rank one spherical tensor. In general, a rank λ spherical tensor, $T_\mu^{(\lambda)}$, is defined as the set of $2\lambda + 1$ operators ($-\lambda \leq \mu \leq \lambda$, where μ varies by integers) which transform under rotation according to the law

$$\mathbf{U}^\dagger \mathbf{T}_\mu^{(\lambda)} \mathbf{U} = \sum_{\mu'} \mathcal{R}_{\mu,\mu'}^{(\lambda)} \mathbf{T}_{\mu'}^{(\lambda)}, \tag{C.13}$$

where the rotation matrix elements are defined as $\mathcal{R}_{m,m'}^{(\lambda)} = \langle lm|\mathbf{U}|lm'\rangle$. As earlier, this law gives rise to the following algebraic relations among the tensor components:

$$[\mathbf{L}_\pm, \mathbf{T}_\mu^{(\lambda)}] = C_{\lambda,\mu}^\pm \mathbf{T}_{\mu\pm1}^{(\lambda)} \tag{C.14}$$

and

$$[\mathbf{L}_z, \mathbf{T}_\mu^{(\lambda)}] = \mu \mathbf{T}_\mu^{(\lambda)}, \tag{C.15}$$

where

$$C^{\pm}_{\lambda,\mu} = \sqrt{\lambda(\lambda+1) - \mu(\mu \pm 1)}. \tag{C.16}$$

We can now derive the Wigner–Eckart theorem, which is the generalization of Eq. (C.9) to tensor operators. Let us consider the matrix elements of the commutators in Eq. (C.14) between two angular momentum eigenstates

$$\langle n', L'M'|[\mathbf{L}_{\pm}, \mathbf{T}^{(\lambda)}_{\mu}]|n, LM\rangle = \langle n', L'M'|C^{\pm}_{\lambda,\mu}\mathbf{T}^{(\lambda)}_{\mu\pm1}|n, LM\rangle, \tag{C.17}$$

we apply in the l.h.s. the properties of the operators $\mathbf{L}_{\pm}$, so to obtain

$$\begin{aligned} C^{\mp}_{L',M'}\langle n', L'M' \mp 1|\mathbf{T}^{(\lambda)}_{\mu}|n, LM\rangle = C^{\pm}_{L,M}\langle n', L'M'|\mathbf{T}^{(\lambda)}_{\mu}|n, LM \pm 1\rangle \\ + C^{\pm1}_{\lambda,\mu}\langle n', L'M'|\mathbf{T}^{(\lambda)}_{\mu\pm1}|n, LM\rangle. \end{aligned}$$

This equation is formally identical, including the same coefficients, to the recurrence equation of the Clebsch–Gordan coefficients, Eq. (C.6). In other words, the matrix elements $\langle n', L'M'|\mathbf{T}^{(\lambda)}_{\mu\pm1}|n, LM\rangle$ fulfill the same equation as the Clebsch–Gordan $\langle LM\lambda\mu|(L\lambda)L'M'\rangle$. It follows that the two sets of matrix elements must be equal to within a constant independent of M, M', μ, hence

$$\langle n', L'M'|\mathbf{T}^{(\lambda)}_{\mu}|n, LM\rangle = \frac{\langle L'M'\lambda\mu|(L'\lambda)LM\rangle}{\sqrt{2L+1}}\langle L'||\mathbf{T}^{(\lambda)}||L\rangle. \tag{C.18}$$

The factor $\sqrt{2L+1}$ is introduced for convenience. This theorem is useful because the angular momentum properties of the matrix elements of the tensor operators are inside the Clebsch–Gordan coefficients, namely $M' - M = \mu$ and $|L - L'| \le \lambda \le L + L'$.

The Wigner–Eckart theorem is frequently used in quantum mechanics, for the calculation of the matrix elements of transition operators. A relevant application concerns electromagnetic transitions in the dipole approximation. The application of this theorem to the calculation of the matrix elements of $Y_{1\mu}$ between electronic levels yields

$$\langle n'l'm'|Y_{1\mu}|nlm\rangle = \frac{\langle l'm'1\mu|(l'1)lm\rangle}{\sqrt{2l+1}}\langle n'l'||Y_1||nl\rangle, \tag{C.19}$$

where the selection rules $m' - m = \mu$ and $|l' - l| = 1$ (dipole transitions) clearly appear.

C.3 Projection Theorem

Let us consider the vector operator $\vec{\mathbf{V}}$, and calculate the matrix element

$$\begin{aligned}
&\langle n', LM|\vec{\mathbf{L}} \cdot \vec{\mathbf{V}}|n, LM\rangle \\
&= \langle n', LM|(\mathbf{L_z V_z} - \mathbf{L_+ V_-} - \mathbf{L_- V_+})|n, LM\rangle \\
&= M\langle n', LM|\mathbf{V_z}|n, LM\rangle - C^-_{LM}\langle n', LM-1|\mathbf{V_-}|n, LM\rangle \\
&\quad - C^+_{LM}\langle n', LM+1|\mathbf{V_+}|n, LM\rangle,
\end{aligned} \tag{C.20}$$

where $\mathbf{V_z}, \mathbf{V_-}, \mathbf{V_+}$ are the components of a rank one spherical tensor, and n represent the additional quantum numbers needed to complete the definition of the state. According to the Wigner–Eckart theorem, these three matrix elements are proportional to the same reduced matrix element $\langle n', L||\mathbf{V}||n, L\rangle$, and hence the r.h.s. is equal to the product of $\langle n', L||\mathbf{V}||n, L\rangle$ times a combination of Clebsch-Gordan coefficients dependent on L and M. But, since the operator $\vec{\mathbf{L}} \cdot \vec{\mathbf{V}}$ on the l.h.s. is a scalar operator such a combination cannot depend on M. Therefore

$$\langle n', LM|\vec{\mathbf{L}} \cdot \vec{\mathbf{V}}|n, LM\rangle = C_L\langle n', L||\mathbf{V}||n, L\rangle. \tag{C.21}$$

In particular, for $\mathbf{V} = \mathbf{L}$ and $n = n'$, one has

$$\langle n, LM|\mathbf{L}^2|n, LM\rangle = C_L\langle n, L||\mathbf{L}||n, L\rangle = L(L+1), \tag{C.22}$$

from which the projection theorem follows:

$$\frac{\langle n', LM'|\mathbf{V}_i|n, LM\rangle}{\langle n', LM'|\mathbf{L}_i|n, LM\rangle} = \frac{\langle n', L|\mathbf{V}|n, L\rangle}{\langle n, L|\mathbf{L}|n, L\rangle} = \frac{\langle n', L|\vec{\mathbf{L}} \cdot \vec{\mathbf{V}}|n, L\rangle}{L(L+1)}. \tag{C.23}$$

An application of this theorem is the level splitting in the Zeeman effect, i.e. the calculation of the first-order correction to the $\mathbf{H_B}$ potential, induced on hydrogenoid atoms, by a constant magnetic field B, directed along the z-axis. This correction is

$$\langle \left(l\frac{1}{2}\right) JM|\mathbf{H_B}|\left(l\frac{1}{2}\right) JM\rangle = \mu\langle \left(l\frac{1}{2}\right) JM|(\mathbf{J_z} + \mathbf{S_z})|\left(l\frac{1}{2}\right) JM\rangle \tag{C.24}$$

$$= \mu\left(M + \langle \left(l\frac{1}{2}\right) JM|\mathbf{S_z}|\left(l\frac{1}{2}\right) JM\rangle\right), \tag{C.25}$$

where μ is the atomic magnetic moment. The spin contribution is calculated by applying the projection theorem, which allows the replacement of $\mathbf{S_z}$ with $\mathbf{S} \cdot \mathbf{J}$

$$
\begin{aligned}
&\left\langle \left(l\frac{1}{2}\right) JM|\mathbf{H_B}|\left(l\frac{1}{2}\right) JM\right\rangle \\
&= \mu\left(M + \frac{\left\langle \left(l\frac{1}{2}\right) JM|\mathbf{S}\cdot\mathbf{J}|\left(l\frac{1}{2}\right) JM\right\rangle}{J(J+1)} \cdot \langle JM|\mathbf{J_z}|JM\rangle \right) \\
&= \mu M\left(1 + \frac{\left\langle \left(l\frac{1}{2}\right) JM|\mathbf{S}\cdot\mathbf{J}|\left(l\frac{1}{2}\right) JM\right\rangle}{J(J+1)} \right) \\
&= \mu M\left(1 + \frac{\left\langle \left(l\frac{1}{2}\right) JM|(\mathbf{J}^2+\mathbf{S}^2-\mathbf{l}^2)/2|\left(l\frac{1}{2}\right) JM\right\rangle}{J(J+1)} \right) \\
&= \mu M\left[1 + \frac{J(J+1)+S(S+1)-l(l+1)}{2J(J+1)}\right] \\
&= \mu M\left(1 \pm \frac{1}{2l+1}\right),
\end{aligned}
\tag{C.26}
$$

being $J = l \pm 1/2$.

Appendix D
Mathematical Complements

D.1 Fourier Transform and Convolution Theorem

The Fourier transform of a real or complex function $f(x)$ is defined as

$$F(k) = \int_{-\infty}^{+\infty} dx\, e^{-ikx} f(x), \ (k \in \mathbb{R}), \tag{D.1}$$

where function $f(x)$ is the Fourier anti-transform

$$f(x) = \frac{1}{2\pi} \int_{-\infty}^{+\infty} dk\, e^{ikx} F(k). \tag{D.2}$$

Extended to n variables, the Fourier transform of a function is

$$F(k_1, k_2, \ldots, k_n) = \int_{-\infty}^{+\infty} dx_1 dx_2 ... dx_n\, e^{-i(k_1x_1+k_2x_2+\cdots+k_nx_n)} f(x_1, x_2, \ldots, x_n), \tag{D.3}$$

and the Fourier anti-transform of a n variables function is

$$f(x_1, x_2, \ldots, x_n) = \left(\frac{1}{2\pi}\right)^n \int_{-\infty}^{+\infty} dk_1 dk_2 ... dk_n e^{i(k_1x_1+k_2x_2+\cdots+k_nx_n)} F(k_1, k_2, \ldots, k_n). \tag{D.4}$$

The convolution of two given functions $f(x)$ and $g(x)$ is a third function $h(x)$ defined by the following integral:

$$h(x) = f(x) * g(x) = \int_{-\infty}^{+\infty} dt f(t) g(x-t). \tag{D.5}$$

U. Lombardo et al., *Quantum Mechanics*, UNITEXT for Physics,
https://doi.org/10.1007/978-981-97-1670-8

The convolution theorem states that: *provided the existence of the Fourier transforms of two given functions, $f(x)$ and $g(x)$, then the Fourier transform of their convolution is equal to the product of the Fourier transforms of $f(x)$ and $g(x)$*, i.e.

$$H(k) = \int_{-\infty}^{+\infty} \mathrm{d}x\, \mathrm{e}^{-ikx} h(x) = F(k)G(k), \tag{D.6}$$

where $F(k)$, $G(k)$, and $H(k)$ are the Fourier transforms of the functions $f(x)$, $g(x)$, and $h(x)$, respectively.

D.2 The Green Function Formalism

The formalism of Green's function is adopted in several branches of quantum theory. We have introduced such a function in Chap. 13, where it was used for solving the Lippmann–Schwinger equation. Here we briefly discuss its main properties. Let us consider the differential equation

$$\mathbf{O}_x u(x) = f(x), \tag{D.7}$$

where $\mathbf{O}_x$ is a differential operator in the x variable. The Green function $G(x.y)$, associated to the operator $\mathbf{O}_x$, is defined as the solution of the following equation:

$$\mathbf{O}_x G(x, y) = \delta(x - y). \tag{D.8}$$

By plugging Green's function into Eq. (D.7), we get

$$\mathbf{O}_x u(x) = \int \mathrm{d}y \delta(x - y) f(y) = \int \mathrm{d}y \mathbf{O}_x G(x, y) f(y), \tag{D.9}$$

and, hence, Eq. (D.7) can be re-written as

$$\mathbf{O}_x \left[u(x) - \int \mathrm{d}y G(x, y) f(y) \right] = 0. \tag{D.10}$$

As a result, we obtain the solution $u(x)$ of the differential equation Eq. (D.7) in terms of Green's function, i.e.

$$u(x) = \int \mathrm{d}y G(x, y) f(y) + h(x). \tag{D.11}$$

The function $h(x)$ is determined by the boundary conditions of the tackled problem, by solving the equation $\mathbf{O}_x h(x) = 0$.

Using the Dirac formalism, we can re-write Eq. (D.7) as

$$\mathbf{O}_x|u\rangle = |f\rangle. \tag{D.12}$$

If the inverse of the operator $\mathbf{O}_x$ exists, it will coincide with the Green operator, namely $\mathbf{O}_x^{-1} \equiv \mathbf{G}$, and hence Eq. (D.12) is written as

$$|u\rangle = \mathbf{G}|f\rangle. \tag{D.13}$$

In the 3D coordinate representation, this takes the form

$$u(\vec{x}) = \int \mathrm{d}^3x' \langle\vec{x}|\mathbf{G}|\vec{x}\,'\rangle f(\vec{x}\,'), \tag{D.14}$$

where

$$\langle\vec{x}|\mathbf{G}|\vec{x}\,'\rangle = G(\vec{x}, \vec{x}'). \tag{D.15}$$

D.3 The Residue Theorem

A complex function $f : A \subset \mathbb{C} \to \mathbb{C}$ is said to be analytic in A, if for each point z_0 belonging to the domain A, the limit of the incremental ratio:

$$\lim_{z \to z_0} \frac{f(z) - f(z_0)}{z - z_0} = f'(z_0) \tag{D.16}$$

exists and is finite. A given point z_1 represents a pole of the n-th order for an analytic function $f(z)$ if there exists one and only one integer n for which

$$\lim_{z \to z_1} (z - z_1)^n f(z) = \alpha \neq 0 \ \ (\alpha \in \mathbb{C}), \tag{D.17}$$

If $n = 1$, the function $f(z)$ has a simple pole around z_1. Given $f(z)$ to be analytic within an interval centered around the pole z_1, we define the residue of $f(z)$ in z_1 the coefficient c_{-1} obtained from the following integral:

$$\mathrm{Res}(f(z), z_1) = \frac{1}{2\pi\mathrm{i}} \oint_\gamma f(z')\mathrm{d}z' = c_{-1}, \tag{D.18}$$

where γ represents a closed path in the counterclockwise direction within the analytical domain, e.g. a circumference. The residue around a simple pole is calculated as follows:

$$c_{-1} = \lim_{z \to z_1} (z - z_1) f(z). \tag{D.19}$$

For a pole of the n-th order,

$$c_{-1} = \lim_{z \to z_1} \frac{1}{(n-1)!} \frac{\mathrm{d}^{n-1}}{\mathrm{d}z^{n-1}} [(z - z_1)^n f(z)]\,. \tag{D.20}$$

The residue theorem states that: *if a function $f(z)$ is analytic within a closed domain $\bar{D}$, with the exception of a finite number of poles $z_1, z_2, \ldots, z_n \in D$, then the integral of $f(z)$ along the frontier ∂D of D is equal to the sum of its residues around the points z_j, $j = 1, \ldots, n$, times $2\pi i$*, that is

$$\oint_{\partial D} f(z)\mathrm{d}z = 2\pi\mathrm{i} \sum_{j=1}^{n} \mathrm{Res}(f(z), z_j) = 2\pi\mathrm{i} \sum_{j=1}^{n} c_{-1}^{(j)}. \tag{D.21}$$

An application of the residue theorem is the calculation of the Green function of a free particle with energy $E = \dfrac{p^2}{2m}$ (see Sect. 13.2)

$$G_0(\vec{r}, \vec{r}\,') = \frac{1}{(2\pi\hbar)^3} \int \mathrm{d}^3 p' \frac{\mathrm{e}^{\mathrm{i}\frac{\vec{p}\,'}{\hbar}\cdot(\vec{r}-\vec{r}\,')}}{E \pm \mathrm{i}\varepsilon - \dfrac{p'^2}{2m}}. \tag{D.22}$$

Hereafter, we choose the positive sign, corresponding to outgoing spherical waves and introduce the complex momentum $p = \sqrt{2mE} + \mathrm{i}\varepsilon$, setting $\hbar = 1$ and $|\vec{r} - \vec{r}\,'| = s$. After integrating over the angular part, we are left with

$$G_0(\vec{r}, \vec{r}\,') = \frac{2m}{(2\pi)^3} \frac{2\pi}{\mathrm{i}s} \int_{-\infty}^{\infty} p'\mathrm{d}p' \frac{\mathrm{e}^{\mathrm{i}p's}}{p^2 - p'^2}. \tag{D.23}$$

The integral on p' will we performed using the residue theorem. For this purpose, we analytically extend the variable p' to the complex plane, and close the integration domain, $-R < p' < R$, with the semi-circle γ of radius R upon the upper half-plane, as shown in Fig. D.1.

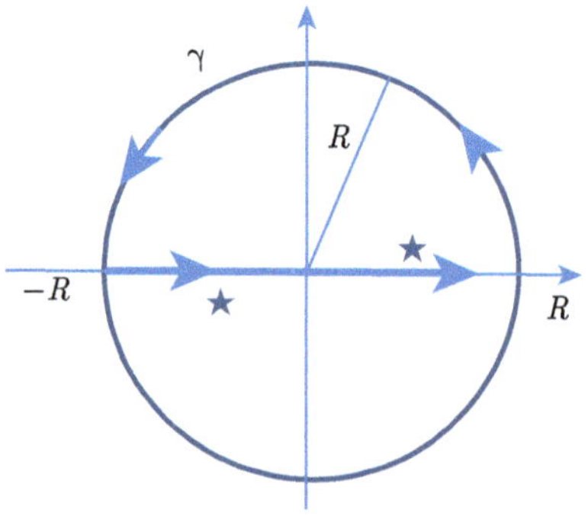

Fig. D.1 Integration contour in the complex p′ plane. The stars represent the two poles of Green's function

By Taking into account that the line integral on γ vanishes for $R \to \infty$, due to the exponential decay of the integrand in upper half-plane, where the imaginary part of p' is greater than zero, we can identify our Green's function with the contour integral of Fig. D.1

$$G_0(\vec{r}, \vec{r}\,') = \frac{2m}{(2\pi)^3} \frac{2\pi}{\mathrm{i}s} \oint p' \mathrm{d}p' \frac{\mathrm{e}^{\mathrm{i}p's}}{p^2 - p'^2}, \tag{D.24}$$

Since we have transformed the integration domain in a closed path, we can use the residue theorem to calculate the integral Eq. (D.24) with pole at $p' = p$ enclosed within γ path. Thus, the residue is

$$c_{-1} = \lim_{p' \to p} (p' - p) \frac{p' \mathrm{e}^{\mathrm{i}sp'}}{p^2 - p'^2} = -\frac{\mathrm{e}^{\mathrm{i}sp}}{2}. \tag{D.25}$$

Finally, plugging this result into Eq. (D.24), we get

$$G_0(\vec{r} - \vec{r}\,') = -\frac{m}{2\pi\hbar^2} \frac{\mathrm{e}^{\mathrm{i}p|\vec{r} - \vec{r}\,'|}}{|\vec{r} - \vec{r}\,'|}. \tag{D.26}$$

Physical Constants

Constant	Symbol	Value
Speed of Light	c	$2.99792458 \times 10^{8}\ \frac{\mathrm{m}}{\mathrm{s}}$
Planck's Constant	h	$6.6260755 \times 10^{-34}\ \mathrm{J \cdot s}$
	$\hbar = \frac{h}{2\pi}$	$1.05457266 \times 10^{-34}\ \mathrm{J \cdot s} = 6.5821220 \times 10^{-22}\ \mathrm{MeV \cdot s}$
	$\hbar c$	$197.327053\ \mathrm{MeV \cdot fm}$
Mass of the Electron	$m_{\mathrm{e}}c^2$	$0.51099906\ \mathrm{MeV}$
Mass of the Neutron	$m_{\mathrm{n}}c^2$	$939.56563\ \mathrm{MeV}$
Mass of the Proton	$m_{\mathrm{p}}c^2$	$938.27231\ \mathrm{MeV}$
Elementary Charge	e	$1.602176634 \cdot 10^{-19}\ \mathrm{C}$
Dielectric Constant	ε_0	$8.854187817 \cdot 10^{-12}\ \frac{\mathrm{F}}{\mathrm{m}}$
Permeability of Vacuum	μ_0	$4\pi \times 10^{-7}\ \frac{\mathrm{N}}{\mathrm{A}^2} = 1.25663706144 \times 10^{-6}\ \frac{\mathrm{N}}{\mathrm{A}^2}$
Fine Structure Constant	$\alpha = \frac{1}{4\pi\varepsilon_0\hbar c}$	$1/137.0359895$
Bohr Radius	$a_0 = \frac{\hbar c}{\alpha m_{\mathrm{e}}c^2}$	$5.29177249 \times 10^{-11}\ \mathrm{m} = 0.529177249\ \mathrm{Å}$
Bohr Magneton	$\mu_{\mathrm{B}} = \frac{e\hbar}{2m_e}$	$5.78838263 \times 10^{-11}\ \mathrm{MeV \cdot T^{-1}}$
Nuclear Magneton	$\mu_{\mathrm{N}} = \frac{e\hbar}{2m_{\mathrm{p}}}$	$3.15245166 \times 10^{-14}\ \mathrm{MeV \cdot T^{-1}}$
Electron Magnetic Moment	μ_{e}	$1.001159652193\ \mu_{\mathrm{B}}$

Constant	Symbol	Value
Neutron Magnetic Moment	μ_n	$-1.91304275\,\mu_N$
Proton Magnetic Moment	μ_p	$2.792847386\,\mu_N$
Avogadro's Number	N_A	$6.0221367 \times 10^{23}\,\text{mol}^{-1}$
Boltzmann's Constant	k_B	$1.380658 \times 10^{-23}\,\text{J} \cdot \text{K}^{-1} = 8.617385 \times 10^{-11}\,\text{MeV} \cdot \text{K}^{-1}$
Gravitational Constant	G	$6.67259 \times 10^{-11}\,\frac{\text{N} \cdot \text{m}^2}{\text{kg}}$
Solar Mass	$M_\odot$	$(1.98847 \pm 0.00007) \times 10^{30}\,\text{kg}$
Solar Radius	$R_\odot$	$(696.342 \pm 650)\,\text{km}$

Units of Measurement

Length

$$1\,\text{Angstrom} = 1\,\text{Å} = 10^{-10}\,\text{m}$$
$$1\,\text{fermi} = 1\,\text{fm} = 10^{-15}\,\text{m}$$

Duration

$$1\,\text{second} = 1\,\text{sec} = 10^{6}\,\text{microsec} = 10^{9}\,\text{nanosec}$$

Energy

$$1\,\text{electron volt} = 1\,\text{eV} = 1.602176634 \cdot 10^{-19}\,\text{J}$$

Cross Section (Area)

$$1\,\text{barn} = 1\,\text{b} = 10^{-24}\,\text{cm}^2 = 10^{2}\,\text{fm}^2$$

Mass

$$1\,\text{atomic mass unit} = 1\,\text{amu} = \frac{1}{12} M({}^{12}_{6}\text{C}) = 931.494013\,\frac{\text{MeV}}{c^2}$$

Temperature

$$1\,\text{MeV} = 1.16 \cdot 10^{10}\,\text{Kelvin}$$

References

Dirac, P.A.M.: The Principles of Quantum Mechanics, 4th edn. Clarendon Press Oxford (1958)
Feynman, R.: The Feynman Lectures on Physics: Quantum Mechanics. New Millennium (2012)
Gottfried, K., Yan, T.-M.: Quantum Mechanics: Fundamentals, 2nd edn. Springer (2004)
Landau, L., Lifshitz, E.: Mecanique Quantique. MIR, Moscou (1967)
Messiah, A.: Quantum Mechanics. North-Holland Publ., Amsterdam (1965)
Preston, M.A.: Physics of the Nucleus. Addison-Wesley (1965)
Ring, P., Schuck, P.: The Nuclear Many-Body Problem. Springer (2000)
Sakurai, J.J.: Modern Quantum Mechanics, Revised Edition. Addison-Wesley Pub. (1994)
Schwinger, J.: Quantum Mechanics: Symbolism of Atomic Measurements. Springer (2011)

Index

U. Lombardo et al., *Quantum Mechanics*, UNITEXT for Physics,
https://doi.org/10.1007/978-981-97-1670-8

FSC
www.fsc.org
MIX
Papier aus verantwortungsvollen Quellen
Paper from responsible sources
FSC® C105338

If you have any concerns about our products,
you can contact us on
ProductSafety@springernature.com

In case Publisher is established outside the EU,
the EU authorized representative is:
Springer Nature Customer Service Center GmbH
Europaplatz 3, 69115 Heidelberg, Germany

Printed by Libri Plureos GmbH
in Hamburg, Germany